STUDENT SOLUTIONS MANUAL
David Robichaud
California State University, Fullerton

CHEMISTRY

THIRD EDITION

John Olmsted III
California State University, Fullerton

Gregory M. Williams
California State University, Fullerton

JOHN WILEY & SONS, INC.

COVER PHOTO: ©Roderick Chen/SUPERSTOCK, Inc.

To order books or for customer service call 1-800-CALL-WILEY (225-5945).

ISBN 0-471-03512-2

Printed in the United States of America.

10 9 8 7 6 5 4 3 2 1

Printed and bound Victor Graphics, Inc.

Table of Contents

1.1   Think about political issues that involve chemical reactions. Examples include the following:
(a) What are appropriate standards for "clean" air and water?
(b) What should be the regulations for testing new drugs before they are approved for human use?
(c) How can toxic waste sites best be cleaned up?
(d) Should chlorofluorocarbons be banned because of potential damage to the ozone layer?

1.3   Think about what a pharmacist does that involves chemistry. A pharmacist deals in chemical substances, most of which are toxic when incorrectly administered. Among the chemistry-related skills that a pharmacist uses are the following:
(a) Weighing, volume measurements and unit conversions;
(b) Knowing chemical compatibility of different drugs;
(c) Identifying similar or chemically equivalent drugs; and
(d) Protecting drugs from degradation due to exposure to adverse conditions.

1.5   Your criticism should be based on the experimental nature of chemistry. When experimental results fail to match theoretical expectations, the experiment may be flawed or the theory may be incorrect. The chemist should redo the experiment, adjusting conditions if necessary, to determine whether or not the results are correct. If repeated measurements show that the results consistently differ from what theory predicts, then the chemist should examine how to revise the theory to accommodate the results.

1.7   You must commit to memory the correspondence between names and symbols of various elements, but remembering them is simplified by the fact that most English names and elemental symbols are related.

        (a) H;   (b) He;   (c) Hf;   (d) N;   (e) Ne;   and (f) Nb.

1.9   Associating an element's name with its symbol requires memorization of both names and symbols. The examples in this problem all begin with the letter "A."

        (a) arsenic; (b) argon; (c) aluminum; (d) americium; (e) silver; (f) gold; (g) astatine; and (h) actinium.

1.11   To convert a molecular picture into a molecular formula, count atoms of each type and consult (or recall) the color scheme used for the elements. See Figure 1-4 of your textbook for the color scheme used in this and many other texts.

        (a) Br_2; (b) HCl; (c) C_2H_5I; (d) C_3H_6O; (e) $C_3H_6O_2$; and (f) $C_7H_5N_3O_6$

1.13   In writing a chemical formula, remember to use elemental symbols and subscripts for the number of atoms:

        (a) CCl_4;   (b) H_2O_2;   (c) P_4O_{10};   and (d) Fe_2S_3.

1.15   The next element after the end of a row is the first element in the succeeding row: cesium.

1.17   Elements in the same column of the periodic table share similar chemical properties. For sulfur, these are its vertical neighbors, oxygen (O) and selenium (Se). The horizontal neighbors are phosphorus (P) and chlorine (Cl).

1.19   Metals from group 1 react in a 1:1 ratio with Br from Group 17:   Li, Na, K, Rb, Cs
        Other metals that also react in a 1:1 ratio with bromine are: Cu, Ag, and Au.

1.21   Consult the periodic table: lithium, Li; beryllium, Be; boron, B; carbon, C; oxygen, O; fluorine, F; and neon, Ne.

1.23   Remember that a pure substance contains a single chemical compound and a solution is a homogeneous mixture of substances.
        (a) Pure substance;
        (b) Solution of various substances in water;
        (c) Heterogeneous mixture (indicated by its opacity);
        (d) Solution of nitrogen, oxygen, and other gases;
        (e) Heterogeneous mixture of gases and particulate matter; and
        (f) Heterogeneous (indicated by the presence of grains).

1.25   Remember that gases and liquids both flow, but gases are much less dense than liquids.
        (a) liquid;   (b) solid;   (c) solid;   and   (d) gas.

1.27   In a chemical transformation, one substance is converted into another. In a physical transformation the substance remains the same, only the state (gas, liquid, solid) changes.
        (a) Water vapor deposits as ice, a physical transformation;
        (b) Liquid water evaporates, a physical transformation; and
        (c) A flame is a chemical transformation.

1.29   A mixture contains more than one substance, while a compound is a single substance containing more than one element.
        (a) A mixture of water and dissolved substances;
        (b) A pure compound, H_2O;
        (c) A mixture of water and one or more solids;
        (d) A single element (He);
        (e) A mixture of isopropanol and water; and
        (f) A mixture of a solvent and various pigments.

1.31   Scientific notation expresses any number as a value between 1 and 10 times a power of ten. Trailing zeros are retained only when they are significant.
        (a) 1.00000×10^5;   (b) 1.0×10^4;   (c) 4.00×10^{-4};   (d) 3×10^{-4}; and   (e) 2.753×10^2.

1.33   To do unit conversions, multiply by a ratio that cancels the unwanted unit(s). Refer to Table 1-3 for the SI base units.

(a) 432 kg = 4.32 x 10^2 kg;

(b) $624 \, \text{ps}\left(\dfrac{10^{-12} \, \text{s}}{1 \, \text{ps}}\right) = 6.24 \times 10^{-10}$ s;

(c) $1024 \, \text{ng}\left(\dfrac{10^{-9} \, \text{g}}{1 \, \text{ng}}\right)\left(\dfrac{10^{-3} \, \text{kg}}{1 \, \text{g}}\right) = 1.024 \times 10^{-9}$ kg;

(d) $93{,}000 \, \text{km}\left(\dfrac{10^{3} \, \text{m}}{1 \, \text{km}}\right) = 9.300 \times 10^{7}$ m;

(e) $1 \, \text{day}\left(\dfrac{24 \, \text{hr}}{1 \, \text{day}}\right)\left(\dfrac{60 \, \text{min}}{1 \, \text{hr}}\right)\left(\dfrac{60 \, \text{sec}}{1 \, \text{min}}\right) = 8.6400 \times 10^{4}$ s (assuming exactly 1 day); and

(f) $0.0426 \, \text{in}\left(\dfrac{2.54 \, \text{cm}}{1 \, \text{in}}\right)\left(\dfrac{1 \, \text{m}}{100 \, \text{cm}}\right) = 1.08 \times 10^{-3}$ m.

1.35 This is a unit conversion problem involving summation and unusual units. First convert all masses into kg, then put them into the same power of ten and add the masses.

Mass of diamonds $= 5.0 \times 10^{-1} \, \text{carat}\left(\dfrac{3.168 \, \text{grains}}{1 \, \text{carat}}\right)\left(\dfrac{1 \, \text{g}}{15.4 \, \text{grains}}\right)\left(\dfrac{10^{-3} \, \text{kg}}{1 \, \text{g}}\right) = 1.0 \times 10^{-4}$ kg;

$7.00 \, \text{g}\left(\dfrac{10^{-3} \, \text{kg}}{1 \, \text{g}}\right) = 7.00 \times 10^{-3}$ kg;

total mass = (7.00 x 10^{-3} kg) + (0.10 x 10^{-3} kg) = 7.10 x 10^{-3} kg.

1.37 Because the quart is a volume measure, density must be used to convert from volume to mass. The density of water is 1.00 g/cm^3 (Table 1-4). Assume exactly one quart.

$1 \, \text{quart}\left(\dfrac{1 \, \text{L}}{1.057 \, \text{quart}}\right)\left(\dfrac{10^{3} \, \text{mL}}{1 \, \text{L}}\right)\left(\dfrac{1 \, \text{cm}^{3}}{1 \, \text{mL}}\right)\left(\dfrac{1.00 \, \text{g}}{1 \, \text{cm}^{3}}\right) = 9.46 \times 10^{2}$ g

1.39 Density is mass divided by volume, and the volume of a block is $V = l \, w \, h$.
V = (15.5 cm)(4.6 cm)(1.75 cm) = 1.25 x 10^2 cm^3;

$\rho = \dfrac{m}{V} = \dfrac{98.456 \, \text{g}}{1.25 \times 10^{2} \, \text{cm}^{3}} = 0.79$ g/cm^3

This result has two significant figures because one dimension is known to only two significant figures.

1.41 To convert from mass to volume, divide by density. See Table 1-4 for densities.

$V = \dfrac{m}{\rho} = 15.4 \, \text{g}\left(\dfrac{1 \, \text{cm}^{3}}{2.70 \, \text{g}}\right) = 5.70$ cm^3

1.43 The question asks for the density of water expressed in SI units (kg/m^3). Begin by analyzing the given information. The mass of the container is given before and after the water has been added. Thus, the mass of water can be obtained from the difference between the masses of the filled and empty container:

m = 270.064 g – 93.054 g = 177.010 g H_2O;

In SI units,

$$m = 177.010 \text{ g}\left(\frac{10^{-3} \text{ kg}}{1 \text{ g}}\right) = 0.177010 \text{ kg}$$

Before computing the volume it is a good idea to convert the units from inches to meters:

$$r = 0.875 \text{ in}\left(\frac{2.54 \text{ cm}}{1 \text{ in}}\right)\left(\frac{10^{-2} \text{ m}}{1 \text{ cm}}\right) = 0.022225 \text{ m};$$

$$h = 4.500 \text{ in}\left(\frac{2.54 \text{ cm}}{1 \text{ in}}\right)\left(\frac{10^{-2} \text{ m}}{1 \text{ cm}}\right) = 0.1143 \text{ m};$$

$V = \pi r^2 h$;

$V = \pi(0.022225 \text{ m})^2(0.1143 \text{ cm}) = 1.7737 \times 10^{-4} \text{ m}^3$;

The density is obtained by dividing the mass of water by the volume.

$$\rho = \frac{m}{V} = \frac{0.177010 \text{ kg}}{1.7737 \times 10^{-4} \text{ m}^3} = 9.98 \times 10^2 \text{ kg/m}^3; \quad (\text{i.e., } 0.998 \text{ g/cm}^3)$$

This result is rounded to three significant figures because the radius of the cylinder is known to only three significant figures.

1.45   The question asks for a comparison of the masses of two objects of different densities and shapes. For each object, $m = \rho V$.

The sphere of Au has ρ = 19.3 g cm^{-3}, $V = \left(\frac{4}{3}\right)\pi r^3$, and r = diameter/2 = 1.00 cm;

$$V = \frac{4\pi(1.00 \text{ cm})^3}{3} = 4.19 \text{ cm}^3;$$

$$m(\text{gold}) = 4.19 \text{ cm}^3\left(\frac{19.3 \text{ g}}{1 \text{ cm}^3}\right) = 80.9 \text{ g};$$

The cube of Pb has ρ = 11.34 g cm^{-3} and $V = l\,w\,h = (2.00 \text{ cm})^3 = 8.00 \text{ cm}^3$;

$$m(\text{lead}) = 8.00 \text{ cm}^3\left(\frac{11.34 \text{ g}}{1 \text{ cm}^3}\right) = 90.7 \text{ g}$$

The lead cube has more mass than the gold sphere.

1.47   Volume and mass are related through density: $V = \dfrac{m}{\rho}$.

$$V = 36.5\,\text{g}\left(\frac{1\,\text{mL}}{3.12\,\text{g}}\right) = 11.7\,\text{mL}.$$

1.49 Consult the periodic table and/or the alphabetical listing at the end of the text to answer questions about elemental symbols. Eight elements have symbols beginning with "T": Ti (titanium), Tc (technetium), Te (tellurium), Ta (tantalum), Tl (thallium), Tb (terbium), Tm (thulium), and Th (thorium).

1.51 Both molecular fluorine and molecular chlorine are diatomic (two atoms per molecule) species. To have a total of 20 molecules in a 3:1 ratio there should be 5 molecules of chlorine (shaded) and 15 molecules of fluorine (white).

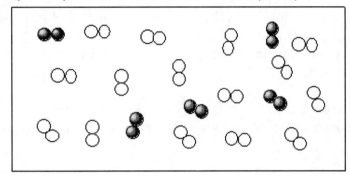

1.53 "Do not match" means that the letters in the symbol are not related to the English name of the element. See Table 1-1 in your textbook.
Lead, Pb.

1.55 Although it may not at first appear to be, this is a unit conversion problem. The speed of the athlete in SI units is:

$$\text{speed} = \left(\frac{100\,\text{yards}}{10.17\,\text{s}}\right)\left(\frac{0.9144\,\text{m}}{1\,\text{yard}}\right) = 8.991\,\text{m/s}\ \ \text{(assuming exactly 100 yards)}$$

The time of the 100-meter dash can be found using the definition of speed:

$$\text{speed} = \frac{\text{distance}}{\text{time}},$$

$$\text{time} = 100\,\text{m}\left(\frac{1\,\text{s}}{8.991\,\text{m}}\right) = 11.12\,\text{s}\ \ \text{(assuming exactly 100 meters)}$$

The result has four significant figures, because the time has four significant figures.

1.57   The periodic table is organized with elements that have similar chemical properties in the same column. Gold shares Group 11 with silver and copper. All three are shiny, non-reactive, soft, and good conductors of heat and electricity.

1.59   The periodic table is organized with elements that have similar chemical properties in the same column. Sodium shares Group 1 with lithium, potassium, rubidium, cesium, and francium. All are soft, highly reactive metals (they react vigorously with water, for example) that form 1:1 salts with elements from Group 17.

1.61   The key to this question is to recognize how solids, liquids, and gases are organized at the molecular level.
A solid has a rigid, constant shape because its molecules are in fixed, regular arrangements. Thus B matches D.
A liquid is dense yet able to flow easily, because its molecules, though close together, are not in fixed positions. Thus C matches E.
A gas has low density and flows easily because its molecules are separated by much empty space. Thus A matches F.

1.63   Kelvin-Celsius conversions involve addition/subtraction rather than multiplication/division, because the two scales have the same unit size but different zero points.

$$K = °C + 273.15;$$
$$T = -11.5 °C + 273.15 = 261.7 K.$$

1.65   The significant figures in a result of multiplication/division is determined by the number containing the least number of significant figures.

(a) $\dfrac{(6.531 \times 10^{13})(6.02 \times 10^{23})}{(435)(2.000)} = \dfrac{3.93 \times 10^{37}}{870.} = 4.52 \times 10^{34};$

(b) $\dfrac{4.476 + (3.44)(5.6223) + 5.666}{(4.3)(7 \times 10^{4})} = \dfrac{4.476 + 19.3 + 5.666}{3 \times 10^{5}} = \dfrac{29.5}{3 \times 10^{5}} = 1 \times 10^{-4}.$

1.67   The value of an extensive property changes with the amount of the substance, while the value of an intensive property is independent of the amount of the substance.
Intensive: (a), (c), and (e);
Extensive: (b) and (d).

1.69   Determine a molecular formula from a model by counting the atoms of each kind and referring to the color code in use for the elements.

(a) $CHClF_2$; (b) CH_2O_2; (c) BrF_3; and (d) C_4H_{10}

1.71 Do unit conversions by multiplying by the appropriate ratios.

(a) $454 \, in^3 \left(\dfrac{2.54 \, cm}{1 \, in} \right)^3 \left(\dfrac{10^{-2} \, m}{1 \, cm} \right)^3 = 7.44 \times 10^{-3} \, m^3$;

(b) $\left(\dfrac{35 \, miles}{1 \, hr} \right) \left(\dfrac{1.609 \, km}{1 \, mile} \right) \left(\dfrac{10^3 \, m}{1 \, km} \right) \left(\dfrac{1 \, hr}{60 \, min} \right) \left(\dfrac{1 \, min}{60 \, s} \right) = 16 \, m \, s^{-1}$;

(c) $\left[6 \, ft \left(\dfrac{12 \, in}{1 \, ft} \right) + 9 \, in \right] \left(\dfrac{2.54 \, cm}{1 \, in} \right) \left(\dfrac{10^{-2} \, m}{1 \, cm} \right) = 2.1 \, m$; and

(d) $227 \, lb \left(\dfrac{1 \, kg}{2.205 \, lb} \right) = 1.03 \times 10^2 \, kg$

1.73 Make use of the periodic table to match these correctly.
(a) Alkaline earth metals are in Group 2: Sr;
(b) Elements in the same column as Al have similar chemical properties: In;
(c) Elements from Columns 16 and 17 react with K: O and Br, respectively;
(d) Transition metals lie in the d block: Co;
(e) Noble gases are in Group 18: Ne; and
(f) Actinides are found in the $n = 5$ f-block: Pu.

1.75 To draw molecular models, make use of the color-coded, scaled atoms shown in Figure 1-4 in your textbook.

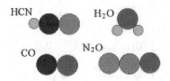

1.77 Time is related to distance through time = distance/speed.

$$time = 2786 \, mile \left(\dfrac{1.609 \, km}{1 \, mile} \right) \left(\dfrac{1 \, hr}{685 \, km} \right) \left(\dfrac{60 \, min}{1 \, hr} \right) = 393 \, min$$

1.79 Physical properties do not relate to other substances, while chemical properties do.
Physical properties: appearance, melting point, softness, density.
Chemical properties: reaction with chlorine and reaction with water.

1.81 Consult the periodic table to find examples of various classes of elements.
(a) Halogens are found in Group 17: fluorine (F), chlorine (Cl), bromine (Br), iodine (I);
(b) Alkaline earth metals are found in Group 2: beryllium (Be), magnesium (Mg), calcium (Ca), strontium (Sr), barium (Ba);
(c) Actinides are those elements in the $n = 5$ f-block, for example uranium (U), and berkelium (Bk);

(d) Noble gases are in Group 18: helium (He), neon (Ne), argon (Ar), krypton (Kr), xenon (Xe) and radon (Rn).

1.83   Distance is related to speed through the equation, speed = distance/time. Thus, distance = (speed)(time). The speed of light is 2.9979×10^8 m/s.
distance traveled in one year =

$$\left(\frac{2.9979 \times 10^8 \, m}{1 \, s}\right)\left(\frac{10^{-3} \, km}{1 \, m}\right)\left(\frac{60 \, s}{1 \, min}\right)\left(\frac{60 \, min}{1 \, hr}\right)\left(\frac{24 \, hr}{1 \, day}\right)\left(\frac{365.24 \, day}{1 \, yr}\right) = 9.4604 \times 10^{12} \, km..$$

1.85   Each atom contributes a distance equal to its diameter, so the number of atoms is the total distance divided by the diameter of an atom. A unit conversion is required between inches and picometers.

$$\# = 1.0 \, in\left(\frac{2.54 \, cm}{1 \, in}\right)\left(\frac{1 \, m}{100 \, cm}\right)\left(\frac{10^{12} \, pm}{1 \, m}\right)\left(\frac{1 \, atom}{200 \, pm}\right) = 1.3 \times 10^8 \, atoms.$$

1.87   This is a unit conversion problem. Multiply by dimensional ratios to convert units.

In feet:

$$20{,}000 \, leagues\left(\frac{3 \, mi}{1 \, league}\right)\left(\frac{10 \, cable}{1 \, mi}\right)\left(\frac{100 \, fathoms}{1 \, cable}\right)\left(\frac{6 \, ft}{1 \, fathom}\right) = 3.6 \times 10^8 \, ft$$

In kilometers:

$$3.6 \times 10^8 \, ft\left(\frac{0.3048 \, m}{1 \, ft}\right)\left(\frac{10^{-3} \, km}{1 \, m}\right) = 1.1 \times 10^5 \, km$$

1.89   This is a unit conversion problem. Multiply by dimensional ratios to convert units.

$$100 \, yr\left(\frac{365.24 \, day}{1 \, yr}\right)\left(\frac{24 \, hr}{1 \, day}\right)\left(\frac{60 \, min}{1 \, hr}\right)\left(\frac{60 \, s}{1 \, min}\right) = 3.1557 \times 10^9 \, s.$$

The number of days in a year is not exact, so the result has the same number of significant figures as the conversion factor for days/yr.

1.91   Your list might include questions about which you have a special interest. Our list includes pollution questions, such as: how to reverse the processes that form the ozone hole; health questions, such as: how to design new drugs to combat bacterial infections and how important for our health are trace elements in our diets; biochemical questions, such as: what molecular processes take place when we smell an odor; and technological questions, such as: how to design materials that are high-temperature superconductors.

1.93   Relative densities can be determined by observing whether materials float or sink.
(a) In a bottle of oil and vinegar, oil floats on the vinegar; thus vinegar has the higher density.
(b) Table salt added to water sinks; thus table salt has the higher density.

(c) Aluminum is much lighter in weight than iron; thus iron has the higher density.

2.1 Molecular pictures must show the structures of individual particles (e.g., atoms, molecules, etc.) and the differences between phases. All these particles have monatomic units. A gas is mostly empty space, a liquid is tightly packed but not entirely regular, and a solid has a regular repeating structure.

(a) gaseous helium  (b) solid tungsten (c) Liquid gallium

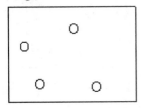 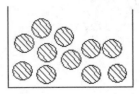

2.3 There are five features of atomic theory. They are:
(1) All matter is composed of tiny particles called atoms. In this reaction, all starting materials and products are made from atoms of carbon and oxygen.
(2) All atoms of a given element have identical chemical properties. All oxygen atoms are in diatomic molecules in the starting materials and CO molecules in the products; likewise, C atoms behave all in the same way.
(3) Atoms of different elements have distinct properties. During the course of the reaction, carbon atoms and oxygen atoms do different things. Also, oxygen is a diatomic gas whereas carbon is a solid.
(4) Atoms form chemical compounds by combining in whole number ratios. C and O combine in a 1:1 ratio to give CO molecules.
(5) In chemical reactions, atoms change the ways they are combined, but they are neither created nor destroyed. There are the same number of oxygen and carbon atoms in the reactants and products. However, they are bonded differently in the products (carbon bonded to oxygen) than in the reactants (oxygen bonded to oxygen, carbon to carbon).

2.5 The law of conservation of mass applies to the entire system, not to any individual item within that system. The system in this problem is the magnesium strip *and* the surrounding air. When magnesium burns in air, magnesium atoms combine with oxygen from the air to form magnesium oxide (the solid residue). The mass of the residue is therefore the total mass of the magnesium strip plus the mass of the oxygen that reacted with the magnesium. Therefore, the mass of the solid increases, but the mass of solid plus gas (the system) remains constant.

2.7 Molecular pictures must show the structures of individual molecules and the differences between phases. All of these pictures should contain diatomic molecules of bromine. A solid has a regular repeating structure, a liquid is tightly packed but not entirely regular, and a gas is mostly empty space.

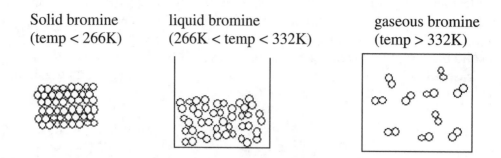

Solid bromine (temp < 266K)   liquid bromine (266K < temp < 332K)   gaseous bromine (temp > 332K)

2.9   The molecules that give roses their aroma continuously evaporate from the surface of the flower. Once in the gas phase, they collide countless times with other gas molecules, moving slowly away from the rose until, when they reach a nose, they are sensed by the olfactory sensors.

2.11   "Dynamic equilibrium" refers to the fact that molecules continue to be transformed from one phase or form to another even though no net change is taking place. In this example, iodine molecules sublime from the crystals at the bottom of the flask into the gas phase, where their presence imparts a pale violet color to the gas. Some of these molecules then condense on the surfaces of the flask, forming crystals. At any time, there are a constant number of molecules in the gas phase, but some molecules are subliming into the gas from the crystals while equal numbers of molecules are condensing onto the crystals.

2.13   Charge per electron and mass per electron can be found in Table 2-1. Because the total charge of the beam is simply the sum of the charges of the individual electrons, to determine the number of electrons, divide the total charge by the charge of a single electron:

(a) $\#_{\text{electrons}} = \dfrac{\text{total charge}}{\text{charge per e}^-} = \dfrac{-1.00 \times 10^{-6}\,\text{C}}{-1.6022 \times 10^{-19}\,\text{C/e}^-} = 6.24 \times 10^{12}\,\text{electrons}$

(b) To determine the combined mass of all of the electrons, multiply the number of electrons by the mass of a single electron:

$m_{\text{electrons}} = (\#_{\text{electrons}})(m_{\text{electron}}) = 6.24 \times 10^{12}\,\text{electrons} \left(\dfrac{9.1094 \times 10^{-31}\,\text{kg}}{1\,\text{electron}} \right) = 5.68 \times 10^{-18}\,\text{kg}$

2.15   In the Millikan experiment, gravitational force was balanced by electrical force for some of the charged particles. Changing the polarity would reverse the direction of the electrical force for all of the charged particles. Negatively-charged particles would be attracted downward, but now positively-charged particles would be attracted upward. Under these conditions, the proper electrical field would cause some of the positively-charged particles to be suspended in space. Because each of these particles had lost one or more electrons, the charge on the electron could be determined from these observations.

2.17   Charge per proton and mass per proton can be found in Table 2-1.

$$\#_{protons} = 1.5\,g\left(\frac{10^{-3}\,kg}{1\,g}\right)\left(\frac{1\,proton}{1.6726 \times 10^{-27}\,kg}\right) = 9.0 \times 10^{23}\ protons$$

To determine the charge, multiply the number of protons by the charge per proton:

$$\text{Charge} = (9.0 \times 10^{23}\,protons)(1.6022 \times 10^{-19}\,C/p) = 1.4 \times 10^{5}\,C$$

2.19   Use the properties of electrons in Table 2-1 to answer this question.
(a) The charge of a proton is equal in magnitude to that of an electron. Therefore, to achieve neutrality (a charge of zero), there must be two electrons to balance the two protons.
(b) There are two ways to determine the fraction of a He atom's mass due to its electrons:
   (1)   Use the proton, neutron, and electron masses to calculate the total mass of the He atom and the mass of two electrons,

$$m_{total} = 2(1.6726 \times 10^{-27}\,kg) + 2(1.6749 \times 10^{-27}\,kg) + 2(9.1094 \times 10^{-31}\,kg)$$

$$m_{total} = 6.6968 \times 10^{-27}\,kg;$$

$$m_{2electron} = 2(9.1094 \times 10^{-31}\,kg) = 1.8219 \times 10^{-30}\,kg$$

To determine the fraction, divide the mass of two electrons by the mass of the He atom:

$$\text{electron mass fraction} = \frac{m_{2electron}}{m_{total}} = \frac{1.82188 \times 10^{-30}\,kg}{6.6968 \times 10^{-27}\,kg} = 2.7205 \times 10^{-4}$$

   (2)   The second method is to use the atomic mass of He (4.0026g/mol) to determine the mass of a He atom:

$$m_{total} = \left(\frac{4.0026\,g}{1\,mol}\right)\left(\frac{1\,kg}{1000\,g}\right)\left(\frac{1\,mol}{6.022 \times 10^{23}\,atom}\right) = 6.647 \times 10^{-27}\,kg$$

Again divide the mass of two electrons by the mass of the He atom to determine the mass fraction:

$$\text{electron mass fraction} = \frac{m_{2electron}}{m_{total}} = \frac{1.82188 \times 10^{-30}\,kg}{6.647 \times 10^{-27}\,kg} = 2.741 \times 10^{-4}$$

These two results differ because, as described in Chapter 21 of your textbook, the mass of an atom is not exactly equal to the sum of the masses of its constituent protons, neutrons, and electrons.

2.21   The left superscript in an isotopic symbol is the mass number, A, which is the sum of protons and neutrons; the left subscript is the atomic number, Z, which is the number of protons, and a right superscript indicates (protons – electrons), when this quantity is non-zero.
(a) 8 protons, 8 neutrons, 10 electrons; (b) 5 p, 6 n, 5 e; (c) 25 p, 30 n, 22 e; (d) 17 p, 18 n, 18 e; and (e) 17 p, 20 n, 16 e.

2.23   An isotopic symbol has the elemental symbol prefaced by a superscript giving the mass number, A (the sum of protons and neutrons), and a subscript giving the atomic number, Z (the number of protons). Use the periodic table in your text to determine which elemental symbol corresponds to the atomic number.

(a) # protons = Z = 26 (iron, Fe). $A = 26 + 30 = 56$. $^{56}_{26}$Fe;

(b) From the periodic table, the atomic number for uranium is 92; $^{236}_{92}$U;

(c) Argon (Ar) has 18 protons, and 20 neutrons. $A = 18 + 20 = 38$; $^{38}_{18}$Ar;

(d) $Z = 9$. $A = 9 + 10 = 19$; The element with 9 protons is fluorine (F); $^{19}_{9}$F.

2.25   A mass spectrum should have the x-axis as the mass number and the y-axis as the intensity (or relative abundance). There are two isotopes, so the mass spectrum has two peaks. The mass-11 peak is 4 times as intense as the mass-10 peak.

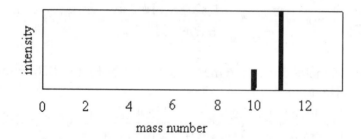

2.27   A pie chart is a circular chart with each piece of the pie proportional in size to the percentage of the isotope that it represents.

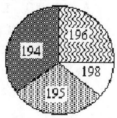

2.29 Cations are positively charged, anions are negatively charged, and neutral species carry no charge. Charge is designated by the right superscript (no superscript indicates a neutral species).

cations: Cl_2^+, CO^+, Cr^{3+}; anions: Cl^-, $Cr_2O_7^{2-}$; and neutrals: C, CCl_4, CO_2.

2.31 The chemical formula for a charged species has a right superscript charge.

(a) $OH^- + H^+ = H_2O$; (b) $Na - e^- = Na^+$; (c) $HCl - H^+ = Cl^-$; and (d) $O + 2e^- = O^{2-}$.

2.33 A molecular picture for a chemical process should follow the five features of atomic theory. You should use labeling, coloring (used here), or shading to differentiate the different atom types.

Hydroxide + hydrogen ion → water

HCl losing a hydrogen ion → chloride ion + hydrogen ion

2.35 Elements in Groups 1 and 2 easily lose one and two electron(s), respectively. Those in Groups 16 and 17 easily gain two and one electron(s), respectively.

(a) Rb^+; (b) F^-; and (c) Ba^{2+}.

2.37 Cations and anions combine to form neutral compounds. In any neutral compound, the total amount of positive charge must match the total amount of negative charge. The compounds are: RbF and BaF_2.

2.39 A molecular picture of a solution of a salt contains cations, anions, and water molecules. The total positive charge on the cations should be equal to the total negative charge on the anions. Thus there should be equal numbers of Na^+ and I^- ions.

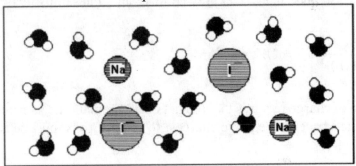

2.41 Any ionic compound must be electrically neutral. Remember that Group 16 and 17 elements form –2 and –1 ions, respectively. The charges on the ions are Al = +3, O = -2, and F = -1. Two Al ions balance the charge of three O ions and three F ions balance the charge of one Al ion.
Aluminum oxide: Al_2O_3
Aluminum fluoride: AlF_3.

2.43 Conservation of energy states that energy is neither created nor destroyed. It can only be transformed from one form into another.
(a) Total energy, which is conserved, is the sum of potential, kinetic, and thermal energy. On the tree, an apple possesses some gravitational potential energy. As an apple falls, that potential energy is converted into kinetic energy of motion.

(b) When an apple hits the ground, the impact transfers energy to molecules in the earth and in the apple. As a result, there is a slight increase in temperature; the kinetic energy of the apple has been converted into thermal energy.

2.45 Speed and kinetic energy are related through the equation, $E_{kinetic} = \frac{1}{2} mu^2$. Note, $1 J = 1 kg\ m^2/s^2$.
$$E_{kinetic} = \frac{(9.1094 \times 10^{-31} kg)(4.55 \times 10^5 m/s)^2}{2} = 9.43 \times 10^{-20} J$$

2.47 Speed and kinetic energy are related through the equation, $E_{kinetic} = \frac{1}{2} mu^2$. Note, $1 J = 1 kg\ m^2/s^2$.
$$E_{kinetic} = \frac{(1.6749 \times 10^{-27} kg)(4.55 \times 10^5 m/s)^2}{2} = 1.73 \times 10^{-16} J$$

2.49 Energy comes in various forms: radiant, kinetic, potential, and thermal.
(a) Radiant energy is consumed and thermal energy is produced.
(b) Gravitational potential energy is consumed and kinetic energy is produced.
(c) Chemical potential energy is consumed and thermal energy is produced.

2.51 Speed and kinetic energy are related through the equation, $u = \sqrt{\dfrac{2E_{kinetic}}{m}}$. The joule has units kg m²/s², so mass must be expressed in kg and u has units m/s.
m (Table 2-1) = 1.6749 x 10⁻²⁷ kg;
$$u = \sqrt{\frac{2(3.75 \times 10^{-23} J)}{1.6749 \times 10^{-27} kg}} = 2.12 \times 10^2 m/s.$$

2.53 Conservation of charge requires that the net charge remain the same during any process.
(a) The products have six negative charges, so six electrons must be added to the neutral N_2 molecule.

(b) An oxalate ion has two negative charges, so two electrons must be removed during decomposition.

(c) The products have equal numbers of positive and negative charge, so charge is conserved.

2.55   Because the ratio of nitrogen to oxygen is 4:1, your picture should show eight nitrogen molecules and two oxygen molecules, randomly distributed in the space available.

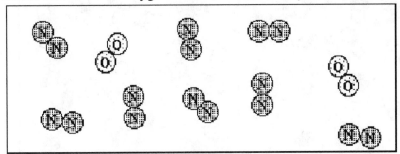

2.57   A mass spectrum contains a peak at each isotopic mass, with peak heights proportional to the isotopic abundances (given in the pie chart). For tin there are a total of 10 peaks, 3 of which are very small and hard to see (peaks at masses 112, 114, and 115), the rest are easily seen.

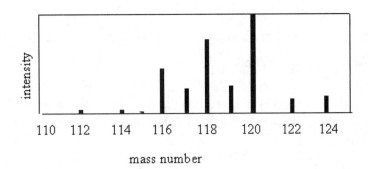

2.59   The upper left superscript in an isotopic symbol represents the sum of protons and neutrons. The number of protons can be deduced from the elemental symbol, with the help of a periodic table.

(a) 43 protons, 56 neutrons, 43 electrons; (b) 26 p, 26 n, 26 e; (c) 54 p, 79 n, 54 e; and (d) 53 p, 78 n, 53e.

2.61   Kinetic energy is given by $E_{kinetic} = \frac{1}{2} mu^2$. , but to obtain energy in joules, we must convert to SI units.

$$2250 \text{ lb} \left(\frac{0.45359 \text{ kg}}{1 \text{ lb}} \right) = 1.021 \times 10^3 \text{ kg};$$

$$\left(\frac{57.5 \text{ mi}}{1 \text{ hr}} \right) \left(\frac{1.609344 \text{ km}}{1 \text{ mi}} \right) \left(\frac{10^3 \text{ m}}{1 \text{ km}} \right) \left(\frac{1 \text{ hr}}{60 \text{ min}} \right) \left(\frac{1 \text{ min}}{60 \text{ s}} \right) = 25.7 \text{ m/s};$$

$$E_{kinetic} = \frac{(1.021 \times 10^3 \text{ kg})(25.7 \text{ m/s})^2}{2} = 3.37 \times 10^5 \text{ kg m}^2/\text{s}^2 = 3.37 \times 10^5 \text{ J}.$$

2.63   A salt is composed of cations and anions, which exist as individual ions in solution.

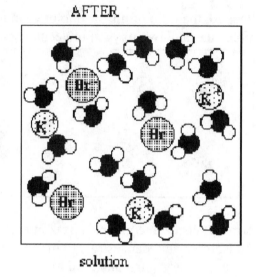

2.65   Each of these isobars has mass number 40, so the upper left superscript is 40.
$^{40}_{17}$Cl, $^{40}_{18}$Ar, $^{40}_{19}$K, $^{40}_{20}$Ca, and $^{40}_{21}$Sc.

2.67   An atomic mass spectrum contains a peak at each isotopic mass, and the peak heights are proportional to the isotopic abundances. The ratio for masses 24:25:26 of magnesium is approximately 10:1:1.

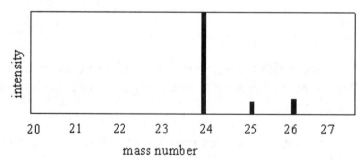

2.69　The atomic number of an element, which equals the number of protons in its nucleus, can be found from the atomic symbol and a periodic table. The right-hand superscript in an ionic symbol represents (protons – electrons).
(a) 11 protons, 10 electrons; (b) 7 p, 10 e; (c) 22 p, 18 e; and (d) 53 p, 54 e.

2.71　Charge must be conserved, and every chemical compound must be electrically neutral. Thus, two chlorine atoms gain an electron for each calcium atom that loses two electrons.

(a) Ca^{2+}, Cl^-;　(b) $CaCl_2$;

(c)

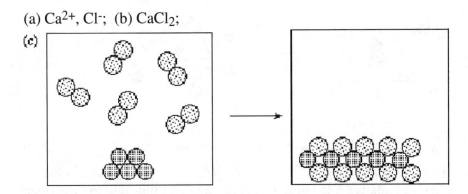

2.73　In a crystalline solid, the ions lie in a regular array, in this case cubic. In the liquid phase, the ions are still close together but the arrangement is no longer entirely regular.

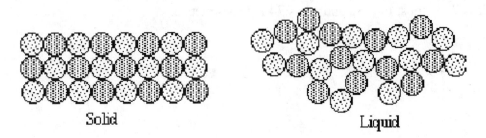

Solid　　　　　　　　　Liquid

2.75　Diatomic bromine in the liquid phase will have molecules close together with little space between them, in the gas phase there is a lot of space between molecules.

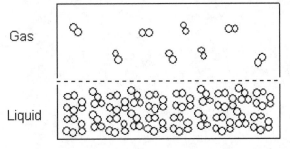

Gas

Liquid

2.77　Only the elements with atomic numbers that are a multiple of 4 can have exactly 1.25 times as many neutrons as protons: $^{45}_{20}Ca$, $^{54}_{24}Cr$, $^{63}_{28}Ni$, $^{72}_{32}Ge$, $^{81}_{36}Kr$, and $^{90}_{40}Zr$.

2.79   The molecular picture is not correct because atoms are not conserved. Molecular pictures of reactions must show that atoms of each element are conserved. There must be equal numbers of atoms of each element on the two sides of the arrow. The question asks you to adjust the right-hand side to make the figure correct. On the left-hand side there are eight big atoms and 12 small atoms, so the right-hand side must show 2 unreacted small molecules:

2.81   There are two isotopes of chlorine, so Cl_2 can be ^{35}Cl–^{35}Cl, ^{35}Cl–^{37}Cl, ^{37}Cl–^{35}Cl, or ^{37}Cl–^{37}Cl. These have mass 70, 72, 72, and 74, so there are three peaks in the mass spectrum. To find the relative intensities of the three peaks, multiply the fractional isotopic abundances:
mass 70, $(0.7577)(0.7577) = 0.5741$;
mass 72, $(0.7577)(0.2423) + (0.2423)(0.7577) = 0.3672$; and
mass 74, $(0.2423)(0.2423) = 0.0587$.
Mass 70 is most intense and mass 74 is least intense.

3.1 Determine the chemical formula from a ball-and-stick model by counting balls of each color and consulting the color code for the elements (Figure 1-4 of your textbook). (a) CH_4; (b) C_2H_4; (c) C_2H_6O; (d) HBr; (e) PCl_3; (f) CH_4N_2O; and (g) C_2H_5I.

3.3 Structural formulas look like ball-and-stick structures but with lines in place of sticks and elemental symbols in place of balls.

(a) (b) 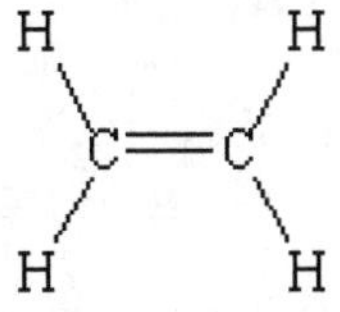 (c) 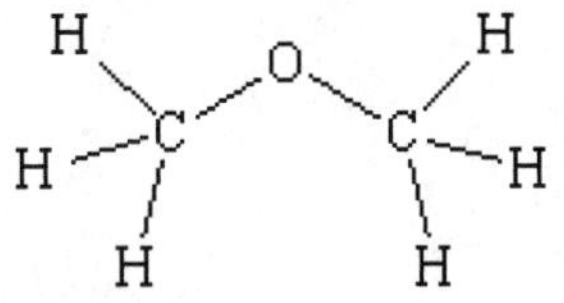 (d)

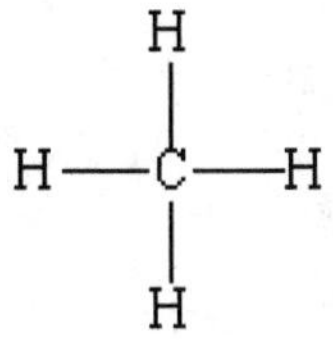

(e) 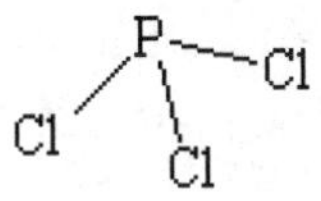 (f) 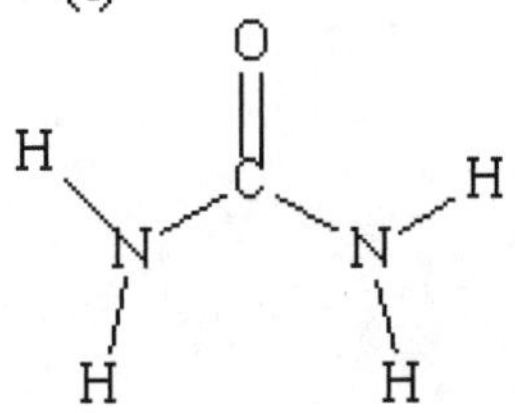 (g)

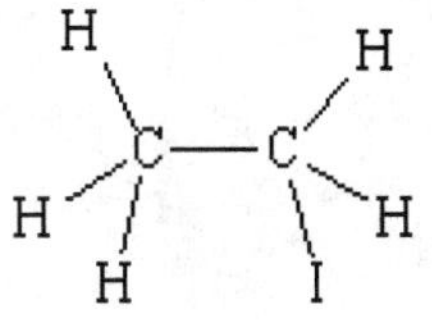

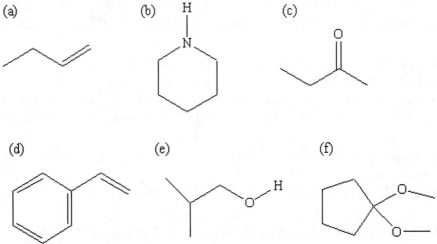

3.5 To convert a structural formula into a line structure, remove all –H connections to carbon atoms and remove all C atom labels.

(a) (b) H (c)

(d) (e) (f)

3.7   To convert a line structure into a structural formula, place a C atom at the end of each line and at each line intersection, then add enough –H connections to give each C atom 4 connections.

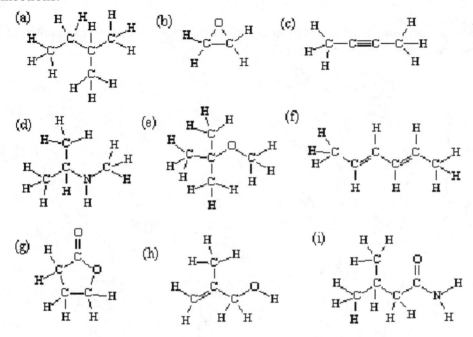

3.9   Determine a chemical formula from a space-filling model by counting atoms of each color and using the color code (Figure 1-4 of your textbook) to identify the elements. Name binary compounds using the rules given in Section 3-2 of your textbook.
(a) CO_2, carbon dioxide; (b) HCl, hydrogen chloride; and (c) CCl_4, carbon tetrachloride.

3.11   Determine a chemical formula from a ball-and-stick model by counting atoms of each color and using the color code (Figure 1-4 of your textbook) to identify the elements. Name binary compounds using the rules given in Section 3-2 of your textbook.
(a) H_2S, hydrogen sulfide; (b) SF_4, sulfur tetrafluoride; and (c) HF, hydrogen fluoride.

3.13   Identify the compound as binary or carbon-based. Then apply the rules given in Section 3.2 of your textbook.
(a) CH_4; (b) HF; (c) CaH_2; (d) PCl_3; (e) N_2O_5; (f) SF_6; (g) BF_3.

3.15   Identify the compound as binary or carbon-based. Then apply the rules given in Section 3.2 of your textbook.
(a) disulfur dichloride; (b) iodine heptafluoride; (c) hydrogen bromide;
(d) dinitrogen trioxide; (e) silicon carbide; (f) methanol.

3.17   Ionic compounds can be identified from the presence of either metals from Group 1 or Group 2 of the periodic table or polyatomic ions. Any compound is neutral, so the formula for an ionic compound must contain equal numbers of positive and negative charges.
(a) not ionic, HF;

(b) Group 2 metal (+2) and Group 17 element (-1), ionic, CaF_2;

(c) polyatomic anion (-2) and +3 metal, ionic, $Al_2(SO_4)_3$;

(d) ammonium cation (+1) and Group 16 element (-2), ionic, $(NH_4)_2S$;

(e) not ionic, SO_2; and

(f) not ionic, CCl_4.

3.19 Identify the compound as ionic, binary or carbon-based. Then apply the rules given in Section 3.2 of your textbook. Ionic compounds can be identified from the presence of either metals from Group 1 or Group 2 of the periodic table or polyatomic ions.
(a) not ionic, carbon-based, dichloromethane;
(b) not ionic, binary, carbon dioxide;
(c) Group 2 metal, ionic, calcium oxide;
(d) polyatomic anion, ionic, potassium carbonate;
(e) not ionic, binary, phosphorus tribromide;
(f) not ionic, binary, hydrogen fluoride; and
(g) polyatomic anion, ionic, sodium hydrogen phosphate.

3.21 To determine a chemical formula from the name of a compound requires knowledge of the polyatomic anions and charges. The compound must be electrically neutral.
(a) Na is a +1 cation and SO_4 is a -2 anion, Na_2SO_4;
(b) K is a +1 cation and S is a -2 anion, K_2S;
(c) K is a +1 cation and H_2PO_4 is a -1 anion, KH_2PO_4;
(d) Co is a +2 cation, F is a -1 anion, $CoF_2 \cdot 4H_2O$;
(e) Pb is a +4 metal and O is a -2 anion, PbO_2;
(f) Na is a +1 cation and HCO_3 is a -1 anion, $NaHCO_3$; and
(g) Li is a +1 cation and BrO_4 is a -1 anion, $LiBrO_4$.

3.23 Follow the rules for naming ionic compounds (cation first, then anion), using Roman numerals for cation charge only when multiple possibilities exist.
(a) calcium chloride hexahydrate; (b) iron(II) ammonium sulfate;
(b) potassium carbonate; (d) tin(II) chloride dihydrate; (e) sodium hypochlorite;
(a) silver sulfate; (g) copper(II) sulfate; (h) potassium dihydrogen phosphate;
(i) sodium nitrate; (j) calcium sulfite; and (k) potassium permanganate.

3.25 Mass-mole-number conversions require the use of masses in grams, molar masses, and/or the Avogadro constant. To determine the number of moles, convert the mass into grams and divide by the molar mass.

(a) $n = \dfrac{m}{MM} = 7.85\,g\left(\dfrac{1\,mol}{55.85\,g}\right) = 0.141$ mol;

(b) $n = \dfrac{m}{MM} = 65.5\,\mu g\left(\dfrac{10^{-6}\,g}{1\,\mu g}\right)\left(\dfrac{1\,mol}{12.01\,g}\right) = 5.45 \times 10^{-6}$ mol;

(c) $n = \dfrac{m}{MM} = 4.68\,mg\left(\dfrac{10^{-3}\,g}{1\,mg}\right)\left(\dfrac{1\,mol}{28.09\,g}\right) = 1.67 \times 10^{-4}$ mol; and

(d) $n = \dfrac{m}{MM} = 1.46\,\text{ton}\left(\dfrac{10^3\,\text{kg}}{1\,\text{ton}}\right)\left(\dfrac{10^3\,\text{g}}{1\,\text{kg}}\right)\left(\dfrac{1\,\text{mol}}{26.98\,\text{g}}\right) = 5.41 \times 10^4\,\text{mol}.$

3.27 The molar mass of a naturally occurring element can be calculated by summing the product of the fractional abundance of each isotope times its isotopic molar mass.

^{36}Ar: $35.96755\,\text{g/mol}\left(\dfrac{0.337\%}{100\%}\right) = 0.121\,\text{g/mol};$

^{38}Ar: $37.96272\,\text{g/mol}\left(\dfrac{0.063\%}{100\%}\right) = 0.024\,\text{g/mol};$

^{40}Ar: $39.9624\,\text{g/mol}\left(\dfrac{99.600\%}{100\%}\right) = 39.803\,\text{g/mol};$

Elemental molar mass = 0.121 + 0.024 + 39.803 = 39.948 g/mol.

3.29 To calculate the molar mass of a compound, multiply each elemental molar mass by the number of atoms in the formula and sum over the elements.

(a) CCl_4, $MM = (12.01\,\text{g/mol C}) + 4(35.45\,\text{g/mol Cl}) = 153.8\,\text{g/mol};$
(b) K_2S, $MM = 2(39.10\,\text{g/mol K}) + 32.07\,\text{g/mol S} = 110.27\,\text{g/mol};$
(c) O_3, $MM = 3(16.00\,\text{g/mol O}) = 48.00\,\text{g/mol};$
(d) $LiBr$, $MM = 6.94\,\text{g/mol Li} + 79.90\,\text{g/mol Br} = 86.84\,\text{g/mol};$
(e) $GaAs$, $MM = 69.72\,\text{g/mol Ga} + 74.92\,\text{g/mol As} = 144.64\,\text{g/mol};$ and
(f) $AgNO_3$, $MM = 107.87\,\text{g/mol Ag} + 14.01\,\text{g/mol N} + 3(16.00\,\text{g/mol O}) = 169.88\,\text{g/mol}.$

3.31 Determine the molecular formula from the line drawing, taking into account the "missing" carbon atoms at the ends and vertices of lines and the "missing" hydrogen atoms attached to carbon atoms. To calculate the molar mass, multiply each elemental molar mass by the number of atoms in the formula and sum over the elements.

tyrosine: 9 C atoms, 4 missing H atoms, 7 shown H atoms, 1 N atom, 3 O atoms; $C_9H_{11}NO_3$,
$MM = 9(12.01\,\text{g/mol}) + 11(1.008\,\text{g/mol}) + 1(14.01\,\text{g/mol}) + 3(16.00\,\text{g/mol}) = 181.19\,\text{g/mol};$

tryptophan: 11 C atoms, 5 missing H atoms, 7 shown H atoms, 2 N atoms, 2 O atoms; $C_{11}H_{12}N_2O_2$,
$MM = 11(12.01\,\text{g/mol}) + 12(1.008\,\text{g/mol}) + 2(14.01\,\text{g/mol}) + 2(16.00\,\text{g/mol}) = 204.23\,\text{g/mol};$

glutamic acid: 5 C atoms, 9 shown H atoms, 1 N atom, 4 O atoms; $C_5H_9NO_4$,
$MM = 5(12.01\,\text{g/mol}) + 9(1.008\,\text{g/mol}) + 1(14.01\,\text{g/mol}) + 4(16.00\,\text{g/mol}) = 147.13\,\text{g/mol};$

lysine: 6 C atoms, 14 shown H atoms, 2 N atoms, 2 O atoms; $C_6H_{14}N_2O_2$,

$$MM = 6(12.01\text{g/mol}) + 14(1.008\text{g/mol}) + 2(14.01\text{g/mol}) + 2(16.00\text{g/mol}) = 146.19 \text{ g/mol}.$$

3.33 To calculate the number of atoms in a mass, convert the mass to grams, divide by molar mass to obtain moles and multiply by the Avogadro constant to obtain number.

(a) $\# = 5.86 \text{ mg}\left(\dfrac{10^{-3}\text{g}}{1\text{mg}}\right)\left(\dfrac{1\text{mol}}{9.012\text{g}}\right)\left(\dfrac{6.022 \times 10^{23}\text{atoms}}{1\text{mol}}\right) = 3.92 \times 10^{20} \text{ atoms};$

(b) $\# = 5.86 \text{ mg}\left(\dfrac{10^{-3}\text{g}}{1\text{mg}}\right)\left(\dfrac{1\text{mol}}{30.97\text{g}}\right)\left(\dfrac{6.022 \times 10^{23}\text{atoms}}{1\text{mol}}\right) = 1.14 \times 10^{20} \text{ atoms};$

(c) $\# = 5.86 \text{ mg}\left(\dfrac{10^{-3}\text{g}}{1\text{mg}}\right)\left(\dfrac{1\text{mol}}{91.22\text{g}}\right)\left(\dfrac{6.022 \times 10^{23}\text{atoms}}{1\text{mol}}\right) = 3.87 \times 10^{19} \text{ atoms};$

(d) $\# = 5.86 \text{ mg}\left(\dfrac{10^{-3}\text{g}}{1\text{mg}}\right)\left(\dfrac{1\text{mol}}{238.0\text{g}}\right)\left(\dfrac{6.022 \times 10^{23}\text{atoms}}{1\text{mol}}\right) = 1.48 \times 10^{19} \text{ atoms}.$

3.35 To calculate the mass of some number of molecules of a substance, divide by the Avogadro constant to obtain moles and multiply by molar mass to obtain grams.

(a) CH_4, $MM = 12.01 \text{ g/mol} + 4(1.008 \text{ g/mol}) = 16.04 \text{ g/mol};$

$m = 3.75 \times 10^5 \text{ molecules}\left(\dfrac{1\text{mol}}{6.022 \times 10^{23}\text{molecules}}\right)\left(\dfrac{16.04\text{g}}{1\text{mol}}\right) = 9.99 \times 10^{-18} \text{ g};$

(b) $C_9H_{13}NO_3$,
$MM = 9(12.01 \text{ g/mol}) + 13(1.008 \text{ g/mol}) + 14.01 \text{ g/mol} + 3(16.00\text{g/mol}) = 183.2 \text{ g/mol};$

$m = 2.5 \times 10^9 \text{ molecules}\left(\dfrac{1\text{mol}}{6.022 \times 10^{23}\text{molecules}}\right)\left(\dfrac{183.2\text{g}}{1\text{mol}}\right) = 7.6 \times 10^{-13} \text{ g};$

(c) $C_{55}H_{72}MgN_4O_5$,
$MM = 55(12.01\text{g/mol}) + 72(1.008\text{g/mol}) + 24.305\text{g/mol} + 4(14.01\text{g/mol}) + 5(16.00\text{g/mol})$
$MM = 893.5 \text{ g/mol};$

$m = 1 \text{ molecule}\left(\dfrac{1\text{mol}}{6.022 \times 10^{23}\text{molecules}}\right)\left(\dfrac{893.5\text{g}}{1\text{mol}}\right) = 1.484 \times 10^{-21} \text{ g};$

3.37 All parts of this question involve mass-mole-number conversions. Moles and mass in grams are related through the equation, $n = \dfrac{m}{MM}$. Use the Avogadro constant to convert between number and moles. When masses are not given in grams, unit conversions must be made. The chemical formula states the number of atoms of each element per molecule of substance.
MM of vitamin A = $20(12.01 \text{ g/mol}) + 30(1.008 \text{ g/mol}) + 16.00\text{g/mol} = 286.44 \text{ g/mol}$

$n = 0.75 \text{ mg}\left(\dfrac{10^{-3}\text{g}}{1\text{mg}}\right)\left(\dfrac{1\text{mol}}{286.44\text{g}}\right) = 2.6 \times 10^{-6} \text{ mol of vitamin A}$

$\#_{molecules} = nN_A = 2.6 \times 10^{-6} mol(6.022 \times 10^{23} molecules/mol) = 1.6 \times 10^{18}$ molecules;

There are 30 H atoms for every molecule of vitamin A:

\# atoms = (atoms/molecule)(\# molecules)

$$\#_H = 1.6 \times 10^{18} molecules\left(\frac{30 \text{ atom H}}{1 \text{ molecules}}\right) = 4.8 \times 10^{19} \text{ atoms of H in 0.75mg of vitamin A;}$$

$$m_H = \frac{\#H}{N_A}MM = 4.8 \times 10^{19} \text{ atoms}\left(\frac{1 \text{ mol}}{6.022 \times 10^{23} \text{ atoms}}\right)\left(\frac{1.008 \text{ g}}{1 \text{ mol}}\right) = 8.0 \times 10^{-5} \text{ g or}$$

$m_H = 8.0 \times 10^{-2}$ mg

3.39   To calculate mass percent composition, which involves mass ratios, work with one mole of substance. Take the ratio of the mass of each element that one mole contains to the mass of one mole (molar mass).

CaO: MM = 40.08 g/mol + 16.00 g/mol = 56.08 g/mol;

$$\% \text{ Ca} = \frac{40.08 \text{ g/mol Ca}}{56.08 \text{ g/mol CaO}} \times 100\% = 71.47 \%;$$

$$\% \text{ O} = \frac{16.00 \text{ g/mol O}}{56.08 \text{ g/mol CaO}} \times 100\% = 28.53 \%;$$

SiO_2: MM = 28.09 g/mol + 2(16.00 g/mol) = 60.09 g/mol;

$$\% \text{ Si} = \frac{28.09 \text{ g/mol Si}}{60.09 \text{ g/mol SiO}_2} \times 100\% = 46.75 \%;$$

$$\% \text{ O} = \frac{2(16.00 \text{ g/mol O})}{60.09 \text{ g/mol SiO}_2} \times 100\% = 53.25 \%;$$

Al_2O_3: MM = 2(26.98 g/mol) + 3(16.00 g/mol) = 101.96 g/mol;

$$\% \text{ Al} = \frac{2(26.98 \text{ g/mol Al})}{101.96 \text{ g/mol Al}_2O_3} \times 100\% = 52.92 \%;$$

$$\% \text{ O} = \frac{3(16.00 \text{ g/mol O})}{101.96 \text{ g/mol Al}_2O_3} \times 100\% = 47.08 \%;$$

Fe_2O_3: MM = 2(55.85 g/mol) + 3(16.00 g/mol) = 159.70 g/mol;

$$\% \text{ Fe} = \frac{2(55.85 \text{ g/mol Fe})}{159.70 \text{ g/mol Fe}_2O_3} \times 100\% = 69.94 \%;$$

$$\% \text{ O} = \frac{3(16.00 \text{ g/mol O})}{159.70 \text{ g/mol Fe}_2O_3} \times 100\% = 30.06 \%.$$

3.41 When working with mass percent compositions, it is convenient to consider 100 g of the substance. Divide the mass of each element by its elemental molar mass to obtain relative amounts of each element in the substance.

$$C: \left(\frac{74.0\,g}{100\,g} \right) \left(\frac{1\,mol}{12.01\,g} \right) = 6.16\ mol\ C/100\ g;$$

$$H: \left(\frac{8.65\,g}{100\,g} \right) \left(\frac{1\,mol}{1.008\,g} \right) = 8.58\ mol\ H/100\ g;$$

$$N: \left(\frac{17.35\,g}{100\,g} \right) \left(\frac{1\,mol}{14.01\,g} \right) = 1.24\ mol\ N/100\ g;$$

To obtain the empirical formula divide each amount by the smallest among them (nitrogen with 1.24 mol/100g) and round to whole numbers. This tells us the number of each element relative to one nitrogen atom in the compound:

$$C: \left(\frac{6.16\,mol\,C}{100\,g} \right) \left(\frac{100\,g}{1.24\,mol\,N} \right) = 4.97\ mol\ C/mol\ N,\ round\ to\ 5\ C/N;$$

$$H: \left(\frac{8.58\,mol\,H}{100\,g} \right) \left(\frac{100\,g}{1.24\,mol\,N} \right) = 6.92\ mol\ H/mol\ N,\ round\ to\ 7\ H/N;$$

Empirical formula: C_5H_7N, empirical MM = 81.12 g/mol;

To convert an empirical formula to a molecular formula, multiply the subscripts in the empirical formula by the ratio of the compound's molar mass to its empirical formula mass.

$$MM/empirical\ MM = \left(\frac{162\,g/mol}{81.12\,g/mol} \right) = 2;\ Molecular\ formula:\ C_{10}H_{14}N_2.$$

3.43 The problem statement identifies this as an empirical formula problem. Use the combustion data to determine the mass percent composition of the burned sample, then use elemental molar masses to find the empirical formula. Assume that the only source of C in CO_2 and the only source of H in H_2O is from the unknown compound.

$$m\ C = 11.8\ g\ CO_2 \left(\frac{1\,mol\,CO_2}{44.01\,g\,CO_2} \right) \left(\frac{1\,mol\,C}{1\,mol\,CO_2} \right) \left(\frac{12.01\,g\,C}{1\,mol\,C} \right) = 3.22\ g\ C;$$

$$\%\ C = \frac{3.22\,g}{5.00\,g} \times 100\% = 64.4\ \%;$$

$$m\ H = 2.42\ g\ H_2O \left(\frac{1\,mol}{18.02\,g} \right) \left(\frac{2\,mol\,H}{1\,mol\,H_2O} \right) \left(\frac{1.008\,g}{1\,mol} \right) = 0.271\ g;$$

$$\% \text{ H} = \frac{0.271 \text{ g}}{5.00 \text{ g}} \times 100\% = 5.42 \text{ \%};$$

Since the unknown contains only C, H, and Fe, the % composition of Fe must be the remainder:

$$\% \text{ Fe} = 100 \text{ \%} - (64.4 \text{ \%} + 5.42 \text{ \%}) = 30.2 \text{ \%};$$

$$\text{C: } 64.4 \text{ g}\left(\frac{1 \text{ mol}}{12.01 \text{ g}}\right) = 5.36 \text{ mol C};$$

$$\text{H: } 5.42 \text{ g}\left(\frac{1 \text{ mol}}{1.008 \text{ g}}\right) = 5.38 \text{ mol H};$$

$$\text{Fe: } 30.2 \text{ g}\left(\frac{1 \text{ mol}}{55.85 \text{ g}}\right) = 0.541 \text{ mol Fe};$$

Divide each by the smallest among them, 0.541 mol Fe:

$$\frac{5.36 \text{ mol C}}{0.541 \text{ mol Fe}} = 9.91 \text{ C/Fe, round to 10};$$

$$\frac{5.38 \text{ mol H}}{0.541 \text{ mol Fe}} = 9.94 \text{ H/Fe, round to 10};$$

The empirical formula is $C_{10}H_{10}Fe$.

3.45   There is much interesting information provided about allicin, but the question asks you to determine its molecular formula, for which the essential data are the combustion analysis and the approximate molar mass. Assume that the only source of C in CO_2, H in H_2O, and S in SO_2 is from the unknown compound.

$$m \text{ C} = 8.13 \text{ mg CO}_2\left(\frac{1 \text{ mol CO}_2}{44.01 \text{ g CO}_2}\right)\left(\frac{1 \text{ mol C}}{1 \text{ mol CO}_2}\right)\left(\frac{12.01 \text{ g C}}{1 \text{ mol C}}\right) = 2.219 \text{ mg};$$

$$\% \text{ C} = \frac{2.219 \text{ mg}}{5.00 \text{ mg}} \times 100\% = 44.37 \text{ \%};$$

$$m \text{ H} = 2.76 \text{ mg H}_2\text{O}\left(\frac{1 \text{ mol}}{18.02 \text{ g}}\right)\left(\frac{2 \text{ mol H}}{1 \text{ mol H}_2\text{O}}\right)\left(\frac{1.008 \text{ g}}{1 \text{ mol}}\right) = 0.3088 \text{ mg};$$

$$\% \text{ H} = \frac{0.3088 \text{ mg}}{5.00 \text{ mg}} \times 100\% = 6.18 \text{ \%};$$

$$m \text{ S} = 3.95 \text{ mg SO}_2\left(\frac{1 \text{ mol}}{64.07 \text{ g}}\right)\left(\frac{1 \text{ mol S}}{1 \text{ mol SO}_2}\right)\left(\frac{32.07 \text{ g}}{1 \text{ mol}}\right) = 1.977 \text{ mg};$$

$$\% \text{ S} = \frac{1.977 \text{ mg}}{5.00 \text{ mg}} \times 100\% = 39.54 \text{ \%};$$

(note, an additional significant figure is used in %C and %S to minimize round-off errors)

Determine the mass percent of O by subtracting the other mass percents from 100%:
% O = 100 % – (44.37 % + 6.18 % + 39.54 %) = 9.91 %;

Next assume you have 100g of the compound and determine the number of moles of each element that are present:

C: $44.37 \text{ g}\left(\dfrac{1 \text{ mol}}{12.01 \text{ g}}\right) = 3.69$ mol C;

H: $6.18 \text{ g}\left(\dfrac{1 \text{ mol}}{1.008 \text{ g}}\right) = 6.13$ mol H;

S: $39.54 \text{ g}\left(\dfrac{1 \text{ mol}}{32.07 \text{ g}}\right) = 1.23$ mol S;

O: $9.91 \text{ g}\left(\dfrac{1 \text{ mol}}{16.00 \text{ g}}\right) = 0.619$ mol O

Divide each by the smallest among them, 0.619 mol O, to determine the relative abundance:

$\dfrac{3.69 \text{ mol C}}{0.619 \text{ mol O}} = 5.96$ C/O, round to 6;

$\dfrac{6.13 \text{ mol H}}{0.619 \text{ mol O}} = 9.90$ H/O, round to 10;

$\dfrac{1.23 \text{ mol S}}{0.619 \text{ mol O}} = 1.99$ S/O, round to 2;

The empirical formula is $C_6H_{10}OS_2$, with an empirical molar mass of:
MM = 6(12.01 g/mol) + 10(1.008 g/mol) + 16.00g/mol + 2(32.07 g/mol) = 162 g/mol. This is nearly equal to the experimental molar mass, so the molecular formula is the same as the empirical formula.

3.47   The solution process is $MgCl_{2(s)} \rightarrow Mg^{2+}_{(aq)} + 2 \, Cl^-_{(aq)}$. Each mole of solid generates one mole of magnesium cations and two moles of chloride anions.

(a) Molarity is found using the equation, $M = \dfrac{n}{V}$.

MM_{MgCl_2} = 24.31 g/mol + 2(35.45 g/mol) = 95.21 g/mol.

$n(Mg^{2+}) = 4.68 \text{ g MgCl}_2\left(\dfrac{1 \text{ mol}}{95.21 \text{ g}}\right)\left(\dfrac{1 \text{ mol Mg}^{2+}}{1 \text{ mol MgCl}_2}\right) = 4.915 \times 10^{-2}$ moles;

$n(Cl^-) = 4.68 \text{ g MgCl}_2\left(\dfrac{1 \text{ mol}}{95.21 \text{ g}}\right)\left(\dfrac{2 \text{ mol Cl}^-}{1 \text{ mol MgCl}_2}\right) = 9.831 \times 10^{-2}$ moles.

Divide by the volume in liters to obtain the concentrations:

$$V = 1.50 \times 10^2 \text{ mL} \left(\frac{10^{-3} \text{ L}}{1 \text{ mL}} \right) = 0.150 \text{ L};$$

$$M(\text{Mg}^{2+}) = \frac{4.915 \times 10^{-2} \text{ mol}}{0.150 \text{ L}} = 0.328 \text{ M}$$

$$M(\text{Cl}^-) = \frac{9.831 \times 10^{-2} \text{ mol}}{0.150 \text{ L}} = 0.655 \text{ M}$$

(answers have three significant figures because the mass and volume are only known to three significant figures)

(b) Molecular pictures of solutions must illustrate the relative numbers of ions of each type present in the solution. Your picture must show twice as many chloride ions as magnesium ions. There are many more molecules of water than of either of the ions.

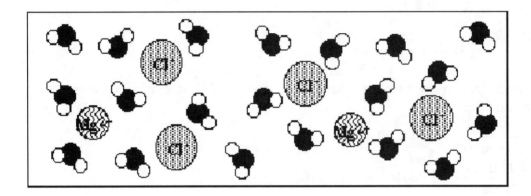

3.49   The solution process is $\text{KOH(s)} \rightarrow \text{K}^+\text{(aq)} + \text{OH}^-\text{(aq)}$. Each mole of solid generates one mole of each ion.

(a) Molarity is found using the equation, $M = \dfrac{n}{V}$.

$$M = \frac{n}{V} = \left(\frac{4.75 \text{ g}}{275 \text{ mL}} \right) \left(\frac{1 \text{ mol}}{56.11 \text{ g}} \right) \left(\frac{10^3 \text{ mL}}{1 \text{ L}} \right) = 0.308 \text{ M} = [\text{K}^+] = [\text{OH}^-];$$

(b) In a dilution, the number of moles of solute remains constant while volume increases, so $M_1 V_1 = M_2 V_2$. The starting volume is 25.00 mL and the final volume is 100.00 mL

$$M_2 = \frac{M_1 V_1}{V_2} = \frac{(0.308 \text{ M})(25.00 \text{ mL})}{100. \text{ mL}} = 0.0770 \text{ M} = [\text{K}^+] = [\text{OH}^-];$$

(c) Molecular pictures of solutions must illustrate the relative numbers of ions of each type present in the solution. Because of the four-fold dilution, the solution in (b) is 1/4 as concentrated as the solution in (a). We omit solvent molecules for clarity, but remember that there are many more molecules of water than of either of the ions.

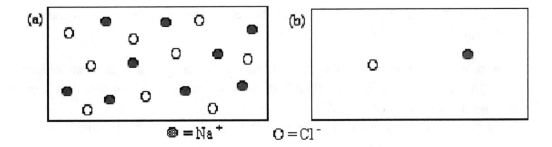

$\bigoplus = Na^+$ $\bigcirc = Cl^-$

3.51   This is a dilution type problem. Begin by deriving the expression to solve for the volume:

$M_i V_i = M_f V_f$, or $V_i = \dfrac{M_f V_f}{M_i}$

$V_i = \dfrac{(0.125\,M)(0.500\,L)}{12.1\,M} = 0.00517\,L$ or 5.17 mL

3.53   In each of the following, remember that concentration is $M = \dfrac{n}{V}$. Begin each problem by identifying the major ionic species present.

(a) Na_2CO_3 contains a Group 1 metal ion, Na^+, and a polyatomic anion, CO_3^{2-}. Therefore this is an ionic compound with major ionic species: Na^+, CO_3^{2-}.

Determine the number of moles in 4.55 g of Na_2CO_3 and use stoichiometric ratios to determine the number of moles of each ion:

$MM(Na_2CO_3) = 2(23.99\text{g/mol}) + 12.01\text{ g/mol} + 3(16.00\text{g/mol}) = 105.99$ g/mol;

$n(Na_2CO_3) = 4.55\text{ g}\left(\dfrac{1\,mol}{105.99\,g}\right) = 0.0429$ mol;

$n(CO_3^{2-}) = 0.0429\text{ mol }Na_2CO_3\left(\dfrac{1\,mol\,CO_3^{2-}}{1\,mol\,Na_2CO_3}\right) = 0.0429\text{ mol }CO_3^{2-}$

Remember there are twice as many moles of Na^+ as for Na_2CO_3:

$n(Na^+) = 0.0429\text{ mol }Na_2CO_3\left(\dfrac{2\,mol\,Na^+}{1\,mol\,Na_2CO_3}\right) = 0.0858\text{ mol }Na^+$

Now determine the molarities by dividing the moles by the volume of the solution:

$V = 245\text{ mL}\left(\dfrac{10^{-3}\,L}{1\,mL}\right) = 0.245$ L;

$[CO_3^{2-}] = \left(\dfrac{0.0429\,mol}{0.245\,L}\right) = 0.175$ M

$[Na^+] = \left(\dfrac{0.0858\,mol}{0.245\,L}\right) = 0.350$ M

(b) NH_4Cl contains a polyatomic ion and is therefore an ionic compound with major ionic species: NH_4^+, Cl^-.

Determine the number of moles in 27.45 mg of the salt. Note that all stoichiometric ratios are 1 for this compound; therefore, the number of moles of each ion is equal to the number of moles of the salt.

$MM(NH_4Cl)$ 14.01 g/mol + 4(1.008 g/mol) + 35.45 g/mol = 53.49 g/mol

$$n(NH_4Cl) = 27.45 \text{ mg} \left(\frac{10^{-3} \text{ g}}{1 \text{ mg}} \right) \left(\frac{1 \text{ mol}}{53.49 \text{ g}} \right) = 5.132 \times 10^{-4} \text{ mol} = n(NH_4^+) = n(Cl^-)$$

Determine the molarities of each ion by dividing the moles by the volume of the solution:

$$[NH_4^+] = [Cl^-] = \frac{5.132 \times 10^{-4} \text{ mol}}{1.55 \times 10^{-2} \text{ L}} = 0.0331 \text{ M}$$

(c) potassium sulfate : K_2SO_4

This compound contains both a Group 1 metal ion and a polyatomic ion. The major ionic species are: K^+, SO_4^{2-}

Determine the number of moles in 1.85 kg of the salt and use stoichiometric ratios to determine the number of moles of each ion:

$$n(K_2SO_4) = 1.85 \text{ kg} \left(\frac{10^3 \text{ g}}{1 \text{ kg}} \right) \left(\frac{1 \text{ mol}}{174.258 \text{ g}} \right) = 10.6 \text{ mol}$$

$$n(SO_4^{2-}) = 10.6 \text{ mol } K_2SO_4 \left(\frac{1 \text{ mol } SO_4^{2-}}{1 \text{ mol } K_2SO_4} \right) = 10.6 \text{ mol};$$

$$n(K^+) = 10.6 \text{ mol } K_2SO_4 \left(\frac{2 \text{ mol } K^+}{1 \text{ mol } K_2SO_4} \right) = 21.2 \text{ mol};$$

Obtain the molarities by dividing the moles of each ion by the volume of the solution:

$$[SO_4^{2-}] = \frac{10.6 \text{ mol}}{5.75 \times 10^3 \text{ L}} = 1.84 \times 10^{-3} \text{ M}$$

Similar to part (a), K^+ has twice the concentration as the anion.

$$[K^+] = \frac{21.2 \text{ mol}}{5.75 \times 10^3 \text{ L}} = 3.69 \times 10^{-3} \text{ M}$$

3.55 Begin by determining the chemical reaction.

$$Ag^+ + Cl^- \rightarrow AgCl.$$

From the chemical equation we can see that for every mole of Ag^+ a mole of AgCl will be formed. Therefore, calculate the number of moles of Ag^+ present and then obtain the mass of AgCl formed.

$$n(AgCl) = n(Ag^+) = M\,V = 595\,mL\left(\frac{10^{-3}\,L}{1\,mL}\right)\left(\frac{1.75 \times 10^{-2}\,mol\,Ag^+}{1\,L}\right) = 1.04 \times 10^{-2}\,mol;$$

$$MM(AgCl) = 107.87\,g/mol + 35.45\,g/mol = 143.32\,g/mol;$$

$$m = n\,MM = 1.04 \times 10^{-2}\,mol\left(\frac{143.32\,g}{1\,mol}\right) = 1.49\,g.$$

3.57   When working empirical formula problems that involve multiple analyses, it is convenient to determine the mass percent composition from the analyses and then determine the empirical formula from the mass percent composition. Each analysis has its own mass of starting sample.

$$m\,C = 4.34\,g\,CO_2\left(\frac{1\,mol}{44.01\,g}\right)\left(\frac{1\,mol\,C}{1\,mol\,CO_2}\right)\left(\frac{12.01\,g}{1\,mol}\right) = 1.184\,g;$$

$$\%\,C = \frac{1.184\,g}{4.00\,g}\times100\% = 29.6\,\%;$$

$$m\,Cl = 0.334\,g\,AgCl\left(\frac{1\,mol}{143.32\,g\,AgCl}\right)\left(\frac{1\,mol\,Cl}{1\,mol\,AgCl}\right)\left(\frac{35.45\,g\,Cl}{1\,mol}\right) = 0.0826\,g;$$

$$\%\,Cl = \frac{0.0826\,g}{0.125\,g}\times100\% = 66.1\,\%;$$

$$\%\,H = 100\,\% - (29.6\,\% + 66.1\,\%) = 4.3\,\%;$$

$$C:\ 29.6\,g\left(\frac{1\,mol}{12.01\,g}\right) = 2.46\,mol\,C;$$

$$H:\ 4.3\,g\left(\frac{1\,mol}{1.008\,g}\right) = 4.3\,mol\,H;$$

$$Cl:\ 66.1\,g\left(\frac{1\,mol}{35.45\,g}\right) = 1.86\,mol\,Cl;$$

Divide each by the smallest among them, 1.86 mol Cl:

$$\frac{2.46\,mol\,C}{1.86\,mol\,Cl} = 1.32\,C/Cl;$$

$$\frac{4.3\,mol\,H}{1.86\,mol\,Cl} = 2.3\,H/Cl;$$

Multiply all values by 3 to get round numbers: 3(1.32) = 4; 3(2.3) = 6.9, round to 7; The empirical formula is $C_4H_7Cl_3$.

3.59   If 0.302 g of pure Hg metal was obtained from the decomposition, then the difference between that and the original cinnabar mass is the mass of sulfur.
$$m_S = m_{cinnabar} - m_{Hg} = 0.350\,g - 0.302\,g = 0.048\,g\ sulfur$$

$$n_S = \frac{m_S}{MM_S} = 0.048 \text{ g}\left(\frac{1 \text{ mol}}{32.066 \text{ g}}\right) = 0.0015 \text{ mol}$$

$$n_{Hg} = \frac{m_{Hg}}{MM_{Hg}} = 0.302 \text{ g}\left(\frac{1 \text{ mol}}{200.59 \text{ g}}\right) = 0.0015 \text{ mol}$$

Since both have the same number of moles, the empirical formula of cinnabar is HgS.

3.61   To obtain a chemical formula, convert a line structure into a structural formula by placing a C atom at the end of each line and at each line intersection. Then add enough –H connections to give each C atom 4 connections. From the structural formula, determine the chemical formula by counting and calculate MM from numbers of atoms and atomic molar masses.

abscisic acid
$C_{15}H_{20}O_4$ $MM = 264.31$ g/mol

indole acetic acid
$C_{10}H_9NO_2$ $MM = 175.18$ g/mol

zeatin
$C_{10}H_{13}N_5O$
$MM = 219.25$ g/mol

abscisic acid, $C_{15}H_{20}O_4$,
$MM = 15(12.01 \text{ g/mol}) + 20(1.008 \text{ g/mol}) + 4(16.00 \text{ g/mol}) = 264.31$ g/mol;
indole acetic acid, $C_{10}H_9NO_2$,
$MM = 10(12.01 \text{ g/mol}) + 9(1.008 \text{ g/mol}) + 14.01 \text{ g/mol} + 2(16.00 \text{ g/mol}) = 175.18$ g/mol;
zeatin, $C_{10}H_{13}N_5O$,
$MM = 10(12.01 \text{ g/mol}) + 13(1.008 \text{ g/mol}) + 5(14.01 \text{ g/mol}) + 16.00 \text{ g/mol} = 219.25$ g/mol;

3.63   To write correct chemical formulas, you must know the compositions and charges of polyatomic ions. The overall formula of a compound must be electrically neutral.
(a) $NaNO_2$ and $NaNO_3$; (b) K_2CO_3 and $KHCO_3$; (c) FeO and Fe_2O_3; and (d) I_2 and I^-.

3.65   The chemical formula of a substance provides all the information needed to compute its molar characteristics.
(a) $MM = 2(26.98 \text{ g/mol}) + 3[32.07 \text{ g/mol} + 4(16.00 \text{ g/mol})] = 342.17$ g/mol;

(b) $n = \dfrac{m}{MM} = 25.0 \text{ g}\left(\dfrac{1 \text{ mol}}{342.17 \text{ g}}\right) = 7.31 \times 10^{-2}$ mol;

(c) To determine the percent composition, work with one mole of substance. Take the ratio of the mass of each element that one mole contains to the mass of one mole (molar mass).

$$\% \text{ Al} = \frac{2(26.98 \text{ g/mol})}{342.17 \text{ g/mol}} \times 100\% = 15.77 \%;$$

$$\% \, S = \frac{3(32.07 \text{ g/mol})}{342.17 \text{ g/mol}} \times 100\% = 28.12 \, \%;$$

$$\% \, O = \frac{12(16.00 \text{ g/mol})}{342.17 \text{ g/mol}} \times 100\% = 56.11 \, \%;$$

(d) $1.00 \text{ mol O} \left(\frac{1 \text{ mol Al}_2(SO_4)_3}{12 \text{ mol O}} \right) \left(\frac{342.17 \text{ g}}{1 \text{ mol}} \right) = 28.5 \text{ g.}$

3.67　This problem poses a set of mole-mass-number conversion questions. Sevin is $C_{12}H_{11}NO_2$,

$MM = 12(12.01 \text{ g/mol}) + 11(1.008 \text{ g/mol}) + 14.01 \text{ g/mol} + 2(16.00 \text{ g/mol}) = 201.22 \text{ g/mol.}$

(a) $n_C = (\text{C atoms/molecule}) = \dfrac{m}{MM} = 8.3 \text{ g} \left(\dfrac{1 \text{ mol}}{201.22 \text{ g}} \right) \left(\dfrac{12.01 \text{ mol C}}{1 \text{ molecule sevin}} \right) = 0.49 \text{ mol};$

(b) $m_O = 4.5 \text{ g} \left(\dfrac{1 \text{ mol}}{201.22 \text{ g}} \right) \left(\dfrac{2 \text{ mol O}}{1 \text{ mol sevin}} \right) \left(\dfrac{16.00 \text{ g}}{1 \text{ mol}} \right) = 0.72 \text{ g};$

(c) $n = 75 \text{ mL} \left(\dfrac{1.00 \text{ g}}{1 \text{ mL}} \right) \left(\dfrac{0.010 \text{ g}}{100 \text{ g}} \right) \left(\dfrac{1 \text{ mol}}{201.22 \text{ g}} \right) = 3.7 \times 10^{-5} \text{ mol};$

$\# = n \, N_A = (3.7 \times 10^{-5} \text{ mol})(6.022 \times 10^{23} \text{ molecules/mol}) = 2.2 \times 10^{19} \text{ molecules};$

(d) 15 gallons of spray requires 15 mL of insecticide. The insecticide contains 3.7×10^{-5} mol per 75 mL, so

$$n = 15 \text{ mL} \left(\frac{3.7 \times 10^{-5} \text{ mol}}{75 \text{ mL}} \right) = 7.4 \times 10^{-6} \text{ mol}$$

3.69　Use the Avogadro constant to convert from moles to number of atoms. Multiply the total number of atoms by the atomic diameter to find the total length.

$$l = 1.000 \text{ mol} \left(\frac{6.022 \times 10^{23} \text{ atoms}}{1 \text{ mol}} \right) \left(\frac{127.8 \text{ pm}}{1 \text{ atom}} \right) \left(\frac{10^{-12} \text{ m}}{1 \text{ pm}} \right) = 7.696 \times 10^{13} \text{ m.}$$

3.71　Chemical names follow the standard rules for chemical nomenclature.

(a) carbon dioxide; (b) potassium nitrate; (c) sodium chloride;
(d) sodium hydrogen carbonate; (e) sodium carbonate; (f) sodium hydroxide;
(g) calcium oxide; and (h) magnesium hydroxide.

3.73　To calculate mass percent composition, which involves mass ratios, work with one mole of substance. Take the ratio of the mass of each element that one mole contains to the mass of one mole (molar mass).

Fe_2SiO_4:
$MM = 2(55.85 \text{ g/mol}) + 28.09 \text{ g/mol} + 4(16.00 \text{ g/mol}) = 203.79 \text{ g/mol};$

$\% \, Fe = \dfrac{2(55.85 \text{ g/mol})}{203.79 \text{ g/mol}} \times 100\% = 54.81 \, \%;$

$$\% \text{ Si} = \frac{28.09 \text{ g/mol}}{203.79 \text{ g/mol}} \times 100\% = 13.78 \text{ \%};$$

$$\% \text{ O} = \frac{4(16.00 \text{ g/mol})}{203.79 \text{ g/mol}} \times 100\% = 31.40 \text{ \%};$$

$NaAlSi_3O_8$:
$$MM = 22.99 \text{ g/mol} + 26.98 \text{ g/mol} + 3(28.09 \text{ g/mol}) + 8(16.00 \text{ g/mol}) = 262.24 \text{ g/mol};$$

$$\% \text{ Na} = \frac{22.99 \text{ g/mol}}{262.24 \text{ g/mol}} \times 100\% = 8.767 \text{ \%};$$

$$\% \text{ Al} = \frac{26.98 \text{ g/mol}}{262.24 \text{ g/mol}} \times 100\% = 10.29 \text{ \%};$$

$$\% \text{ Si} = \frac{3(28.09 \text{ g/mol})}{262.24 \text{ g/mol}} \times 100\% = 32.13 \text{ \%};$$

$$\% \text{ O} = \frac{4(16.00 \text{ g/mol})}{262.24 \text{ g/mol}} \times 100\% = 24.41 \text{ \%};$$

$Al_2Si_2O_5(OH)_4$:
$$MM = 2(26.98 \text{ g/mol}) + 2(28.09 \text{ g/mol}) + 5(16.00 \text{ g/mol}) + 4(16.00 \text{ g/mol} + 1.008 \text{ g/mol})$$
$$MM = 258.17 \text{ g/mol};$$

$$\% \text{ Al} = \frac{2(26.98 \text{ g/mol})}{258.17 \text{ g/mol}} \times 100\% = 20.90 \text{ \%};$$

$$\% \text{ Si} = \frac{2(28.09 \text{ g/mol})}{258.17 \text{ g/mol}} \times 100\% = 21.76 \text{ \%};$$

$$\% \text{ O} = \frac{9(16.00 \text{ g/mol})}{258.17 \text{ g/mol}} \times 100\% = 55.78 \text{ \%};$$

$$\% \text{ H} = \frac{4(1.008 \text{ g/mol})}{258.17 \text{ g/mol}} \times 100\% = 1.562 \text{ \%};$$

$MgSi_4O_{10}(OH)_8$:
$$MM = 24.31 \text{ g/mol} + 4(28.09 \text{ g/mol}) + 10(16.00 \text{ g/mol}) + 8(16.00 \text{ g/mol} + 1.008 \text{ g/mol}) =$$
$$MM = 432.73 \text{ g/mol};$$

$$\% \text{ Mg} = \frac{24.31 \text{ g/mol}}{432.73 \text{ g/mol}} \times 100\% = 5.62 \text{ \%};$$

$$\% \text{ Si} = \frac{4(28.09 \text{ g/mol})}{432.73 \text{ g/mol}} \times 100\% = 25.97 \text{ \%};$$

$$\% \text{ O} = \frac{18(16.00 \text{ g/mol})}{432.73 \text{ g/mol}} \times 100\% = 66.55 \text{ \%};$$

$$\% \text{ H} = \frac{8(1.008 \text{ g/mol})}{432.73 \text{ g/mol}} \times 100\% = 1.86 \text{ \%};$$

3.75 When salts dissolve in water, they dissociate into cations and anions. Molecular substances retain their structure in solution, and H_2O is always present as a species in aqueous solutions.

(a) H_2O, NH_4^+, SO_4^{2-}; (b) H_2O, CO_2; (c) H_2O, Na^+, F^-; (d) H_2O, K^+, CO_3^{2-};

(e) H_2O, Na^+, HSO_4^-; and (f) H_2O, Cl_2.

3.77 This is a mass-mole-mass problem.

NH_4NO_3:

$MM = 2(14.01 \text{ g/mol}) + 4(1.008 \text{ g/mol}) + 3(16.00 \text{ g/mol}) = 80.05 \text{ g/mol}$;

$$1.00 \text{ kg N}\left(\frac{1 \text{ mol}}{14.01 \text{ g}}\right)\left(\frac{1 \text{ mol NH}_4\text{NO}_3}{2 \text{ mol N}}\right)\left(\frac{80.05 \text{ g}}{1 \text{ mol}}\right) = 2.86 \text{ kg};$$

$(NH_4)_2SO_4$:

$MM = 2(14.01 \text{ g/mol}) + 8(1.008 \text{ g/mol}) + 32.07 \text{ g/mol} + 4(16.00 \text{ g/mol})\ 132.15 \text{ g/mol}$;

$$1.00 \text{ kg N}\left(\frac{1 \text{ mol}}{14.01 \text{ g}}\right)\left(\frac{1 \text{ mol }(\text{NH}_4)_2\text{SO}_4}{2 \text{ mol N}}\right)\left(\frac{132.15 \text{g}}{1 \text{ mol}}\right) = 4.72 \text{ kg};$$

$(NH_2)_2CO$:

$MM = 2(14.01 \text{ g/mol}) + 4(1.008 \text{ g/mol}) + 12.01 \text{ g/mol} + 16.00 \text{ g/mol} = 60.06 \text{ g/mol}$;

$$1.00 \text{ kg N}\left(\frac{1 \text{ mol}}{14.01 \text{ g}}\right)\left(\frac{1 \text{ mol }(\text{NH}_2)_2\text{CO}}{2 \text{ mol N}}\right)\left(\frac{60.06 \text{g}}{1 \text{ mol}}\right) = 2.14 \text{ kg};$$

$(NH_4)_2HPO_4$:

$MM = 2(14.01 \text{ g/mol}) + 9(1.008 \text{ g/mol}) + 30.97 \text{ g/mol} + 4(16.00 \text{ g/mol})\ 132.06 \text{ g/mol}$;

$$1.00 \text{ kg N}\left(\frac{1 \text{ mol}}{14.01 \text{ g}}\right)\left(\frac{1 \text{ mol }(\text{NH}_4)_2\text{HPO}_4}{2 \text{ mol N}}\right)\left(\frac{132.06 \text{g}}{1 \text{ mol}}\right) = 4.71 \text{ kg};$$

3.79 Use nomenclature rules to determine a name from a chemical formula.

(a) ammonium chloride; (b) xenon tetrafluoride; (c) iron(III) oxide; (d) sulfur dioxide; and (e) potassium perchlorate.

3.81 The question asks about mole-mass-number quantities. Although there is much interesting biomedical information provided, the only relevant data are the chemical formula of Verapamil, $C_{27}H_{38}O_4N_2$, and the amount of Verapamil per tablet, 120.0 mg.

(a) $MM = 27(12.01 \text{ g/mol}) + 38(1.008 \text{ g/mol}) + 4(16.00 \text{ g/mol}) + 2(14.01 \text{ g/mol})$

$MM = 454.59 \text{ g/mol}$;

(b) $n = \dfrac{m}{MM} = 120.0 \text{ mg}\left(\dfrac{10^{-3}\text{g}}{1 \text{ mg}}\right)\left(\dfrac{1 \text{ mol}}{454.59 \text{ g}}\right) = 2.640 \times 10^{-4} \text{ mol}$;

(c) # atoms $= n\, N_A = 2.640 \times 10^{-4} \text{ mol}\left(\dfrac{6.022 \times 10^{23}\text{ molec. verap.}}{1 \text{ mol}}\right)\left(\dfrac{2 \text{ atoms N}}{1 \text{ molec. verap.}}\right)$

atoms $= 3.179 \times 10^{20}$ atoms N.

3.83 This is an empirical formula problem involving combustion analysis. Do a mass analysis of the combustion data to obtain the mass percent composition of the compound. Determine the mass percent oxygen by difference. Once the mass percentages are known, use elemental molar masses to determine relative amounts of the elements.

$$\% \, N = \frac{0.0158 \, g \, N}{0.183 \, g \, cmpd} \times 100\% = 8.63 \, \%;$$

$$m \, C = 0.372 \, g \left(\frac{1 \, mol}{44.01 g}\right)\left(\frac{1 \, mol \, C}{1 \, mol \, CO_2}\right)\left(\frac{12.01 \, g}{1 \, mol}\right) = 0.1015 \, g; \ \ \text{(leave an extra digit to avoid}$$

round-off error)

$$\% \, C = \frac{0.1015 \, g \, C}{0.137 \, g} \times 100\% = 74.1 \, \%;$$

$$m \, H = 0.0910 \, g\left(\frac{1 \, mol}{18.02 g}\right)\left(\frac{2 \, mol \, H}{1 \, mol \, H_2O}\right)\left(\frac{1.008 g}{1 \, mol}\right) = 1.018 \times 10^{-2} \, g;$$

$$\% \, H = \frac{1.018 \times 10^{-2} \, g \, C}{0.137 \, g} \times 100\% = 7.43 \, \%;$$

$$\% \, O = 100 \, \% - (8.63 \, \% + 74.1 \, \% + 7.43 \, \%) = 9.84 \, \%;$$

Next, determine the number of moles of each element assuming 100g of the substance:

$$C: \ 74.1 g\left(\frac{1 \, mol}{12.01 \, g}\right) = 6.17 \, mol \, C;$$

$$H: \ 7.43 \, g\left(\frac{1 \, mol}{1.008 \, g}\right) = 7.37 \, mol \, H;$$

$$N: \ 8.63 \, g\left(\frac{1 \, mol}{14.01 \, g}\right) = 0.616 \, mol \, N;$$

$$O: \ 9.84 \, g\left(\frac{1 \, mol}{16.00 \, g}\right) = 0.615 \, mol \, O;$$

Divide each by the smallest among them, 0.615 mol O, to determine the relative amounts of each element:

$$\frac{6.17 \, mol \, C}{0.615 \, mol \, O} = 10.03 \, C/O, \text{ round to 10};$$

$$\frac{7.37 \, mol \, H}{0.615 \, mol \, O} = 11.98 \, H/O, \text{ round to 12};$$

$$\frac{0.616 \, mol \, N}{0.615 \, mol \, O} = 1.002 \, N/O, \text{ round to 1};$$

The empirical formula of quinine is $C_{10}H_{12}ON$, with molar mass:

$MM = 10(12.01 \, g/mol) + 12(1.008 \, g/mol) + 16.00 \, g/mol + 14.01 \, g/mol = 162 \, g/mol;$

Since the MM is known to be between 300 and 350 g/mol, we see that twice our empirical molar mass fits (i.e., 324 g/mol):

Molecular formula: $C_{20}H_{24}O_2N_2$.

3.85 Use the definition of mass percentage composition to compute the molar mass:

$$\% \, Co = \frac{(\text{atoms Co/molec})(MM_{Co})}{MM_{molecule}} \times 100\%$$

$$MM_{molecule} = 1 \text{ atoms Co/molec} \left(\frac{58.93 g}{1 \, mol} \right)\left(\frac{100\%}{4.34\%} \right) = 1.36 \text{ x } 10^3 \text{ g/mol}.$$

3.87 "Same number of atoms" also means "same number of moles," so work with moles:

$$n_{Li} = n_{Pt} \text{ and } n = \frac{m}{MM} \, ;$$

$$n_{Li} = 5.75 \, g \left(\frac{1 \, mol}{195.08 \, g} \right)\left(\frac{1 \text{ atom Li}}{1 \text{ atom Pt}} \right)\left(\frac{6.94 \, g}{1 \, mol} \right) = 0.205 \text{ g}.$$

3.89 Moles and mass in grams are related through the equation, $n = \dfrac{m}{MM}$. When masses are not given in grams, unit conversions must be made.

MM (H_3PO_4) = 3(1.008 g/mol) + 30.97 g/mol + 4(16.00g/mol) = 97.99 g/mol;

1 lb = 453.6 g;

$$n = 26.19 \text{ x } 10^7 \, lb \left(\frac{453.6 \, g}{1 \, lb} \right)\left(\frac{1 \, mol}{97.99 \, g} \right) = 1.212 \text{ x } 10^9 \text{ mol};$$

The chemical formula shows that there is one mole of P in every mole of H_3PO_4. The problem states that 15% of the annual production of H_3PO_4 comes from elemental P:

$$n(P) = 1.212 \text{ x } 10^9 \, mol \left(\frac{15\%}{100\%} \right) = 1.8 \text{ x } 10^8 \text{ mol};$$

$$m(P) = n \, MM = 1.8 \text{ x } 10^8 \, mol \left(\frac{30.97 \, g}{1 \, mol} \right)\left(\frac{10^{-3} kg}{1 \, g} \right) = 5.6 \text{ x } 10^6 \text{ kg of P consumed}$$

3.91 To determine a molecular formula, mass percent composition and an approximate molar mass must be known. The data given are insufficient to calculate mass percent composition. To calculate the mass percentages of carbon and hydrogen, the mass of compound that was burned must be known. Unless the percentages of C and H together total 100 %, further information about the other elements present in the compound is also needed.

3.93 A molar concentration is moles in one liter of liquid. First use density to calculate mass/L, then use MM to convert to mol/L:

$$\text{mass/L} = \left(\frac{1.0 \, g}{1 \, mL} \right)\left(\frac{10^3 \, mL}{1 \, L} \right) = 1.0 \text{ x } 10^3 \text{ g/L};$$

MM = 2(1.008 g/mol) + 15.999 g/mol = 18.015 g/mol.

$$\text{molar concentration} = \left(\frac{1.0 \text{ x } 10^3 \, g}{1 \, L} \right)\left(\frac{1 \, mol}{18.015 \, g} \right) = 56 \text{ mol/L}$$

3.95   To find the mass percent composition of each element find the mass of each element in one mole of the compound and divide that by the mass of one mole of the compound. The masses of the elements in 1 mol of $C_{10}H_{16}N_5O_{13}P_3$ follow:

$$10 \text{ mol C} \left(\frac{12.011 \text{ g}}{1 \text{ mol}} \right) = 120.11 \text{ g}$$

$$16 \text{ mol H} \left(\frac{1.008 \text{ g}}{1 \text{ mol}} \right) = 16.13 \text{ g}$$

$$5 \text{ mol N} \left(\frac{14.007 \text{ g}}{1 \text{ mol}} \right) = 70.035 \text{ g}$$

$$13 \text{ mol O} \left(\frac{15.999 \text{ g}}{1 \text{ mol}} \right) = 207.987 \text{ g}$$

$$3 \text{ mol P} \left(\frac{30.974 \text{ g}}{1 \text{ mol}} \right) = 92.922 \text{ g}$$

Mass of 1 mol of $C_{10}H_{16}N_5O_{13}P_3 = 507.18$ g

$$\text{Mass percentage} = \frac{\text{mass of element}}{\text{mass of compound}} \times 100\%$$

C: $\dfrac{120.11 \text{ g}}{507.18 \text{ g}} \times 100\% = 23.682\% \text{ C}$

H: $\dfrac{16.13 \text{ g}}{507.18 \text{ g}} \times 100\% = 3.180\% \text{ H}$

N: $\dfrac{70.035 \text{ g}}{507.18 \text{ g}} \times 100\% = 13.809\% \text{ N}$

O: $\dfrac{207.987 \text{ g}}{507.18 \text{ g}} \times 100\% = 41.009\% \text{ O}$

P: $\dfrac{92.922 \text{ g}}{507.18 \text{ g}} \times 100\% = 18.321\% \text{ P}$

Summing the percentages gives 100%, which confirms the work.

3.97   This problem asks for mass-mole calculations on turquoise.
(a)
$MM_{\text{turquoise}} = 63.55$ g/mol Cu + 6(26.98 g/mol Al) + 4(30.97 g/mol P) + 28(16.00 g/mol O) + 16(1.01 g/mol H) = 813.47 g/mol turquoise

$$7.25 \text{ g turquoise} \left(\frac{1 \text{ mol}}{813.47 \text{ g}} \right) \left(\frac{6 \text{ mol Al}}{1 \text{mol turquoise}} \right) \left(\frac{26.98 \text{ g}}{1 \text{ mol}} \right) = 1.44 \text{ g Al}$$

(b) there are 4 phosphate ions per every 28 O atoms present:

$$5.50 \times 10^{-3} \text{ g O} \left(\frac{1 \text{mol}}{16.00 \text{ g}} \right) \left(\frac{4 \text{ mol PO}_4^{3-}}{28 \text{ mol O}} \right) \left(\frac{6.022 \times 10^{23} \text{ ions}}{1 \text{ mol}} \right) = 2.96 \times 10^{19} \text{ ions}$$

(c) charge on OH = -1, charge on PO_4 = -3, charge on Al = +3
thus, charge on Cu must balance 8(-1)+4(-3)+6(+3) = -2
charge on Cu = +2

3.99   This problem contains a series of mole-mass-number problems.
$MM_{Al_2(SO_4)_3}$ = 2(26.98g/mol) + 3(32.07g/mol) + 30(16.00g/mol) + 36(1.01g/mol)
$MM_{Al_2(SO_4)_3}$ = 666.53 g/mol

(a) $0.570 \text{ moles} \left(\dfrac{3 \text{ mol S}}{1 \text{ mol Al}_2(SO_4)_3} \right) \left(\dfrac{32.07g}{1 \text{ mol}} \right) = 54.8 \text{ g}$

(b) $\# = 5.1g \left(\dfrac{1 \text{ mol}}{666.53 \text{ g}} \right) \left(\dfrac{18 \text{ mol H}_2O}{1 \text{ mol Al}_2(SO_4)_3 \bullet 18H_2O} \right) \left(\dfrac{6.022 \times 10^{23} \text{ molecules}}{1 \text{ mol}} \right)$
$\# = 8.3 \times 10^{22}$ molecules H_2O

(c) $12.5 \text{ mol O} \left(\dfrac{3 \text{ mol SO}_4^{2-}}{30 \text{ mol O}} \right) = 1.25 \text{ mol SO}_4^{2-}$

(d) $\left(\dfrac{1.05 \text{ g}}{1 \text{ mL}} \right) \left(\dfrac{1.25 \text{ g}}{100 \text{ g}} \right) \left(\dfrac{1 \text{ mol}}{26.98 \text{ g}} \right) \left(\dfrac{1000 \text{ mL}}{1 \text{ L}} \right) = 0.486 \text{ M Al}^{3+}$

3.101 (a) $MM = 2(1.01 \text{ g/mol H}) + 32.07 \text{ g/mol S} + 4(16.00 \text{ g/mol O}) = 98.09 \text{ g/mol H}_2SO_4$
$1.75 \text{ g/ml} \left(\dfrac{1000 \text{ mL}}{1 \text{ L}} \right) \left(\dfrac{80.\%}{100\%} \right) \left(\dfrac{1 \text{ mol}}{98.09 \text{ g}} \right) = 14 \text{ M}$

(b) This is a dilution problem. Thus, $M_iV_i = M_fV_f$ or
$V_i = \dfrac{(0.65 \text{ M})(2.50 \text{ L})}{14.3 \text{ M}} = 0.11 \text{ L or } 110 \text{ mL}$

4.1 A balanced chemical equation must have equal numbers of atoms of each element on each side of the arrow. Balance each element in turn, beginning with those that appear in only one reactant and product, by adjusting stoichiometric coefficients. Generally, H and O are balanced last. In each case we should start first by determining the number of atoms on each side of the chemical equation.

(a) $NH_4NO_3 \rightarrow N_2O + H_2O$

$2N + 3O + 4H \rightarrow 2N + 2O + 2H$

There are two nitrogen atoms in the reactant and two in the products, thus nitrogen is already balanced. Balance the H's by changing the stoichiometric coefficient of water from 1 to 2 and determine the new amounts of each atom on both sides of the equation:

$NH_4NO_3 \rightarrow N_2O + \textbf{2}\, H_2O$

$2N + 3O + 4H \rightarrow 2N + 3O + 4H$

Notice that now all three atoms are balanced, thus the equation is balanced.

(b) $P_4O_{10} + H_2O \rightarrow H_3PO_4$

$4P + 11O + 2H \rightarrow P + 4O + 3H$

Start by balancing the P since it occurs in only one reactant and one product. There are 4 P on the reactant side and 1 P on the product side. Hence, balance P by putting a coefficient of 4 in front of the H_3PO_4:

$P_4O_{10} + H_2O \rightarrow \textbf{4}\, H_3PO_4$

$4P + 11O + 2H \rightarrow 4P + 16O + 12H$

Next balance H by giving H_2O a coefficient of 6 (each water contributes 2H and 12 H's are required on the reactant side to balance the 12H on the product side):

$P_4O_{10} + \textbf{6}\, H_2O \rightarrow 4\, H_3PO_4$

$4P + 16O + 12H \rightarrow 4P + 16O + 12H$

Note that now all atoms are balanced on both sides of the equation, thus the equation is balanced.

(c) $HIO_3 \rightarrow I_2O_5 + H_2O$

$I + H + 3O \rightarrow 2I + 2H + 6O$

Notice that there are half as many of each atom on the reactant side as on the product side. Therefore this equation can be balanced by simply increasing the HIO_3 coefficient from 1 to 2.

$\textbf{2}\, HIO_3 \rightarrow I_2O_5 + H_2O$

$2I + 2H + 6O \rightarrow 2I + 2H + 6O$

(d) $As + Cl_2 \rightarrow AsCl_5$

$1As + 2Cl \rightarrow 1As + 5Cl$

As is already balanced. There are 2 Cl on the reactant side and 5 on the product. To balance the number Cl we need to change the Cl_2 coefficient from 1 to 5/2:

$$As + \tfrac{5}{2} Cl_2 \rightarrow AsCl_5$$

$$1As + 5Cl \rightarrow 1As + 5Cl$$

Notice that both the As and Cl are now balanced. However, we do not want fractions in a chemical equation. Therefore, multiply **all** coefficients by 2:

$$\mathbf{2(As + \tfrac{5}{2} Cl_2 \rightarrow AsCl_5)}$$

$$\mathbf{2\,As + 5\,Cl_2 \rightarrow 2\,AsCl_5}$$

$$2\,As + 10\,Cl \rightarrow 2\,As + 10\,Cl$$

Hence the equation is balanced.

4.3   Molecular pictures must show the correct number of molecules undergoing the reaction. In 4.1d, 2 atoms of As react with 5 molecules of Cl_2 to form 2 molecules of $AsCl_5$. Remember that when drawing molecular pictures you must differentiate between the different atom types by color, labeling, or shading (here we use shading).

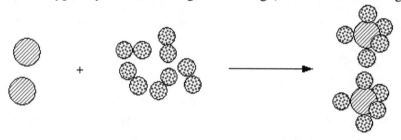

4.5   A balanced chemical equation must have equal numbers of atoms of each element on each side of the arrow. Balance each element in turn, beginning with those that appear in only one reactant and product, by adjusting stoichiometric coefficients. Generally, H and O are balanced last. When balancing the equation start by determining the number of atoms on each side of the chemical equation.

(a) We should start first by determining the chemical formula of each compound of the reactants and products (you may want to review Chapter 3).
       molecular hydrogen : H_2
       carbon monoxide: CO
       methanol : CH_3OH

The chemical equation (unbalanced) is thus:

$$H_2 + CO \rightarrow CH_3OH$$

$$2H + 1C + 1O \rightarrow 4H + 1C + 1O$$

Both carbon and oxygen are balanced. There are 4H on the product side and 2H on the reactant. Change the coefficient of H_2 from 1 to 2:

$$\mathbf{2H_2 + CO \rightarrow CH_3OH}$$

$$4H + 1C + 1O \rightarrow 4H + 1C + 1O$$

Since all atoms are now balanced this is the balanced equation.

(b) $CaO + C \rightarrow CO + CaC_2$

$1Ca + 1O + 1C \rightarrow 1Ca + 1O + 3C$

Note that both Ca and O are balanced. Multiply the coefficient of C by 3:

$CaO + \mathbf{3}\,C \rightarrow CO + CaC_2$

$1Ca + 1O + 3C \rightarrow 1Ca + 1O + 3C$

Double-checking the amount of each atom on both sides of the equation shows that this is the balanced equation.

(c) $C_2H_4 + O_2 + HCl \rightarrow C_2H_4Cl_2 + H_2O$

$2C + 5H + 2O + Cl \rightarrow 2C + 6H + 1O + 2Cl$

Carbon is already balanced. Both O and Cl are found in only one reactant and one product. Hence we can choose either to start with; let's choose Cl. Since there is one Cl on the reactant side and 2 on the product side, change the HCl coefficient from 1 to 2.

$C_2H_4 + O_2 + \mathbf{2}\,HCl \rightarrow C_2H_4Cl_2 + H_2O$

$2C + 6H + 2O + 2Cl \rightarrow 2C + 6H + 1O + 2Cl$

Note that this balanced both Cl and H and left C balanced. To balance the O change the coefficient on O_2 from 1 to 1/2:

$C_2H_4 + \frac{1}{2}\,O_2 + 2\,HCl \rightarrow C_2H_4Cl_2 + H_2O$

$2C + 6H + 1O + 2Cl \rightarrow 2C + 6H + 1O + 2Cl$

The reaction is now balanced. Multiply **all** coefficients by 2 to eliminate fractions from the chemical equation:

$\mathbf{2}(C_2H_4 + \frac{1}{2}\,O_2 + 2\,HCl \rightarrow C_2H_4Cl_2 + H_2O)$

$\mathbf{2}\,C_2H_4 + O_2 + \mathbf{4}\,HCl \rightarrow \mathbf{2}\,C_2H_4Cl_2 + \mathbf{2}\,H_2O.$

4.7   Molecular pictures must show the correct number of molecules undergoing the reaction. In 4.5a, 2 molecules of hydrogen react with a molecule of CO to form one molecule of CH_3OH.

4.9   A balanced chemical equation must have equal numbers of atoms of each element on each side of the arrow. Balance each element in turn, beginning with those that appear in only one reactant and product, by adjusting stoichiometric coefficients. Generally, H and O are balanced last. When balancing the equation, start by determining the number of atoms on each side of the chemical equation.

(a) $Ca(OH)_2 + H_3PO_4 \rightarrow H_2O + Ca_3(PO_4)_2$

$1Ca + 6O + 5H + 1P \rightarrow 3Ca + 9O + 2H + 2P$

Start by balancing the Ca by changing the coefficient of $Ca(OH)_2$ from 1 to 3:

$3 Ca(OH)_2 + H_3PO_4 \rightarrow H_2O + Ca_3(PO_4)_2$

$3Ca + 10O + 9H + 1P \rightarrow 3Ca + 9O + 2H + 2P$

Next balance P by multiplying the H_3PO_4 coefficient by 2:

$3 Ca(OH)_2 + 2 H_3PO_4 \rightarrow H_2O + Ca_3(PO_4)_2$

$3Ca + 14O + 12H + 2P \rightarrow 3Ca + 9O + 2H + 2P$

Now balance the H by multiplying the water coefficient by 6 (this also balances the O):

$3 Ca(OH)_2 + 2 H_3PO_4 \rightarrow 6 H_2O + Ca_3(PO_4)_2$

$3Ca + 14O + 12H + 2P \rightarrow 3Ca + 14O + 12H + 2P$

All atoms are balanced, thus this is the balanced equation

(b) $Na_2O_2 + H_2O \rightarrow NaOH + H_2O_2$

$2Na + 3O + 2H \rightarrow 1Na + 3O + 3H$

Start by balancing the Na by changing the NaOH coefficient from 1 to 2:

$Na_2O_2 + H_2O \rightarrow 2 NaOH + H_2O_2$

$2Na + 3O + 2H \rightarrow 2Na + 4O + 4H$

Since H occurs in only one reactant (O occurs in both), balance H first. Balance the H by multiplying the water coefficient by 2:

$Na_2O_2 + 2 H_2O \rightarrow 2 NaOH + H_2O_2$

$2Na + 4O + 4H \rightarrow 2Na + 4O + 4H$

Notice that this also balanced the oxygen, hence, we are done.

(c) $BF_3 + H_2O \rightarrow HF + H_3BO_3$

$1B + 3F + 2H + 1O \rightarrow 1B + 1F + 4H + 3O$

Boron is already balanced. Balance the F by multiplying the HF coefficient by 3:

$BF_3 + H_2O \rightarrow 3 HF + H_3BO_3$

$1B + 3F + 2H + 1O \rightarrow 1B + 3F + 6H + 3O$

Next balance the H by multiplying the water coefficient by 3:

$BF_3 + 3 H_2O \rightarrow 3 HF + H_3BO_3$

$1B + 3F + 6H + 3O \rightarrow 1B + 3F + 6H + 3O$

Since this also balances O we are done.

(d) $NH_3 + CuO \rightarrow Cu + N_2 + H_2O.$

$$1N + 3H + 1Cu + 1O \rightarrow 2N + 2H + 1Cu + 1O$$

Cu is already balanced. To balance the H, multiply the NH_3 by 2 and H_2O by 3:
$2NH_3 + CuO \rightarrow Cu + N_2 + 3H_2O$.
$2N + 6H + 1Cu + 1O \rightarrow 2N + 6H + 1Cu + 3O$

This has also balance N. Finally balance O by giving *both* CuO *and* Cu a coefficient of 3.
$2NH_3 + 3CuO \rightarrow 3Cu + N_2 + 3H_2O$.
$2N + 6H + 3Cu + 3O \rightarrow 2N + 6H + 3Cu + 3O$

4.11   This problem is asking us to calculate the mass of the second reactant that will completely react with 5.00 g of the first. Remember that calculations of amounts in chemistry always center on the mole. Thus, we will need first to determine how many moles there are in the first reactant and then determine how many moles of the second are required to completely react.

(a) balanced eqn: $2 H_2 + CO \rightarrow CH_3OH$;

$$5.00 \text{ g } H_2 \left(\frac{1 \text{ mol}}{2.016 \text{ g}} \right) \left(\frac{1 \text{ mol CO}}{2 \text{ mol } H_2} \right) \left(\frac{28.01 \text{ g}}{1 \text{ mol}} \right) = 34.7 \text{ g CO}$$

(b) balanced eqn: $CaO + 3 C \rightarrow CO + CaC_2$;

$$5.00 \text{ g CaO} \left(\frac{1 \text{ mol}}{56.08 \text{ g}} \right) \left(\frac{3 \text{ mol C}}{1 \text{ mol CaO}} \right) \left(\frac{12.01 \text{ g}}{1 \text{ mol}} \right) = 3.21 \text{ g C}$$

(c) balanced eqn: $2 C_2H_4 + O_2 + 4 HCl \rightarrow 2 C_2H_4Cl_2 + 2 H_2O$;

$$5.00 \text{ g } C_2H_4 \left(\frac{1 \text{ mol}}{28.05 \text{ g}} \right) \left(\frac{1 \text{ mol } O_2}{2 \text{ mol } C_2H_4} \right) \left(\frac{32.00 \text{ g}}{1 \text{ mol}} \right) = 2.85 \text{ g } O_2$$

4.13   You are asked to calculate the mass of sodium iodide required to produce 1.50 kg of iodine. The balanced equation is given in the problem. Remember to convert the mass into grams before dividing by the molar mass:

$$1.50 \text{ kg } I_2 \left(\frac{1000 \text{ g}}{1 \text{ kg}} \right) \left(\frac{1 \text{ mol}}{253.8 \text{ g}} \right) \left(\frac{2 \text{ mol NaI}}{1 \text{ mol } I_2} \right) \left(\frac{149.89 \text{ g}}{1 \text{ mol}} \right) = 1.77 \text{ x } 10^3 \text{ g NaI}$$

4.15   This problem tells you how much of the reactant you have (in kg) and asks you to calculate the amount of product formed. Remember that calculations of amounts in chemistry always center on the mole.

$$1.00 \text{ kg sugar} \left(\frac{10^3 \text{ g}}{1 \text{ kg}} \right) \left(\frac{1 \text{ mol}}{180.2 \text{ g}} \right) \left(\frac{2 \text{ mol } C_2H_5OH}{1 \text{ mol sugar}} \right) \left(\frac{46.07 \text{ g}}{1 \text{ mol}} \right) = 511 \text{ g } C_2H_5OH$$

4.17   To calculate masses for a chemical reaction, balance the equation, then do appropriate mass-mole-mass conversions. The reaction is:

$CCl_4 + HF \rightarrow CCl_2F_2 + HCl$.

$1C + 4Cl + 1H + 1F \rightarrow 1C + 3Cl + 2F + 1H$

Carbon is balanced. Fluorine occurs in only 1 reactant and 1 product, therefore balance it next. Fluorine can be balanced by giving HF a coefficient of 2:

$CCl_4 + \mathbf{2}\,HF \rightarrow CCl_2F_2 + HCl$.

$1C + 4Cl + 2H + 2F \rightarrow 1C + 3Cl + 2F + 1H$

Finally, balance H and Cl by giving HCl a coefficient of 2:

$CCl_4 + 2\,HF \rightarrow CCl_2F_2 + \mathbf{2}\,HCl$.

$1C + 4Cl + 2H + 2F \rightarrow 1C + 4Cl + 2F + 2H$

Balanced eqn: $CCl_4 + 2HF \rightarrow CCl_2F_2 + 2HCl$.

To determine the amount of HF required to completely react with the CCl_4, convert to moles and do the appropriate conversions.

$$175\text{ kg }CCl_4\left(\frac{10^3\text{ g}}{1\text{ kg}}\right)\left(\frac{1\text{ mol}}{153.8\text{ g}}\right)\left(\frac{2\text{ mol HF}}{1\text{ mol }CCl_4}\right)\left(\frac{20.01\text{ g}}{1\text{ mol}}\right)\left(\frac{1\text{ kg}}{10^3\text{ g}}\right) = 45.5\text{ kg HF}$$

Since this reaction is 100% efficient, all of the reactants will be converted in products.

Mass of products formed:

$$175\text{ kg }CCl_4\left(\frac{10^3\text{ g}}{1\text{ kg}}\right)\left(\frac{1\text{ mol}}{153.8\text{ g}}\right)\left(\frac{1\text{ mol }CCl_2F_2}{1\text{ mol }CCl_4}\right)\left(\frac{120.9\text{ g}}{1\text{ mol}}\right)\left(\frac{1\text{ kg}}{10^3\text{ g}}\right) = 138\text{ kg }CCl_2F_2$$

$$175\text{ kg }CCl_4\left(\frac{10^3\text{ g}}{1\text{ kg}}\right)\left(\frac{1\text{ mol}}{153.8\text{ g}}\right)\left(\frac{2\text{ mol HCl}}{1\text{ mol }CCl_4}\right)\left(\frac{36.46\text{ g}}{1\text{ mol}}\right)\left(\frac{1\text{ kg}}{10^3\text{ g}}\right) = 83.0\text{ kg HCl}$$

4.19   In a yield problem, we compare actual amounts with theoretical amounts. Carry out standard mole-mass conversions to find the theoretical amount (FA = fluoroapatite). The reaction (balanced) is:

$Ca_5(PO_4)_3F + 5\,H_2SO_4 + 10\,H_2O \rightarrow 3\,H_3PO_4 + 5\,CaSO_4\cdot 2H_2O + HF$.

The percent yield is: $\%\text{yield} = \dfrac{\text{actual yield}}{\text{theoretical yield}} \times 100\%$

The problem gives the actual yield (400 g), thus we only need to calculate the theoretical yield.

Use stoichiometry to calculate the theoretical yield.

$$1.00 \text{ kg FA}\left(\frac{10^3 \text{ g}}{1 \text{ kg}}\right)\left(\frac{1 \text{ mol}}{504.3 \text{ g}}\right)\left(\frac{3 \text{ mol H}_3\text{PO}_4}{1 \text{ mol FA}}\right)\left(\frac{97.99 \text{ g}}{1 \text{ mol}}\right) = 583 \text{ g H}_3\text{PO}_4$$

Now divide the actual yield by the theoretical yield to obtain the percent yield:

$$\%\text{yield} = \frac{400 \text{ g}}{583 \text{ g}} \times 100\% = 68.6\%$$

4.21   In a multiple-step synthesis, the yield of each step must be multiplied to determine the overall yield. In this case, there are eight steps, each with a yield of 88%, so the overall fractional yield is $(0.88)(0.88)(0.88)(0.88)(0.88)(0.88)(0.88)(0.88) = (0.88)^8 = 0.36$. Use this value and the desired amount of phenobarbital in moles to determine how many moles of toluene are required. Do the usual mole-mass conversions. For phenobarbital, $MM = 12(12.01 \text{ g/mol}) + 12(1.01 \text{ g/mol}) + 2(14.01 \text{ g/mol}) + 3(16.00 \text{ g/mol}) = 232.3 \text{ g/mol}$.

$$\text{Moles toluene} = 25 \text{ kg}\left(\frac{10^3 \text{ g}}{1 \text{ kg}}\right)\left(\frac{1 \text{ mol}}{232.3 \text{ g}}\right)\left(\frac{1 \text{ mol toluene}}{1 \text{ mol phenobarbital}}\right) = 107.7 \text{ mol toluene};$$

$$\text{Moles of toluene required} = \frac{107.7 \text{ mol}}{0.36} = 299 \text{ mol};$$

$$\text{Mass required} = 299 \text{ mol}\left(\frac{92.13 \text{ g}}{1 \text{ mol}}\right)\left(\frac{10^{-3} \text{ kg}}{1 \text{ g}}\right) = 28 \text{ kg}.$$

There are two significant figures because the mass and yield are only known to two significant figures.

4.23   The calculation of a yield requires a theoretical amount and an actual amount. The theoretical amount is calculated as in Problem 4.17:

$$175 \text{ kg CCl}_4\left(\frac{10^3 \text{ g}}{1 \text{ kg}}\right)\left(\frac{1 \text{ mol}}{153.8 \text{ g}}\right)\left(\frac{1 \text{ mol CCl}_2\text{F}_2}{1 \text{ mol CCl}_4}\right)\left(\frac{120.9 \text{ g}}{1 \text{ mol}}\right)\left(\frac{10^{-3} \text{ kg}}{1 \text{ g}}\right) = 137.6 \text{ kg CCl}_2\text{F}_2;$$

(carry an additional significant figure until the calculation is complete)

The actual amount (given) is 105 kg. The $\%\text{yield} = \dfrac{105 \text{ kg}}{137.6 \text{ kg}} \times 100\% = 76.3\%$

To find how much of each reactant is required, divide the desired amount by the percent yield to obtain the theoretical yield and then do the usual stoichiometric calculations:

$$\text{Theoretical yield} = 155 \text{ kg}\left(\frac{100\%}{76.3\%}\right) = 203 \text{ kg};$$

$$203 \text{ kg CCl}_2\text{F}_2\left(\frac{10^3 \text{ g}}{1 \text{ kg}}\right)\left(\frac{1 \text{ mol}}{120.9 \text{ g}}\right)\left(\frac{1 \text{ mol CCl}_4}{1 \text{ mol CCl}_2\text{F}_2}\right)\left(\frac{153.8 \text{ g}}{1 \text{ mol}}\right)\left(\frac{10^{-3} \text{ kg}}{1 \text{ g}}\right) = 258 \text{ kg CCl}_4;$$

$$203 \text{ kg CCl}_2\text{F}_2\left(\frac{10^3 \text{ g}}{1 \text{ kg}}\right)\left(\frac{1 \text{ mol}}{120.9 \text{ g}}\right)\left(\frac{2 \text{ mol HF}}{1 \text{ mol CCl}_2\text{F}_2}\right)\left(\frac{20.01 \text{ g}}{1 \text{ mol}}\right)\left(\frac{10^{-3} \text{ kg}}{1 \text{ g}}\right) = 67.2 \text{ kg HF}.$$

4.25 The problem gives information about the amounts of starting materials, so this is a limiting reactant situation. We must calculate the number of moles of each species, construct a table of amounts, and use the results to determine the final product masses. Starting amounts are in metric tons (1 metric ton = 10^6 g), so it will be convenient to work with 10^6 mol amounts. See the answers to problem 4.5 for the balanced equations.

(a) Calculate the initial amounts:

$$1000 \text{ kg CO} \left(\frac{10^3 \text{ g}}{1 \text{ kg}} \right) \left(\frac{1 \text{ mol}}{28.01 \text{ g}} \right) = 0.0357 \times 10^6 \text{ mol};$$

$$1000 \text{ kg H}_2 \left(\frac{10^3 \text{ g}}{1 \text{ kg}} \right) \left(\frac{1 \text{ mol}}{2.02 \text{ g}} \right) = 0.495 \times 10^6 \text{ mol}$$

Now we set up the table of amounts:

Reaction	2 H$_2$ +	CO →	CH$_3$OH
Amt (10^6 mol)	0.495	0.0357	0
10^6 mol/coeff	0.248	0.0357 (LR)	
Change (10^6 mol)	- 2(0.0357)	-0.0357	+0.0357
Final (10^6 mol)	0.424	0.00	0.0357

The mass that could be produced is

$$0.0357 \times 10^6 \text{ mol} \left(\frac{32.04 \text{ g}}{1 \text{ mol}} \right) \left(\frac{1 \text{ ton}}{10^6 \text{ g}} \right) = 1.14 \text{ metric ton}$$

(b)
Calculate the initial amounts:

$$1000 \text{ kg CaO} \left(\frac{10^3 \text{ g}}{1 \text{ kg}} \right) \left(\frac{1 \text{ mol}}{56.08 \text{ g}} \right) = 0.0178 \times 10^6 \text{ mol};$$

$$1000 \text{ kg C} \left(\frac{10^3 \text{ g}}{1 \text{ kg}} \right) \left(\frac{1 \text{ mol}}{12.01 \text{ g}} \right) = 0.0833 \times 10^6 \text{ mol}$$

Now set up the table of amounts:

Reaction	CaO +	3 C →	CO +	CaC$_2$
Amt (10^6 mol)	0.0178	0.0833	0	0
10^6 mol/coeff	0.0178 (LR)	0.0278		
Change (10^6 mol)	-0.0178	-3(0.0178)	+0.0178	+0.0178
Final (10^6 mol)	0	0.0299	+0.0178	+0.0178

The masses that could be produced are:

$$\text{CO: } 0.0178 \times 10^6 \text{ mol} \left(\frac{28.01 \text{ g}}{1 \text{ mol}} \right) \left(\frac{1 \text{ ton}}{10^6 \text{ g}} \right) = 0.499 \text{ metric ton};$$

CaC_2: $0.0178 \times 10^6 \text{ mol} \left(\dfrac{64.10 \text{ g}}{1 \text{ mol}} \right) \left(\dfrac{1 \text{ ton}}{10^6 \text{ g}} \right) = 1.14$ metric ton.

(c)

Calculate the initial amounts:

$1000 \text{ kg C}_2\text{H}_4 \left(\dfrac{10^3 \text{ g}}{1 \text{ kg}} \right) \left(\dfrac{1 \text{ mol}}{28.05 \text{ g}} \right) = 0.0357 \times 10^6 \text{ mol};$

$1000 \text{ kg O}_2 \left(\dfrac{10^3 \text{ g}}{1 \text{ kg}} \right) \left(\dfrac{1 \text{ mol}}{32.00 \text{ g}} \right) = 0.0313 \times 10^6 \text{ mol};$

$1000 \text{ kg HCl} \left(\dfrac{10^3 \text{ g}}{1 \text{ kg}} \right) \left(\dfrac{1 \text{ mol}}{36.46 \text{ g}} \right) = 0.0274 \times 10^6 \text{ mol}$

Now set up the table of amounts:

Reaction	2 C_2H_4 +	O_2 +	4 HCl $\rightarrow$	2 $C_2H_4Cl_2$	+ 2 H_2O
Amt (10^6 mol)	0.0357	0.0313	0.0274	0	0
10^6 mol/coeff	0.0179	0.0313	0.00685 (LR)		
Change (10^6 mol)	-¾ (0.0274)	-¼ (0.0274)	-.0274	+¾ (0.0274)	+¾ (0.0274)
Final (10^6 mol)	0.0220	0.0245	0	0.0137	0.0137

The masses that could be produced are:

$C_2H_4Cl_2$: $0.0137 \times 10^6 \text{ mol} \left(\dfrac{98.96 \text{ g}}{1 \text{ mol}} \right) \left(\dfrac{1 \text{ ton}}{10^6 \text{ g}} \right) = 1.36$ metric ton;

H_2O: $0.0137 \times 10^6 \text{ mol} \left(\dfrac{18.02 \text{ g}}{1 \text{ mol}} \right) \left(\dfrac{1 \text{ ton}}{10^6 \text{ g}} \right) = 0.247$ metric ton.

4.27 The problem gives information about the amounts of both starting materials, so this is a limiting reactant situation. We must calculate the number of moles of each species, construct a table of amounts, and use the results to determine the mass of the product formed.

Starting amounts are in kilograms (1 kg = 10^3 g), so it will be convenient to work with 10^3 mol amounts. The balanced equation is given in the problem.

Begin by calculating the initial amounts:

$75.0 \text{ kg N}_2 \left(\dfrac{10^3 \text{ g}}{1 \text{ kg}} \right) \left(\dfrac{1 \text{ mol}}{28.02 \text{ g}} \right) = 2.68 \times 10^3 \text{ mol};$

$75.0 \text{ kg H}_2 \left(\dfrac{10^3 \text{ g}}{1 \text{ kg}} \right) \left(\dfrac{1 \text{ mol}}{2.02 \text{ g}} \right) = 37.1 \times 10^3 \text{ mol};$

Next, using the balanced chemical equation, construct an amounts table:

Reaction	N_2 +	3 H_2 $\rightarrow$	2 NH_3
Amt (10^3 mol)	2.68	37.1	0

10^3 mol/coeff	2.68 (LR)	12.4	
Change (10^3 mol)	-2.68	-8.04	+5.36
Final (10^3 mol)	0	29.1	5.36

The mass of ammonia that could be produced is:

$$5.36 \times 10^3 \text{ mol}\left(\frac{17.04\,\text{g}}{1\,\text{mol}}\right)\left(\frac{10^{-3}\,\text{kg}}{1\,\text{g}}\right) = 91.3 \text{ kg.}$$

4.29 To determine which ingredient will run out first, calculate how many cheeseburgers could be made with each ingredient:

INGREDIENT	INVENTORY	AMT/BURGER	# BURGERS
Roll	(12 dozen)(12/dozen)	1	144
Beef	40 lb	2(1/4 lb)	80
Cheese	(2 pkg)(65/pkg)	1	130
Tomato	40	¼	160
Lettuce	(1 kg)(10^3 g/kg)	15 g	66.7

The lettuce will run out first, after 66 burgers have been made.

4.31 The problem gives information about the amounts of both starting materials, so this is a limiting reactant situation. We must calculate the number of moles of each species, construct a table of amounts, and use the results to determine the masses of the product formed and the remaining reactant.
Begin by determining the balanced chemical equation:
$P_4 + O_2 \rightarrow P_4O_{10}$

The P is already balanced. To balance the O, give O_2 a coefficient of 5:
$P_4 + \mathbf{5}O_2 \rightarrow P_4O_{10}$

Next, calculate the initial amounts:
$$3.75 \text{ g } P_4\left(\frac{1\,\text{mol}}{123.9\,\text{g}}\right) = 0.0303 \text{ mol;}$$

$$6.55 \text{ g } O_2\left(\frac{1\,\text{mol}}{32.00\,\text{g}}\right) = 0.205 \text{ mol;}$$

Construct the amounts table using the balanced equation above:

Reaction	P_4 +	$5\,O_2$ →	P_4O_{10}
Amt (mol)	0.0303	0.205	0
mol/coeff	0.0303 (LR)	0.0410	
Change (mol)	-0.0303	-(5)(0.0303)	+0.0303
Final (mol)	0.00	0.0535	0.0303

Now obtain the mass of P_4O_{10} produced, using the information from the amounts table and the molar mass:
$MM = 4(30.97 \text{ g/mol}) + 10(16.00 \text{ g/mol}) = 283.9 \text{ g/mol}$

The mass that could be produced is: $0.0303 \text{ mol} \left(\frac{283.9 \text{ g}}{1 \text{ mol}} \right) = 8.60 \text{ g } P_4O_{10}$ produced.

Finally, determine the mass of O_2 left over:

There would be $0.0535 \text{ mol} \left(\frac{32.00 \text{ g}}{1 \text{ mol}} \right) = 1.7 \text{ g of } O_2$ left unreacted. (2 sig. figs. because the final moles of O_2 has only two sig. figs., although 3 are shown to avoid roundoff errors)

4.33   Identify species based on the type of substance present: Ions for salts and strong acids/bases, molecules for other substances. H_2O is always a major species in aqueous solution.
(a) Salt: NH_4^+, Cl^-, H_2O; (b) Salt: Fe^{2+}, ClO_4^-, H_2O; (c) Salt: Na^+, SO_4^{2-}, H_2O;
(d) Not a salt: Br_2, H_2O; (e) Salt: K^+, Br^-, H_2O.

4.35   Identify species based on the type of substance present: Ions for salts and strong acids/bases, molecules for other substances. H_2O is always a major species in aqueous solution.
(a) Salt: K^+, HPO_4^-, H_2O; (b) Weak acid: CH_3CO_2H, H_2O; (c) Salt: Na^+, $CH_3CO_2^-$, H_2O; (d) Weak base: NH_3, H_2O; (e) Salt: NH_4^+, Cl^-, H_2O.

4.37   To determine whether a precipitate will form, consider the solubility guidelines. If any combination of cations and anions is insoluble, that salt will precipitate.
(a) $AgNO_3$: All nitrates are soluble, so a precipitate forms only if the silver salt is insoluble.   (a) AgCl precipitates; (b, c, and d) No precipitate; (e) AgBr precipitates.

(b) Na_2CO_3: All sodium salts are soluble, so a precipitate forms only if the carbonate salt is insoluble.   (a) No precipitate; (b) $FeCO_3$ precipitates; (c, d, and e) No precipitate.

(c) $Ba(OH)_2$: Both ions form insoluble salts unless their partners convey solubility.
(a, d, and e) No precipitate; (b) $Fe(OH)_2$ precipitates; (c) $BaSO_4$ precipitates.

4.39   A net ionic equation shows which ions combine to give new products. Spectator ions do not appear in the net equation, and charges must be balanced. Start by determining what ions are present in solution. Then use the Solubility Guidelines from your text to determine what precipitates can form.
(a) The ions present are: Ag^+, NO_3^-, K^+, and OH^-. So the possible combinations are: $AgNO_3$, $AgOH$, KOH, and KNO_3. Guideline 2 predicts that salts of potassium and nitrate are soluble. Since $AgOH$ is not soluble by the guidelines, it must be an insoluble salt:

$Ag^+(aq) + OH^-(aq) \rightarrow AgOH(s);$ spectator ions: NO_3^-, K^+

(b) The ions present are: Fe^{3+}, ClO_4^-, NH_4^+, and $C_2O_4^{2-}$. The possible combinations are: $Fe(ClO_4)_3$, $Fe_2(C_2O_4)_3$, NH_4ClO_4, $(NH_4)_2C_2O_4$. Guidelines 1 and 2 predict that all ammonium salts and all perchlorate salts are soluble. $Fe(C_2O_4)_3$ is not covered in guidelines 1,2, or 5 and thus is insoluble by guideline 3:

$$2\ Fe^{3+}{}_{(aq)} + 3\ C_2O_4{}^{2-}{}_{(aq)} \rightarrow Fe_2(C_2O_4)_{3(s)}; \quad \text{spectator ions: } ClO_4^-,\ NH_4^+$$

(c) The ions present are: Pb^{2-}, Br^-, Na^+, and NO_3^-. By guidelines 1 and 2, all salts of sodium and nitrate are soluble. $PbBr_2$ is covered in guideline 4 as an insoluble salt:

$$Pb^{2-}{}_{(aq)} + 2Br^-{}_{(aq)} \rightarrow PbBr_{2(s)}; \qquad \text{spectator ions: } NO_3^-,\ Na^+$$

(d) The ions present are: Ni^{2+}, OH^-, K^+, and $SO_4{}^{2-}$. By guidelines 1 and 3, all salts of potassium and most of sulfate are soluble. Thus, the only possible precipitate is $Ni(OH)_2$, which is insoluble by guideline 3:

$$Ni^{2+}{}_{(aq)} + 2\ OH^-{}_{(aq)} \rightarrow Ni(OH)_{2(s)}; \qquad \text{spectator ions: } K^+,\ SO_4^{2-}$$

4.41　The problem gives information about the amounts of both starting materials, so this is a limiting reactant situation. We must calculate the number of moles of each species, construct a table of amounts, and use the results to determine the final product mass. Start by determining the balanced net ionic reaction using the solubility guidelines.

The ions present are Ag^+, NO_3^-, K^+, and $CO_3{}^{2-}$. Guideline 1 and 2 state that all salts containing potassium and nitrate are soluble. Ag_2CO_3 is not covered in guidelines 1 or 2, and thus by guideline 3 is insoluble:

$$2\ Ag^+{}_{(aq)} + CO_3{}^{2-}{}_{(aq)} \rightarrow Ag_2CO_{3(s)}.$$

Calculations of initial amounts:

For Ag^+, 55.0 ml = 0.0550 L and 5.00×10^{-2} M = 0.0500 mol Ag^+/L

$$0.0550\,L\left(\frac{0.0500\,mol}{1\,L}\right) = 2.75 \times 10^{-3}\,mol\ Ag^+$$

For $CO_3{}^{2-}$, 95.0 mL = 0.0950 L and 3.50×10^{-2} M = 0.0350 mol $CO_3{}^{2-}$/L

$$0.0950\,L\left(\frac{0.0350\,mol}{1\,L}\right) = 3.33 \times 10^{-3}\ mol\ CO_3^{2-}$$

Next set up the amounts table:

Reaction	2 Ag^+ +	CO_3^{2-} $\rightarrow$	$Ag_2CO_{3(s)}$
Start (10^{-3} mol)	2.75	3.33	0
Mol/coeff	2.75/2 =1.38 (LR)	3.33	
Change (10^{-3} mol)	-2.75	-2.75/2	+2.75/2
Final (10^{-3} mol)	0	1.96	1.38

The mass of solid that forms is:

1.38×10^{-3} mol $\left(\dfrac{275.8\,g}{1\,mol}\right) = 0.381$ g. The ions remaining in solution are the excess CO_3^{2-} and the spectator ions, NO_3^- and K^+.

4.43   In each case all are acid-base reactions, where the H^+ is being transferred from the acid to the base.

(a) HCl is a strong acid and forms H_3O^+ and Cl^- ions in solution, $Ca(OH)_2$ is soluble and a strong base, therefore, the reaction is:
$$H_3O^+(aq) + OH^-(aq) \rightarrow H_2O(l)$$
spectator ions: Ca^{2+}, Cl^-

(b) H_3PO_4 is a weak acid and doesn't dissociate in soln. LiOH is a strong base and forms OH^- and Li^+ ions in solution. Reaction:
$$H_3PO_4(aq) + 3OH^-(aq) \rightarrow 3H_2O(l) + PO_4^{3-}(aq)$$
spectator ions: Li^+

(c) NH_3 is a weak base, HNO_3 is a strong acid and forms H_3O^+ and NO_3^- ions. Reaction:
$$NH_3(aq) + H_3O^+(aq) \rightarrow H_2O(l) + NH_4^+(aq)$$
spectator ions: NO_3^-

(d) CH_3CO_2H is a weak acid, KOH is a strong base and forms OH^- and K^+ ions in soln. Reaction:
$$CH_3CO_2H(aq) + OH^-(aq) \rightarrow H_2O(l) + CH_3CO_2^-(aq)$$
spectator ions: K^+

4.45   This problem describes an acid-base reaction. We are asked to determine the mass of base required to completely neutralize the acid. The starting materials are HCl, a strong acid, and $Al(OH)_3$, a weak base. In addition to water, the major species are:
$$H_3O^+, Cl^-, Al(OH)_3,$$
The presence of hydronium ions and a weak base together as major species will result in an acid-base reaction:
The balanced reaction is:
$$Al(OH)_3(s) + 3\ H_3O^+(aq) \rightarrow Al^{3+}(aq) + 6\ H_2O(l).$$

The problem gives information about the amount of acid present. This is a mol-mass conversion problem:

Calculation of initial amount of acid:
For H_3O^+: 0.155 L$\left(\dfrac{0.175\,mol}{1\,L}\right) = 2.71 \times 10^{-2}$ mol H_3O^+;

$MM\ [Al(OH)_3] = 26.98$ g/mol $+ 3(16.00$ g/mol$) + 3(1.008$ g/mol$) = 78.00$ g/mol

2.71×10^{-2} mol $H_3O^+ \left(\dfrac{1\,mol\,Al(OH)_3}{3\,mol\,H_3O^+}\right)\left(\dfrac{78.00\,g}{1\,mol}\right) = 0.705$ g $Al(OH)_3$

4.47 This problem describes an acid-base reaction. We are asked to determine the final concentrations of all the major ions in a final solution. Begin by analyzing the chemistry. The starting materials are HCl $_{(aq)}$, a strong acid, and Ba(OH)$_2$ $_{(aq)}$, a strong base. In addition to water, the major species in solution are:

$$H_3O^+, Cl^-, Ba^{2+}, OH^-$$

The presence of hydronium and hydroxide ions together as major species in solution always results in an acid-base reaction:

$$H_3O^+ + OH^- \rightarrow 2\,H_2O$$

The problem gives information about the amounts of both starting materials, so this is a limiting reactant situation. We must calculate the number of moles of each species, construct a table of amounts, and use the results to determine the final solution concentrations.

Calculations of initial amounts:

For HCl $_{(aq)}$, 100.0 mL = 0.1000 L and 5.00×10^{-2} M = 0.0500 mol HCl/L

$$(0.100\ L)\left(0.0500\,\frac{mol}{L}\right) = 0.00500\ mol\ H_3O^+\ and\ \ 0.00500\ mol\ Cl^-$$

For Ba(OH)$_2$ $_{(aq)}$, 150.0 mL = 0.1500 L and 2.00×10^{-2} M = 0.0200 mol Ba(OH)$_2$/L

$$(0.150\ L)\left(0.0200\,\frac{mol}{L}\right) = 0.00300\ mol\ Ba^{2+}\ and\ \ 0.00600\ mol\ OH^-$$

Now we set up the table of amounts. The acid reaction involves only H_3O^+ and OH^-. The amounts of Ba^{2+} and Cl^- do not change, so we may omit these species from the table. Moreover, the water generated in the proton transfer reaction joins the bulk solvent, so we may ignore the water in our calculations.

The balanced equation shows that the starting materials react in a 1:1 mole ratio, so we can identify the limiting reactant by inspection; the limiting reactant is H_3O^+. Here is the complete table of amounts:

Reaction:	H_3O^+ +	$OH^- \rightarrow$	$2\,H_2O$
Start (mol)	0.00500	0.00600	---
Change	–0.00500	–0.00500	---
Final	0	0.00100	solvent

Before calculating the final concentrations, it is a good idea to organize the final amounts:

$H_3O^+ = 0$ mol $OH^- = 0.00100$ mol $Cl^- = 0.00500$ mol $Ba^{2+} = 0.00300$ mol

The final concentrations are obtained by dividing the number of moles of each species by the final volume of the solution.

$$V_{final} = 100.0 \text{ mL} + 150.0 \text{ mL} = 250.0 \text{ mL} = 0.2500 \text{ L}$$

Here are the final concentrations:

$$[H_3O^+] = 0 \qquad\qquad [OH^-] = \frac{0.00100 \text{ mol}}{0.250 \text{ L}} = 0.00400 \text{ M}$$

$$[Cl^-] = \frac{0.00500 \text{ mol}}{0.2500 \text{ L}} = 0.0200 \text{ M} \qquad [Ba^{2+}] = \frac{0.00300 \text{ mol}}{0.2500 \text{ L}} = 0.0120 \text{ M}$$

4.49   This problem describes an acid-base titration. We are asked to determine the concentration of base used to neutralize the acid. The starting materials are HCl, a strong acid, and KOH, a strong base. In addition to water, the major species are:

$$H_3O^+, Cl^-, K^+, OH^-$$

The presence of hydronium ions and hydroxide ions together as major species will result in an acid-base reaction:
The balanced reaction is:

$$OH^- + H_3O^+ \rightarrow 2 H_2O_{(l)}.$$

Calculation of the amount of acid added:

$$H_3O^+: 27.35 \text{ mL} \left(\frac{10^{-3} \text{ L}}{1 \text{ mL}} \right) \left(\frac{0.1206 \text{ mol}}{1 \text{ L}} \right) = 3.298 \times 10^{-3} \text{ mol}$$

Since this is a completed titration, the moles of acid added to the solution will equal the moles of base in the solution.

$$n(H_3O^+) = 3.298 \times 10^{-3} \text{ mol} = n(OH^-)$$

The concentration of the KOH solution is obtained by dividing the number of moles of base by the volume of the KOH solution.

$$V_{soln} = 0.00500 \text{ L}$$

$$[OH^-] = \frac{3.298 \times 10^{-3} \text{ mol}}{0.00500 \text{ L}} = 0.660 \text{ M}$$

4.51   This problem describes an acid-base titration. We are asked to determine the concentration of base used to neutralize the acid. The starting materials are NaOH, a strong base, and $KHC_8H_4O_4$, the salt of a weak acid. In addition to water, the major species are:

$$Na^+, OH^-, K^+, HC_8H_4O_4^-$$

The presence of hydroxide ions and $HC_8H_4O_4^-$ ions together as major species will result in an acid base reaction.
The balanced reaction is:

$$OH^- + HC_8H_4O_4^- \rightarrow C_8H_4O_4^{2-} + H_2O_{(l)}$$

Calculation of the amount of $KHC_8H_4O_{4(aq)}$:

$$MM = 39.10 \text{ g/mol} + 8(12.01 \text{ g/mol}) + 5(1.01 \text{ g/mol}) + 4(16.00 \text{ g/mol}) = 204.23 \text{ g/mol}$$

$$n = 0.6634 \text{ g} \left(\frac{1 \text{ mol}}{204.23 \text{ g}} \right) = 0.003248 \text{ mol of } HC_8H_4O_4^- \text{(aq)}$$

Since this is a completed titration, the moles of base added to the solution will equal the moles of acid in the solution.

$$n(HC_8H_4O_4^-) = 0.003248 \text{ mol} = n(OH^-)$$

The concentration of the NaOH solution is obtained by dividing the number of moles of base by the volume of the NaOH solution.

$$V_{soln} = 0.03655 \text{ L}$$

$$[OH^-] = \frac{0.003248 \text{ mol}}{0.03655 \text{ L}} = 0.08886 \text{ M}$$

4.53 A redox reaction between a metal and a cation occurs when the metal is higher in the activity series than the cation. The activity series appears as Table 4-3 of your textbook.
(a and b) Cu is below H_2 and Mg in the activity series, so no reaction occurs;
(c) Cu is above Ag in the activity series, so a displacement reaction occurs:
$$Cu(s) + 2 Ag^+(aq) \rightarrow Cu^{2+}(aq) + 2 Ag(s);$$
(d) K is above H_2 in the activity series, so a displacement reaction occurs:
$$2 K(s) + 2 H_2O(l) \rightarrow 2 K^+(aq) + H_2(g) + 2 OH^-(aq)$$

4.55 Formation reactions are easily balanced once the chemical formula of the oxide is known. Remember that O carries a charge of −2 and that the compound must be electrically neutral.
(a) Sr is in Group 2, so it forms a +2 cation and its oxide is SrO:
$$2 Sr + O_2 \rightarrow 2 SrO;$$
(b) Cr(III) has a +3 charge and its oxide is Cr_2O_3:
$$4 Cr + 3 O_2 \rightarrow 2 Cr_2O_3;$$
(c) Sn(IV) has a +4 charge and its oxide is SnO_2:
$$Sn + O_2 \rightarrow SnO_2.$$

4.57 This problem describes a redox reaction. We are asked to determine the mass H_2 that will form from the reaction. Begin by analyzing the chemistry. The starting materials are HCl (aq), a strong acid, and Al metal. In addition to water, the major species in solution are:
$$Al^{3+}, H_3O^+, Cl^-$$

Use Table 4-3 to identify the half-reactions:
$$Al \rightarrow Al^{3+} + 3e^-$$
$$H_3O^+ + e^- \rightarrow \tfrac{1}{2} H_2 + H_2O$$
To balance the number of electrons, multiply the second equation by 3 and add the two equations:
$$3 H_3O^+ + Al \rightarrow Al^{3+} + \tfrac{3}{2} H_2 + 3 H_2O.$$

Multiply all coefficients by 2 to eliminate the fraction and obtain the balanced reaction:
$$6\ H_3O^+ + 2\ Al \rightarrow 2\ Al^{3+} + 3\ H_2 + 6\ H_2O.$$

The problem gives information about the amounts of both starting materials, so this is a limiting reactant situation. We must calculate the number of moles of each species, construct a table of amounts, and use the results to determine the final amounts. Calculations of initial amounts:
(keep an additional significant figure to avoid round off error)

$HCl_{(aq)}$: 8.00 mL = 0.00800 L;
$$n = 0.00800\ L\left(\frac{6.00\ mol}{1\ L}\right) = 4.800 \times 10^{-2}\ mol$$

Al: $n = 0.355\ g\left(\frac{1\ mol}{26.98\ g}\right) = 1.316 \times 10^{-2}\ mol$

Because the problem asks only about the mass of H_2 formed, we do not need to do calculations for the other products.

Reaction	$6\ H_3O^+ +$	$2\ Al \rightarrow$	$3\ H_2$
Amt (10^{-2} mol)	4.800	1.316	0
10^{-2} mol/coeff	0.800	0.658 (LR)	
Change (10^{-2} mol)	-3(1.316)	-1.316	+(³⁄₂)(1.316)
Final (10^{-2} mol)	0.852	0	1.974

Obtain the mass of H_2 formed by multiplying the moles of H_2 from the table by the molar mass:
$$H_2\!:\ m = 1.974 \times 10^{-2}\ mol\left(\frac{2.016\ g}{1\ mol}\right) = 3.98 \times 10^{-2}\ g$$

4.59   The first two parts of this problem ask about an ionic reaction. Identify the species present in the solution and use the products to find the net ionic reaction and the spectator ions. The species present are H_2O, NH_4^+, HCO_3^-, Na^+, and Cl^-. The solid product is $NaHCO_3$.

(a) The net ionic reaction is $Na^+ + HCO_3^- \rightarrow NaHCO_{3(s)}$;
(b) NH_4^+ and Cl^- are spectator ions;

(c) The second two parts of this problem involve stoichiometric calculations. This is a limiting reagent problem as well as a yield problem, requiring a table of amounts to determine the theoretical yield.

Begin by determining the initial amounts of the reactants:

Na^+: $n = 5.00 \times 10^2 \, L \left(\dfrac{6.00 \, mol}{1 \, L} \right) = 3.00 \times 10^3 \, mol$

HCO_3^-: $n = 5.00 \times 10^2 \, L \left(\dfrac{1.50 \, mol}{1 \, L} \right) = 7.50 \times 10^2 \, mol$

The balanced equation shows that the starting materials react in a 1:1 mole ratio, so we can identify the limiting reactant by inspection; the limiting reactant is HCO_3^-. Here is the complete table of amounts:

Reaction	Na^+ +	HCO_3^- $\rightarrow$	$NaHCO_3$
Start (10^2 mol)	30.0	7.50	0
Change (10^2 mol)	-7.50	-7.50	+7.50
Final (10^2 mol)	22.5	0	7.50

The theoretical yield of $NaHCO_3$ is 7.50×10^2 mol.

Now determine the actual yield from the amount of product formed (35.0 kg):
MM ($NaHCO_3$) = 22.99 g/mol + 1.01 g/mol + 12.01 g/mol + 3(16.00 g/mol)
MM = 84.01 g/mol

The actual yield is $35.0 \, kg \left(\dfrac{10^3 \, g}{1 \, kg} \right) \left(\dfrac{1 \, mol}{84.01 \, g} \right) = 4.17 \times 10^2$ mol;

Obtain the percent yield by dividing the actual yield by the theoretical yield:

The percent yield is $\dfrac{4.17 \times 10^2 \, mol}{7.50 \times 10^2 \, mol} \times 100\% = 55.6\%$

(Low because $NaHCO_3$ is relatively soluble, so many ions remain in solution).

(d) To find concentrations in the final solution, first determine how many moles of each ion remain in solution, then divide by the final volume of the solution:
$V_{final} = 5.00 \times 10^2 \, L + 5.00 \times 10^2 \, L = 1.000 \times 10^3 \, L$

The amounts of the two spectator ions are unaffected by the reaction:

$[NH_4^+] = \dfrac{7.50 \times 10^2 \, mol}{1.000 \times 10^3 \, L} = 0.750 \, M$;

$[Cl^-] = \dfrac{30.0 \times 10^2 \, mol}{1.000 \times 10^3 \, L} = 3.00 \, M$;

The amounts of the reactant ions are reduced by the amount that precipitates (Note the final amounts from the amounts table are the theoretical yields; we want to use the actual amounts):

$[Na^+] = \dfrac{30.0 \times 10^2 \, mol - 4.17 \times 10^2 \, mol}{1.000 \times 10^3 \, L} = 2.58 \, M$;

$$[HCO_3^-] = \frac{7.50 \times 10^2 \, mol - 4.17 \times 10^2 \, mol}{1.000 \times 10^3 \, L} = 0.333 \, M.$$

4.61   The formation reactions for oxides are easily balanced by inspection. The number of electrons lost by the metal atom can be determined by assigning two additional electrons to each oxygen atom in the oxide.

$4 \, Ru + 3 \, O_2 \rightarrow 2 \, Ru_2O_3$. Three O atoms gain six electrons, so each Ru loses three.

$Ru + O_2 \rightarrow RuO_2$. Two O atoms gain four electrons, so each Ru loses four.

$2 \, Ru + 3 \, O_2 \rightarrow 2 \, RuO_3$. Three O atoms gain six electrons, so each Ru loses six.

$Ru + 2 \, O_2 \rightarrow RuO_4$. Four O atoms gain eight electrons, so each Ru loses eight.

4.63   This problem describes a redox reaction. We are asked to determine the mass of iron(III) oxide that will form from the reaction. Begin by analyzing the chemistry. The starting materials are FeS_2 and excess O_2 and the products are iron(III) oxide and SO_2. The reaction (unbalanced) is:

$$FeS_2 + O_2 \rightarrow Fe_2O_3 + SO_2$$
$$1Fe + 2 \, S + 2O \rightarrow 2Fe + 1S + 5O$$

Balance the Fe by giving FeS_2 a coefficient of 2:

$$\mathbf{2}FeS_2 + O_2 \rightarrow Fe_2O_3 + SO_2$$
$$2Fe + 4S + 2O \rightarrow 2Fe + 1S + 5O$$

Next, balance the sulfur by changing the coefficient on SO_2 from 1 to 4:

$$2FeS_2 + O_2 \rightarrow Fe_2O_3 + \mathbf{4}SO_2$$
$$2Fe + 4S + 2O \rightarrow 2Fe + 4S + 11O$$

Finally, balance oxygen by giving the O_2 a coefficient of 11/2 and eliminate the fraction by multiplying by 2:

$$4 \, FeS_2 + 11 \, O_2 \rightarrow 2 \, Fe_2O_3 + 8 \, SO_2.$$

Determine the amount of iron(II) pyrite by multiplying the percent composition by the mass of the ore:

$$m_{pyrite} = 175 \, ton\left(\frac{55\%}{100\%}\right) = 96.25 \, ton;$$

Use the mass-mole-mass calculations to obtain the theoretical mass of the iron(III) oxide formed:

$$m_{oxide} = 96.25 \, ton\left(\frac{1 \, mol}{120.2 \, g}\right)\left(\frac{2 \, mol \, Fe_2O_3}{4 \, mol \, FeS_2}\right)\left(\frac{159.7 \, g}{1 \, mol}\right) = 63.9 \, ton;$$

Multiply by the percent efficiency to determine the actual mass recovered:

Mass recovered $= 63.9 \text{ ton} \left(\dfrac{85\%}{100\%} \right) = 54$ metric tons.

4.65 When determining reaction products, first identify the type of substances present to determine what kind of reaction can occur.

(a) Oxygen can oxidize metals to metal oxides:
$2\,Al(s) + 3\,O_2(g) \rightarrow 2\,Al_2O_3(s)$ (redox reaction);

(b) Oxygen can oxidize hydrocarbons to carbon dioxide and water when heated:
$C_3H_8(g) + 5\,O_2(g) \rightarrow 3\,CO_2(g) + 4\,H_2O(g)$ (redox reaction);

(c) Mg displaces hydrogen gas from strong acids such as HBr:
$Mg(s) + 2\,H_3O^+(aq) \rightarrow Mg^{2+}(aq) + H_2(g) + 2\,H_2O(l)$ (redox reaction);

(d) NaOH is a strong base, HCl is a strong acid:
$OH^-(aq) + H_3O^+(aq) \rightarrow 2H_2O(l)$ (acid-base reaction);

(e) Both compounds are salts, so a precipitate will form if any combination is insoluble:
$Pb^{2+}(aq) + CO_3^{2-}(aq) \rightarrow PbCO_3(s)$ (precipitation reaction).

(f) Ca(OH)$_2$ is a strong base, H$_2$SO$_4$ is a strong acid for the first H;
$OH^-(aq) + H_3O^+(aq) \rightarrow 2\,H_2O(l)$ (acid-base reaction)

4.67 This problem describes a redox reaction. We are asked to determine the percent yield for the reaction. Begin by analyzing the chemistry. The starting materials are Xe and F_2, and the product is XeF_4.

The balanced reaction is: $Xe + 2\,F_2 \rightarrow XeF_4$.

The problem gives information about the amount of one of the starting materials (the other is in excess), so this is a simple mass-mole-mass problem. We must calculate the theoretical yield, determine the percent yield, and use the actual yield to determine the final amount of Xe left over.

Calculations of theoretical yield:
$MM\,(XeF_4) = 131.3 \text{ g/mol} + 4(19.0 \text{ g/mol}) = 207.3 \text{ g/mol}$

$n = 5.00 \text{ g} \left(\dfrac{1\,mol}{131.3\,g} \right) \left(\dfrac{1\,mol\,XeF_4}{1\,mol\,Xe} \right) = 3.81 \times 10^{-2} \text{ mol}$

The actual yield of XeF_4 is $4.00 \text{ g} \left(\dfrac{1\,mol}{207.3\,g} \right) = 1.93 \times 10^{-2}$ mol.

Now determine the percent yields using the actual yield and the theoretical yield:
The percent yield is $\dfrac{1.93 \times 10^{-2}\,mol}{3.81 \times 10^{-2}\,mol} \times 100\% = 50.7\%$

The amount of Xe left unreacted is the difference of the starting amount and the actual XeF_4 yield:

$$m = (3.81 \times 10^{-2} \text{ mol} - 1.93 \times 10^{-2} \text{ mol}) \left(\frac{131.3 \text{ g}}{1 \text{ mol}} \right) = 2.47 \text{ g}.$$

4.69   Examine the molecular picture to see what chemical reaction occurs. The reactants are atomic X and atomic Y and the product of the reaction is YX_2 (there are atoms of X left over). Thus the reaction is: $Y + 2X \rightarrow YX_2$.

4.71   A balanced chemical equation has equal numbers of atoms of each element on each side of the arrow. Adjust stoichiometric coefficients to balance each element in turn, beginning with those that appear in only one reactant and product. Generally, balance H and O last.

(a) $H_2 + NO \rightarrow NH_3 + H_2O$

$2H + 1N + 1O \rightarrow 5H + 1N + 1O$

N and O are already balanced. Give H_2 a coefficient of 5/2 to balance H, then multiply through by 2 to clear fractions:

$$5H_2 + 2NO \rightarrow 2NH_3 + 2H_2O$$

(b) $CO + NO \rightarrow N_2 + CO_2$

$1C + 1N + 2O \rightarrow 1C + 2N + 2O$

C is already balanced. Give N_2 a coefficient of ½ to balance N, then multiply through by 2 to clear fractions:

$$2CO + 2NO \rightarrow N_2 + 2CO_2$$

(c) $NH_3 + O_2 \rightarrow N_2O + H_2O$

$3H + 1N + 2O \rightarrow 2H + 2N + 2O$

Give NH_3 a coefficient of 2 to balance N

$$2NH_3 + O_2 \rightarrow N_2O + H_2O$$

$6H + 2N + 2O \rightarrow 2H + 2N + 2O$

Both N and O are now balanced. To balance H, give H_2O a coefficient of 3:

$$2NH_3 + O_2 \rightarrow N_2O + 3H_2O$$

$6H + 2N + 2O \rightarrow 6H + 2N + 4O$

This balances H, but now O is unbalanced. To balance O give O_2 a coefficient of 2 which gives the balanced reaction:

$$2NH_3 + 2O_2 \rightarrow N_2O + 3H_2O$$

$6H + 2N + 4O \rightarrow 6H + 2N + 4O$

(d)   The reaction (unbalanced) is:

$$NO + NH_3 \rightarrow N_2 + H_2O$$

$3H + 2N + 1O \rightarrow 2H + 2N + 1O$

N appears in three reagents, so balance it last. Balance H by giving NH_3 a coefficent of 2 and H_2O a coefficient of 3:

$$NO + 2NH_3 \rightarrow N_2 + 3H_2O$$

$$6 \, H + 3 \, N + 1 \, O \rightarrow 6 \, H + 2 \, N + 3 \, O$$

Next balance O by giving NO a coefficient of 3:
$$3 \, NO + 2 \, NH_3 \rightarrow N_2 + 3 \, H_2O$$
$$6 \, H + 5 \, N + 3 \, O \rightarrow 6 \, H + 2 \, N + 3 \, O$$

Now there are 5 N on the left, so give N_2 a coefficient of 5/2 to balance N and multiply through by 2 to clear fractions and obtain the balanced reaction:
$$6 \, NO + 4 \, NH_3 \rightarrow 5 \, N_2 + 6 \, H_2O$$

(d) The reaction (unbalanced) is:

$$H_2O + NO \rightarrow O_2 + NH_3$$
$$2 \, H + 1 \, N + 2 \, O \rightarrow 3 \, H + 1 \, N + 2 \, O$$

N is already balanced. Balance H by giving NH_3 a coefficient of 2 and H_2O a coefficient of 3:
$$\mathbf{3} \, H_2O + NO \rightarrow O_2 + \mathbf{2}NH_3$$
$$6 \, H + 1 \, N + 4O \rightarrow 6 \, H + 2 \, N + 2 \, O$$

Next give NO a coefficient of 2 to balance N:
$$3 \, H_2O + \mathbf{2}NO \rightarrow O_2 + 2NH_3$$
$$6 \, H + 2 \, N + 5O \rightarrow 6 \, H + 2 \, N + 2 \, O$$

Now there are 5 O on the left, so give O_2 a coefficient of 5/2 to balance O and multiply through by 2 to clear fractions:
$$6 \, H_2O + 4 \, NO \rightarrow 5 \, O_2 + 4 \, NH_3$$

(e) The reaction (unbalanced) is:
$$H_2 + O_2 \rightarrow H_2O$$
$$2 \, H + 2O \rightarrow 2 \, H + 1O$$
Give O_2 a coefficient of ½ to balance O, then multiply through by 2 to clear fractions:
$$2 \, H_2 + O_2 \rightarrow 2 \, H_2O$$

4.73   A convenient synthesis of ionic salts starts with solutions of soluble salts of the cation and anion which, when mixed, form a precipitate of the desired product. Sodium salts are inexpensive soluble sources of anions and chlorides or nitrates are inexpensive soluble sources of cations. Do mole-mass conversions to compute the required masses.

(a) Mix solutions of Na_3PO_4 and $FeCl_3$. The net ionic reaction is:
$$Fe^{3+}(aq) + PO_4^{3-}(aq) \rightarrow FePO_4(aq);$$

$$n_{FePO_4} = \frac{m}{MM} = 2.50 \, kg \left(\frac{10^3 \, g}{1 \, kg} \right) \left(\frac{1 \, mol}{150.82 \, g} \right) = 16.6 \, mol;$$

$$n_{FeCl_3} = n_{Na_3PO_4} = n_{FePO_4};$$

$$m_{FeCl_3} = n\ MM = 16.6\ mol\left(\frac{162.21\ g}{1\ mol}\right)\left(\frac{10^{-3}\ kg}{1\ g}\right) = 2.69\ kg;$$

$$m_{Na_3PO_4} = n\ MM = 16.6\ mol\left(\frac{163.94\ g}{1\ mol}\right)\left(\frac{10^{-3}\ kg}{1\ g}\right) = 2.72\ kg$$

(b) Mix solutions of NaOH and $ZnCl_2$. The net ionic reaction is:
$$Zn^{2+}{}_{(aq)} + 2\ OH^-{}_{(aq)} \rightarrow Zn(OH)_{2(s)};$$

$$n_{Zn(OH)_2} = \frac{m}{MM} = 2.50\ kg\left(\frac{10^3\ g}{1\ kg}\right)\left(\frac{1\ mol}{99.41\ g}\right) = 25.1\ mol;$$

$$n_{ZnCl_2} = n_{Zn(OH)_2}; \qquad n_{NaOH} = 2\ n_{Zn(OH)_2};$$

$$m_{ZnCl_2} = n\ MM = 25.1\ mol\ Zn(OH)_2\left(\frac{1\ mol\ ZnCl_2}{1\ mol\ Zn(OH)_2}\right)\left(\frac{136.3\ g}{1\ mol}\right)\left(\frac{10^{-3}\ kg}{1\ g}\right) = 3.42\ kg;$$

$$m_{NaOH} = n\ MM = 25.1\ mol\ Zn(OH)_2\left(\frac{2\ mol\ NaOH}{1\ mol\ Zn(OH)_2}\right)\left(\frac{40.01\ g}{1\ mol}\right)\left(\frac{10^{-3}\ kg}{1\ g}\right) = 2.01\ kg$$

(c) Mix solutions of Na_2CO_3 and $NiCl_2$. The net ionic reaction is
$$Ni^{2+}{}_{(aq)} + CO_3{}^{2-}{}_{(aq)} \rightarrow NiCO_{3(s)};$$

$$n_{NiCO_3} = \frac{m}{MM} = 2.50\ kg\left(\frac{10^3\ g}{1\ kg}\right)\left(\frac{1\ mol}{118.7\ g}\right) = 21.1\ mol;$$

$$n_{NiCl_2} = n_{Na_2CO_3} = n_{NiCO_3};$$

$$m_{NiCl_2} = n\ MM = 21.1\ mol\left(\frac{129.59\ g}{1\ mol}\right)\left(\frac{10^{-3}\ kg}{1\ g}\right) = 2.73\ kg;$$

$$m_{Na_2CO_3} = n\ MM = 21.1\ mol\left(\frac{105.99\ g}{1\ mol}\right)\left(\frac{10^{-3}\ kg}{1\ g}\right) = 2.24\ kg.$$

4.75   This problem describes a combustion reaction. Begin by analyzing the chemistry. The problem asks about the amount of a product that forms from a given amount of reactant. First balance the chemical equation. The reaction is:

$C_2H_5OH + O_2 \rightarrow CO_2 + H_2O$.
$2C + 6H + 3O \rightarrow 1C + 2H + 3O$

Give CO_2 a coefficient of 2 and H_2O a coefficient of 3 to balance C and H:

$C_2H_5OH + O_2 \rightarrow \mathbf{2}CO_2 + \mathbf{3}H_2O$.
$2C + 6H + 3O \rightarrow 2C + 6H + 7O$

There are seven O on the product side, so give O_2 a coefficient of 3 to balance O:

$C_2H_5OH + 3\ O_2 \rightarrow 2\ CO_2 + 3\ H_2O$.

Use the balanced equation to do the appropriate mass-mol-number calculations:

(a) $n_{H_2O} = 4.6 \text{ g} \left(\dfrac{1 \text{ mol}}{46.07 \text{ g}} \right) \left(\dfrac{3 \text{ mol H}_2\text{O}}{1 \text{ mol C}_2\text{H}_5\text{OH}} \right) = 0.30 \text{ mol H}_2\text{O};$

(b) $\# = n \, N_A = 0.30 \text{ mol} \left(\dfrac{6.022 \times 10^{23} \text{ molecules}}{1 \text{ mol}} \right) = 1.8 \times 10^{23} \text{ molecules};$

(c) $m = n \, MM = 0.30 \text{ mol} \left(\dfrac{18.02 \text{ g}}{1 \text{ mol}} \right) = 5.4 \text{ g.}$

4.77 Molecular pictures and limiting reactant calculations require a balanced chemical equation for the reaction under consideration. For this process, the balanced equation is $N_2 + 3 \, H_2 \rightarrow 2 \, NH_3$.

(a) There are 15 H_2 molecules and 6 N_2 molecules in this picture. Divide number of molecules by coefficient to see which reactant is limiting: $\dfrac{15}{3} = 5$ for H_2, $\dfrac{6}{1} = 6$ for N_2. The smaller number identifies the limiting reactant, H_2.

(b) The new molecular picture must show that all the H_2 has been consumed and NH_3 has been produced, but the number of atoms of each element must be the same as before: 30 atoms of H and 12 atoms of N. Fifteen H_2 molecules react with five N_2 molecules to produce 10 molecules of NH_3:

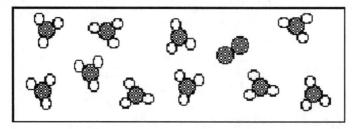

(c) Construct an amounts table for the stoichiometric calculations:

Reaction	N_2 +	$3 \, H_2 \rightarrow$	$2 \, NH_3$
Amt (mol)	6	15	0
Change (mol)	-15/3	-15	+(2/3)(15)
Final (mol)	1	0	10
Mass (g)	28.02	0	170.3

Use the results from the amounts table to calculate the desired masses:

$1.0 \text{ mol N}_2 \left(\dfrac{28.02 \text{ g}}{1 \text{ mol}} \right) = 28.02 \text{ g N}_2 \text{ remaining}$

$10 \text{ mol NH}_3 \left(\dfrac{17.03 \text{ g}}{1 \text{ mol}} \right) = 170.3 \text{ g NH}_3 \text{ produced}$

4.79   Identify species based on the type of substance present: Ions for salts and strong acids/bases, molecules for other substances. H_2O is always a major species in aqueous solution. Reactions that can occur include precipitation, proton transfer (acid-base), and redox.

(a) NH_3 is a weak base, so the species present are NH_3 and H_2O; HCl is a strong acid, so the species present are H_2O, H_3O^+ and Cl^-; an acid-base reaction occurs:

$$NH_3 + H_3O^+ \rightarrow NH_4^+ + H_2O;$$

(b) Both substances are salts, so the species present are the appropriate cations and anions: H_2O, Ca^{2+}, Cl^-, Na^+, and SO_4^{2-}; sulfates are generally insoluble, so a precipitation reaction occurs:

$$Ca^{2+} + SO_4^{2-} \rightarrow CaSO_{4(s)};$$

(c) KOH is a strong base, so the species present are H_2O, K^+, and OH^-; HBr is a strong acid, so the species present are H_2O, H_3O^+, and Br^-; an acid-base reaction occurs:

$$H_3O^+ + OH^- \rightarrow 2 H_2O;$$

(d) HNO_2 is a weak acid, so the species present are H_2O and HNO_2; KOH is a strong base, so the species present are H_2O, K^+, and OH^-; an acid-base reaction occurs:

$$HNO_2 + OH^- \rightarrow NO_2^- + H_2O.$$

4.81   A molecular picture must show the stoichiometry of the reaction and conserve atoms of each element. The net ionic reaction is $H_3O^+ + OH^- \rightarrow 2H_2O$. The starting solutions have equal overall reactant concentrations, $[H_3O^+] = [OH^-]$. Your picture should show this, for example using 3 OH^- ions and 3 H_3O^+ ions. Omit solvent water molecules and spectator ions for clarity:

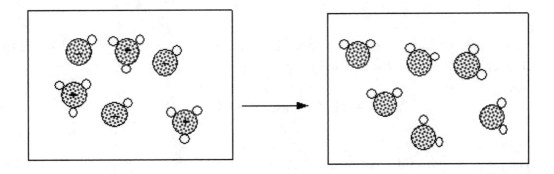

4.83   There is much information about propylene oxide provided in this problem, but all that is needed for the calculations is the balanced chemical equation, the amount of starting material, and molar masses. The stoichiometry of this reaction is 1:1 in all reagents.

MM(propylene oxide) = 3(12.01 g/mol) + 6(1.01 g/mol) + 16.00 g/mol = 58.1 g/mol
MM(propene) = 3(12.01 g/mol) + 6(1.01 g/mol) = 42.1 g/mol
MM(hydroperoxide) = 4(12.01 g/mol) + 10(1.01 g/mol) + 2(16.00 g/mol) = 90.1 g/mol

(a)   g/mol = kg/kmol

$$m_{propyleneoxide} = 75\,kg\left(\frac{1\,kmol}{90.1\,kg}\right)\left(\frac{1\,kmol\ propylene\ oxide}{1\,kmol\ hydroperoxide}\right)\left(\frac{58.1\,kg}{1\,kmol}\right) = 48\,kg$$

(b) $m_{propene} = 75\,kg\left(\frac{1\,kmol}{90.1\,kg}\right)\left(\frac{1\,kmol\ propene}{1\,kmol\ hydroperoxide}\right)\left(\frac{42.1\,kg}{1\,kmol}\right) = 35\,kg$

4.85   A molecular picture must show the stoichiometry of the reaction and conserve atoms of each element. The balanced reaction is $2\,Mg + O_2 \rightarrow 2\,MgO$. The desired picture starts with six Mg atoms and four O_2 molecules. The six Mg atoms react with three O_2 molecules to produce six MgO molecules, leaving one O_2 molecule unreacted. Your figure should show the Mg-containing species as solids:

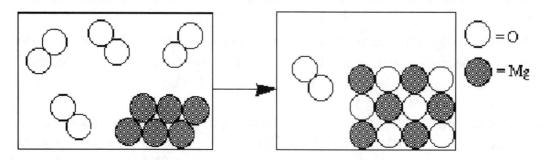

4.87   (a) The reaction is:
$$CaCO_3 + H_3O^+ \rightarrow Ca^{2+} + CO_2 + H_2O.$$
$$1Ca + 1\,C + 4O + 3H \rightarrow 1Ca + 1C + 3O + 2H$$

Carbon and calcium are already balanced. Give H_3O^+ a coefficient of 2 to balance the charges and H_2O a coefficient of 3 to balance H. Then O is also balanced (5 on each side):
$$CaCO_3 + 2\,H_3O^+ \rightarrow Ca^{2+} + CO_2 + 3\,H_2O.$$

(b and c)
The problem gives information about the amounts of both starting materials, so this is a limiting reactant situation. We must calculate the number of moles of each species, construct a table of amounts, and use the results to determine the final amounts.

Calculations of initial amounts:
$CaCO_3$:
$MM = 40.08\,g/mol + 12.01\,g/mol + 3(16.00\,g/mol) = 100.09\,g/mol$
$$n = 5.0\,g\left(\frac{1\,mol}{100.09\,g}\right) = 0.050\,mol$$

H_3O^+:

$n = 0.50$ L (0.10 M) $= 0.050$ mol

Use the balanced equation and the initial amounts to construct the amounts table:

Reaction	$CaCO_3$ +	$2 H_3O^+$ $\rightarrow$	Ca^{2+} +	CO_2
Amt (mol)	0.050	0.050	0	0
mol/coeff	0.050	0.025 (LR)		
Change (mol)	-0.025	-0.050	+0.025	+0.025
Final (mol)	0.025	0	0.025	0.025

Use the final amounts from the table to obtain the desired results:

(b) $m = n\ MM = 0.025$ mol $\left(\dfrac{44.01\,\text{g}}{1\,\text{mol}} \right) = 1.1$ g CO_2 formed;

(c) The spectator ion concentration remains unchanged: $[Cl^-] = 0.10$ M;
   Use the amounts table to calculate the concentrations of the other ionic species (Ca^{2+}) present in the solution:

   $$[Ca^{2+}] = \frac{0.025\,\text{mol}}{0.500\,\text{L}} = 0.050 \text{ M}.$$

4.89   To predict reaction products, first determine what species are present if an aqueous solution is involved, then identify the types of reaction each reactant can undergo.
   (a) Species are Ca, H_3O^+, Cl^-, and H_2O. Ca is a metal that displaces H_2 from acidic solutions (redox):

   $$Ca + 2 H_3O^+ \rightarrow Ca^{2+} + H_2 + 2 H_2O$$

   (b) Li is a Group 1 metal, which forms an oxide, Li_2O (redox):

   $$4 Li + O_2 \rightarrow 2 Li_2O$$

   (c) Species are H_3O^+, Br^-, NH_3, and H_2O. The strong acid reacts with the weak base (acid/base):

   $$NH_3 + H_3O^+ \rightarrow NH_4^+ + H_2O$$

   (d) C_3H_8 is a hydrocarbon and combusts in air (redox):

   $$C_3H_8 + 5O_2 \rightarrow 3CO_2 + 4H_2O$$

4.91   Balance each element in turn by adjusting coefficients, leaving O, which appears in several products, to the end. Because of the odd number of atoms of C and H, it is convenient to give the starting material a coefficient of 2 at the outset:

   $$2 C_3H_5N_3O_9 \text{ (l)} \rightarrow N_2 \text{ (g)} + CO_2 \text{ (g)} + H_2O \text{ (g)} + O_2 \text{ (g)}$$
   $$6C + 10H + 6N + 18\ O \rightarrow 1C + 2H + 2N + 5\ O$$

Give N_2 a coefficient of 3 to balance N, CO_2 a coefficient of 6 to balance C, and H_2O a coefficient of 5 to balance H:

$$2\ C_3H_5N_3O_9\ (l) \rightarrow \mathbf{3}N_2\ (g) + \mathbf{6}CO_2\ (g) + \mathbf{5}H_2O\ (g) + O_2\ (g)$$
$$6C + 10H + 6N + 18\ O \rightarrow 6C + 10H + 6N + 19O$$

Now there are 18 O atoms on the left and 17 in CO_2 and H_2O on the right, so give O_2 a coefficient of 1/2 to balance O and multiply through by 2 to clear fractions:

$$4\ C_3H_5N_3O_9\ (l) \rightarrow 6\ N_2\ (g) + 12\ CO_2\ (g) + 10\ H_2O\ (g) + O_2\ (g)$$

4.93   The problem asks about yields and provides information about the amounts of both starting materials, so it is both a limiting reagent and a yield problem. Use a table of amounts to determine the theoretical yield:

Calculations of initial amounts:

$$N_2:\ 84.0\ kg\left(\frac{1\,g}{10^{-3}\,kg}\right)\left(\frac{1\,mol}{28.02\,g}\right) = 3.00 \times 10^3\ mol$$

$$H_2:\ 24.0\ kg\left(\frac{1\,g}{10^{-3}\,kg}\right)\left(\frac{1\,mol}{2.016\,g}\right) = 1.19 \times 10^4\ mol$$

Reaction	N_2 +	$3\ H_2$ →	$2\ NH_3$
Amt (10^3 mol)	3.00	11.9	0
mol/coeff	3.00 (LR)	3.97	
Change (10^3 mol)	-3.00	-9.0	+(2)(3.00)
Final (10^3 mol)	0	2.9	6.00

Use the table to obtain the theoretical yield of NH_3:

$$6.00 \times 10^3\ mol\left(\frac{17.04\,g}{1\,mol}\right)\left(\frac{10^{-3}\,kg}{1\,g}\right) = 102\ kg$$

The theoretical yield is 102 kg, and the actual yield is 68 kg.

$$\%\ yield = \frac{68\ kg}{102\ kg} \times 100\% = 67\%$$

Repeat the table of amounts using the actual change rather than the theoretical change:

Reaction	N_2 +	$3\ H_2$ →	$2\ NH_3$
Amt (10^3 mol)	3.00	11.9	0
Change (10^3 mol)	-2.00	-6.00	+(2)(2.00)
Final (10^3 mol)	1.00	5.9	4.00
Mass (kg)	28	12	68

The masses of leftover reactants are:

N_2: 1.00×10^3 mol $\left(\dfrac{28.02\,g}{1\,mol}\right)\left(\dfrac{1\,kg}{10^3\,g}\right)$ = 28 kg

H_2: 5.9×10^3 mol $\left(\dfrac{2.016\,g}{1\,mol}\right)\left(\dfrac{1\,kg}{10^3\,g}\right)$ = 12 kg

4.95   The quantity asked for is a volume ratio, which can be related to a mass ratio using densities and to a mole ratio using molar masses. First balance the chemical reaction, leaving H, which appears in both reactants, for last.

N_2H_4 (l) + H_2O_2 (l) → N_2 (g) + H_2O (g).
2N + 6H + 2O → 2N + 2H + O

N is already balanced; give H_2O a coefficient of 2 to balance O.

N_2H_4 (l) + H_2O_2 (l) → N_2 (g) + 2 H_2O (g).
2N + 6H + 2O → 2N + 4H + 2 O

Now there are six H on the left and four on the right; multiply the coefficients of both H_2O_2 and H_2O by 2 to give 8 on each side:

N_2H_4 (l) + 2 H_2O_2 (l) → N_2 (g) + 4 H_2O (g).

The two reactants should be present in a mole ratio of 1 N_2H_4: 2 H_2O_2. Now use

$m = n\,MM$ and $V = \dfrac{m}{\rho}$, giving $V = \dfrac{n\,MM}{\rho}$:

Vol. ratio = 1 mol $N_2H_4\left(\dfrac{32.05\,g}{1\,mol}\right)\left(\dfrac{1\,mL}{1.44\,g}\right)$: 2 mol $H_2O_2\left(\dfrac{34.02\,g}{1\,mol}\right)\left(\dfrac{1\,mL}{1.01\,g}\right)$

Vol. ratio = 22.26:67.37 or 1:3.03

4.97   This problem describes an acid-base titration. We are asked to determine the mass of acid in the tablet.

n_{base} = 0.1045 M(0.02445 L) = 0.002555 mol base at the stoichiometric pt. Remember, at the stoichiometric point moles base = moles acid, therefore
$n_{acid} = n_{base}$ = 0.002555 mol
Converted to grams,

$m_{acid} = n\,MM$ =0.002555 mol$\left(\dfrac{176.13\,g}{1\,mol}\right)\left(\dfrac{10^3\,mg}{1\,g}\right)$ = 450.0 mg

Therefore, since the mass calculated from the titration is not the same as a pure tablet (500.0mg), the tablet is not pure.

The mass of the impurities is the difference of a pure tablet and that of the actual tablet: 500.0 mg - 450.0 mg = 50.0 mg impurities.

$$Mass\% = \frac{50.0 mg}{500.0 mg} \times 100\% = 10.0\% \text{ impurites}$$

4.99 This problem asks about the stoichiometry of a mixture of two salts, $CaSO_4$ and $Ca(H_2PO_4)_2$.

(a) The 2:1 mole ratio indicates the coefficients for the balanced equation. There is no water present during this reaction, so we need not ask about species present:
$2 H_2SO_4 + Ca_3(PO_4)_2 \rightarrow 2 CaSO_4 + Ca(H_2PO_4)_2$.

(b) Because the mixture has fixed stoichiometric proportions of 2:1, it has a combined molar mass of $2 MM_{CaSO_4} + MM_{Ca(H_2PO_4)_2}$:

$MM_{combination} = 2(136.1 \text{ g/mol}) + 234.1 \text{ g/mol} = 506.3 \text{ g/mol}$;

$$n_{combination} = 50.0 \text{ kg} \left(\frac{10^3 \text{ g}}{1 \text{ kg}} \right) \left(\frac{1 \text{ mol}}{506.3 \text{ g}} \right) = 98.76 \text{ mol};$$

$n_{acid} = 2 n_{combination}$; $n_{Ca_3(PO_4)_2} = n_{combination}$;

$$m_{acid} = 98.76 \text{ mol} \left(\frac{2 \text{ mol } H_2SO_4}{1 \text{ mol combination}} \right) \left(\frac{98.09 \text{ g}}{1 \text{ mol}} \right) \left(\frac{10^{-3} \text{ kg}}{1 \text{ g}} \right) = 19.4 \text{ kg};$$

$$m_{Ca_3(PO_4)_2} = 98.76 \text{ mol} \left(\frac{1 \text{ mol } Ca_3(PO_4)_2}{1 \text{ mol combination}} \right) \left(\frac{310.2 \text{ g}}{1 \text{ mol}} \right) \left(\frac{10^{-3} \text{ kg}}{1 \text{ g}} \right) = 30.6 \text{ kg}.$$

(c) Each mole of $Ca_3(PO_4)_2$ yields 2 mol of phosphate ion, so the amount of phosphate ion available is $2 n_{combination} = 2(98.76 \text{ mol}) = 198 \text{ mol}$.

4.101 First identify the chemical reactions. Here, the metal in the sample is dissolved in acid, and AgCl is then precipitated from the solution. The precipitate contains all the silver that was present in the metal sample, so we can calculate the composition of the metal from the amount of silver present in the precipitate.

$$m_{Ag} = (m_{AgCl}) \left(\frac{MM_{Ag}}{MM_{AgCl}} \right) = 0.156 \text{ g} \left(\frac{1 \text{ mol}}{143.4 \text{ g}} \right) \left(\frac{107.9 \text{ g}}{1 \text{ mol}} \right) = 0.1174 \text{ g}; \text{ (use an additional}$$

digit to avoid round-off error)

$$\% \text{ Ag} = (100\%) \left(\frac{m_{Ag}}{m_{total}} \right) = \frac{0.1174 \text{ g}}{0.135 \text{ g}} \times 100\% = 87.0\%;$$

$\% \text{ Cu} = 100\% - 87.0\% = 13.0\%.$

4.103 You are asked to calculate moles for a chemical reaction. To do this requires the use of mole ratios, which in turn requires a balanced chemical equation. Balance the equation, then do appropriate mass-mole-mass conversions:
The reaction is:

$C_5H_{12}O + O_2 \rightarrow H_2O + CO_2$.

$$5C + 12H + 3O \rightarrow 1C + 2H + 3O$$

Balance C by giving CO_2 a coefficient of 5 and H by giving H_2O a coefficient of 6:

$$C_5H_{12}O + O_2 \rightarrow \mathbf{6}\ H_2O + \mathbf{5}\ CO_2.$$
$$5C + 12H + 3O \rightarrow 5C + 12H + 16O$$

This gives 16 O atoms on the right. Give O_2 a coefficient of 15/2 so there are 16 O atoms on the left, then multiply through by 2 to clear fractions:

$$2\ C_5H_{12}O + 15\ O_2 \rightarrow 12\ H_2O + 10\ CO_2.$$

Use the density equation to calculate the mass of the compound: $m = \rho V$

$$3.15\ \text{mL}\ C_5H_{12}O \left(\frac{0.740\ \text{g}}{1\ \text{mL}} \right) \left(\frac{1\ \text{mol}}{88.15\ \text{g}} \right) \left(\frac{10\ \text{mol}\ CO_2}{2\ \text{mol}\ C_5H_{12}O} \right) = 0.132\ \text{mol}\ CO_2$$

4.105 The problem asks about the amount of reactant needed to react completely with a given amount of another reactant. The reaction is $CO_2\ (g) + 2\ KOH\ (s) \rightarrow K_2CO_3(s) + H_2O\ (l)$. Use mole amounts and stoichiometric ratios:

$$n_{CO_2} = (5\ \text{astronaut})(6\ \text{days}) \left(\frac{1.0\ \text{kg}}{(1\ \text{astronaut})\ (1\ \text{day})} \right) \left(\frac{10^3\ \text{g}}{1\ \text{kg}} \right) \left(\frac{1\ \text{mol}}{44.0\ \text{g}} \right) = 6.8 \times 10^2\ \text{mol}$$

$$m_{KOH} = 6.8 \times 10^2\ \text{mol}\ CO_2 \left(\frac{2\ \text{mol}\ KOH}{1\ \text{mol}\ CO_2} \right) \left(\frac{56.1\ \text{g}}{1\ \text{mol}} \right) \left(\frac{1\ \text{kg}}{10^3\ \text{g}} \right) = 76\ \text{kg}\ KOH.$$

The more expensive LiOH is used on the space shuttle because it is 43 kg lighter.

5.1 Molecular speed is related to temperature by the equations, $u = \sqrt{\dfrac{2E_{kinetic}}{m}}$, and

$\overline{E}_{kinetic} = \dfrac{3RT}{2N_A}$, so $u_{avg} = \sqrt{\dfrac{3RT}{mN_A}}$. Thus, molecular speed increases with the square root of kinetic energy, so molecules will reach the detector sooner at higher temperature, because they have greater speed.

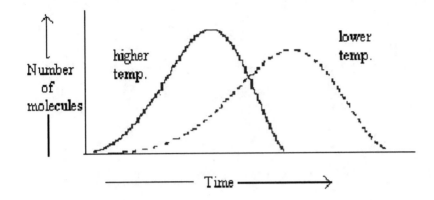

5.3 Molecular speed is related to temperature by the equations, $u = \sqrt{\dfrac{2E_{kinetic}}{m}}$, and

$\overline{E}_{kinetic} = \dfrac{3RT}{2N_A}$, so $u_{avg} = \sqrt{\dfrac{3RT}{mN_A}}$. Each distribution of molecular speeds will have the

same general shape, with the position of the peak depending on $\sqrt{\dfrac{T}{m}}$:

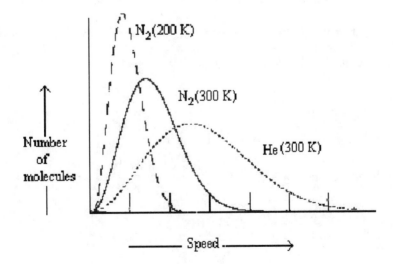

5.5 Average molecular kinetic energy is directly proportional to temperature and is independent of molecular mass: $\overline{E}_{kinetic} = \dfrac{3RT}{2N_A}$. N_2 and He at 300 K have identical kinetic energy distributions, while the distribution for N_2 at 200 K is shifted to lower energy:

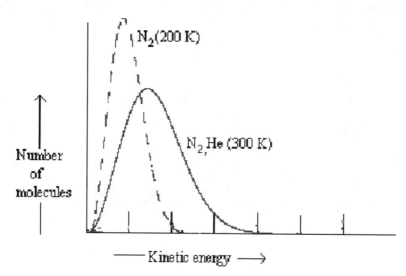

5.7 The most probable speed at any temperature is related to the most probable kinetic energy by the equation $E_{kinetic} = \dfrac{mu^2}{2}$, or $u_{mp} = \sqrt{\dfrac{2E_{kinetic,mp}}{m}}$. According to your textbook, the most probable kinetic energy at 300 K is 4.13 x 10^{-21} J/molecule. Because energy is proportional to temperature, the most probable kinetic energy at any other temperature can be calculated using proportions: $\dfrac{E_{kinetic,2}}{E_{kinetic,1}} = \dfrac{T_2}{T_1}$. Average kinetic energy, which is not the same as most probable kinetic energy because the distribution of energies is asymmetric, is calculated using $E_{kinetic,\ molar} = \dfrac{3RT}{2}$.

(a) He: $m = \left(\dfrac{4.00\,g}{1\,mol}\right)\left(\dfrac{10^{-3}\,kg}{1\,g}\right)\left(\dfrac{1\,mol}{6.022 \times 10^{23}\,atoms}\right) = 6.64 \times 10^{-27}$ kg;

at 900 K, $E_{kinetic,\ mp} = 4.13 \times 10^{-21}$ J $\left(\dfrac{900\,K}{300\,K}\right) = 1.24 \times 10^{-20}$ J.

$u_{mp} = \sqrt{\dfrac{2(1.24 \times 10^{-20}\,J)}{6.64 \times 10^{-27}\,kg}} = 1.93 \times 10^3$ m/s;

$E_{kinetic,\ molar} = \left(\dfrac{3}{2}\right)\left(\dfrac{8.314\,J}{1\,mol\,K}\right)(900K) = 1.12 \times 10^4$ J/mol;

(b) O_2: $m = \left(\dfrac{32.00\,g}{1\,mol}\right)\left(\dfrac{10^{-3}\,kg}{1\,g}\right)\left(\dfrac{1\,mol}{6.022 \times 10^{23}\,molecules}\right) = 5.31 \times 10^{-26}\,kg$;

at 300 K, $E_{kinetic,\,mp} = 4.13 \times 10^{-21}$ J;

$u_{mp} = \sqrt{\dfrac{2(4.13 \times 10^{-21}\,J)}{5.31 \times 10^{-26}\,kg}} = 3.94 \times 10^2$ m/s;

$E_{kinetic,\,molar} = \left(\dfrac{3}{2}\right)\left(\dfrac{8.314\,J}{1\,mol\,K}\right)(300\,K) = 3.74 \times 10^3$ J/mol;

(c) SF_6: $MM = [32.066\,g/mol + 6(18.998\,g/mol)] = 146.1\,g/mol$;

$m = \left(\dfrac{146.1\,g}{1\,mol}\right)\left(\dfrac{10^{-3}\,kg}{1\,g}\right)\left(\dfrac{1\,mol}{6.022 \times 10^{23}\,atoms}\right) = 2.43 \times 10^{-25}\,kg$;

at 900 K, $E_{kinetic,\,mp} = 4.13 \times 10^{-21}\,J\left(\dfrac{900\,K}{300\,K}\right) = 1.24 \times 10^{-20}$ J.

$u_{mp} = \sqrt{\dfrac{2(1.24 \times 10^{-20}\,J)}{2.43 \times 10^{-25}\,kg}} = 3.19 \times 10^2$ m/s;

$E_{kinetic,\,molar} = \left(\dfrac{3}{2}\right)\left(\dfrac{8.314J}{1\,mol\,K}\right)(900K) = 1.12 \times 10^4$ J/mol.

5.9 The size of the force exerted by a single particle is dependent on the mass of the particle and its speed. A particle with a small mass will exert a smaller force than one with a larger mass. Similarly a particle moving at a slower speed will exert a smaller force. Recall that at higher temperatures, particles will have higher speeds.
(a) At equal temperatures, a xenon atom generates a greater force than a neon atom because of its greater mass.
(b) Average speed increases with temperature, so a methane molecule with average speed at 700 K generates a greater force than one with average speed at 390 K.
(c) At equal speeds, an F_2 molecule generates a greater force than an H_2 molecule because of its greater mass.

5.11 The ideal gas is defined by the conditions that both molecular volumes and intermolecular forces are negligible.
(a) At very high pressure, molecules are very close together, so their volumes are significant compared to the volume of their container; the first condition is not met and the gas is not ideal.
(b) At very low temperature, molecules move very slowly, so the forces between molecules, even though small, are sufficient to influence molecular motion; the second condition is not met and the gas is not ideal.

5.13 (a) If the piston is stationary and there is no friction, the forces and pressures on each side must be equal, so the internal pressure is also 1 atm. This pressure is generated by gas molecules colliding with the face of the piston.
(b) When temperature is doubled, the increase in average molecular speed causes an increase in pressure. The piston will move outward, causing the concentration of gas molecules to decrease. This will lead to a lower frequency of collisions with the wall, reducing the pressure. The piston will stop when the reduced rate of collisions counterbalances the increased impulse per collision and the internal pressure is once again 1 atm.

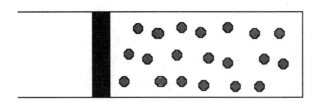

5.15 A pinhole in the top of the tube would let air leak into the space until the internal pressure matched the external, atmospheric pressure. The barometer would then indicate zero pressure, because the height of the mercury column is determined by the pressure *difference* between inside and outside, and this difference would be zero in the presence of a pinhole.

5.17 The SI unit for pressure is the pascal: $1 \text{ atm} = 1.01325 \times 10^5$ Pa, and the atmosphere-torr conversion factor is 1 atm = 760 torr.

(a) $455 \text{ torr}\left(\dfrac{1 \text{ atm}}{760 \text{ torr}}\right)\left(\dfrac{1.01325 \times 10^5 \text{ Pa}}{1 \text{ atm}}\right) = 6.07 \times 10^4 \text{ Pa};$

(b) $2.45 \text{ atm}\left(\dfrac{1.01325 \times 10^5 \text{ Pa}}{1 \text{ atm}}\right) = 2.48 \times 10^5 \text{ Pa};$

(c) $0.46 \text{ torr}\left(\dfrac{1 \text{ atm}}{760 \text{ torr}}\right)\left(\dfrac{1.01325 \times 10^5 \text{ Pa}}{1 \text{ atm}}\right) = 61 \text{ Pa; and}$

(d) $1.33 \times 10^{-3} \text{ atm}\left(\dfrac{1.01325 \times 10^5 \text{ Pa}}{1 \text{ atm}}\right) = 1.35 \times 10^2 \text{ Pa}.$

5.19 The question asks for moles, which can be obtained from *P-V-T* data using the ideal gas equation: $n = \dfrac{PV}{RT}$.

$$n = \frac{PV}{RT} = \frac{(5.00 \text{ atm})(20.0 \text{ L})}{(8.206 \times 10^{-2} \; \frac{\text{L atm}}{\text{mol K}})(298 \text{ K})} = 4.09 \text{ mol.};$$

$$V = \frac{nRT}{P} = \frac{(4.09 \text{ mol})(8.206 \times 10^{-2} \; \frac{\text{L atm}}{\text{mol K}})(298 \text{ K})}{1.00 \text{ atm}} = 1.00 \times 10^2 \text{ L}.$$

5.21 When some variables are held fixed, rearrange the ideal gas equation, $PV = nRT$, to collect fixed values on the right.

(a) n, R, V are constant: $\dfrac{P}{T} = \dfrac{nR}{V} = \text{constant}$; $\dfrac{P_i}{T_i} = \dfrac{P_f}{T_f}$;

(b) $V = \dfrac{nRT}{P}$

(c) n, R, T are constant: $PV = nRT = \text{constant}$; $P_i V_i = P_f V_f$;

5.23 When some conditions change but others remain fixed, rearrange the ideal gas equation so the constant terms are grouped on the right. In this problem, n and P are fixed:

$\dfrac{V}{T} = \dfrac{nR}{P} = \text{constant}$, so $\dfrac{V_i}{T_i} = \dfrac{V_f}{T_f}$.

Begin by converting volume and temperature to the units in R (liters and kelvin):
$V_i = 0.255$ L,
$T_i = 25 + 273.15 = 298$ K,
$T_f = -15 + 273.15 = 258$ K;

$$V_f = \dfrac{V_i T_f}{T_i} = \dfrac{(0.255\ \text{L})(258\ \text{K})}{(298\ \text{K})} = 0.221\ \text{L}$$

5.25 As a gas cools, its molecules move more slowly, so they impart smaller impulses on the walls of their container. This reduces the internal pressure, so the balloon collapses until the increase in gas density inside the balloon brings the internal pressure back up to 1.000 atm.

5.27 The equation is valid only for a gas for which n and T are fixed.
(a) n and T are fixed, so the equation is valid. (b) n can change, so the equation is not valid.
(c) T changes, so the equation is not valid. (d) The equation is not valid for liquids.

5.29 The ideal gas equation, $PV = nRT$, can be used to calculate moles using P-V-T data.

Molar mass is related to moles through $n = \dfrac{m}{MM}$. These equations yield two expressions for n,

$n = \dfrac{PV}{RT} = \dfrac{m}{MM}$, from which $MM = \dfrac{mRT}{PV}$.

Begin by converting the initial data into the units of R:
$T = 25.0 + 273.15 = 298.2$ K,

$P = 262\ \text{torr} \left(\dfrac{1\ \text{atm}}{760\ \text{torr}} \right) = 0.345\ \text{atm}$;

Use the modified ideal gas equation to obtain the molar mass:

$$MM = \frac{(2.55 \text{ g})(8.206 \times 10^{-2} \frac{\text{L atm}}{\text{mol K}})(298.2 \text{ K})}{(0.345 \text{ atm})(1.50 \text{ L})} = 121 \text{ g/mol}$$

Knowing that the compound contains only C (12.0 g/mol), F (19.0 g/mol), and Cl (35.5 g/mol), the formula can be determined by trial and error. In this case, the combination of 1 C, 2 F, and 2 Cl has MM = 12.0 g/mol + 2(19.0 g/mol) + 2(35.5 g/mol) = 121 g/mol, matching the experimental value.
The formula is CF_2Cl_2.

5.31   According to Dalton's law of partial pressures, each gas exerts a pressure equal to the total pressure times its mole fraction. Mole fraction can be found from concentration in ppm:

$$X = \left(\frac{\text{ppm}}{10^6} \right).$$

$$p(NO_2) = 758.4 \text{ torr} \left(\frac{1 \text{ atm}}{760 \text{ torr}} \right) \left(\frac{0.78 \text{ molecules NO}_2}{10^6 \text{ molecules of air}} \right) = 7.8 \times 10^{-7} \text{ atm.}$$

5.33   According to Dalton's law of partial pressures, each gas exerts a pressure equal to its mole fraction times the total pressure: $p = XP_{\text{tot}}$. Standard atmospheric conditions is P_{tot} = 1 atm.

$$p(N_2) = (0.7808)(1 \text{ atm}) \left(\frac{760 \text{ torr}}{1 \text{ atm}} \right) = 593.4 \text{ torr;}$$

$$p(O_2) = (0.2095)(1 \text{ atm}) \left(\frac{760 \text{ torr}}{1 \text{ atm}} \right) = 159.2 \text{ torr;}$$

$$p(Ar) = (9.34 \times 10^{-3})(1 \text{ atm}) \left(\frac{760 \text{ torr}}{1 \text{ atm}} \right) = 7.10 \text{ torr; and}$$

$$p(CO_2) = (3.25 \times 10^{-4})(1 \text{ atm}) \left(\frac{760 \text{ torr}}{1 \text{ atm}} \right) = 2.47 \times 10^{-1} \text{ torr.}$$

5.35   Molecular pictures show the relative numbers of molecules of various substances, which also represents the relative numbers of moles of those substances. The figure contains 8 He atoms and 4 O_2 molecules.

(a) and (b) There are more He atoms, so the pressure due to He and the mole fraction of He are higher.

(c) $X_{\text{He}} = \dfrac{n_{\text{He}}}{n_{\text{tot}}} = \dfrac{8}{12} = 0.67.$

5.37 To calculate partial pressures from a total pressure, mole fractions are required: $p_i = X_i P_{tot}$. Mole fractions can be calculated from the analytical data for the gas sample. The equations needed to solve this problem are:

$$n = \frac{m}{MM}, \qquad X_i = \frac{n_i}{n_{tot}}, \qquad p_i = \frac{n_i RT}{V}$$

$$n(CH_4) = 1.57 \text{ g} \left(\frac{1 \text{ mol}}{16.04 \text{ g}} \right) = 9.79 \times 10^{-2} \text{ mol};$$

$$n(C_2H_6) = 0.41 \text{ g} \left(\frac{1 \text{ mol}}{30.07 \text{ g}} \right) = 1.36 \times 10^{-2} \text{ mol};$$

$$n(C_3H_8) = 0.020 \text{ g} \left(\frac{1 \text{ mol}}{44.09 \text{ g}} \right) = 4.54 \times 10^{-4} \text{ mol};$$

$$n_{tot} = 9.79 \times 10^{-2} \text{ mol} + 1.36 \times 10^{-2} \text{ mol} + 4.54 \times 10^{-4} \text{ mol} = 1.120 \times 10^{-1} \text{ mol};$$

Determine the mole fraction by dividing the moles of each component by the total number of moles:

$$X(CH_4) = \frac{9.79 \times 10^{-2} \text{ mol}}{1.120 \times 10^{-1} \text{ mol}} = 0.874;$$

$$X(C_2H_6) = \frac{1.36 \times 10^{-2} \text{ mol}}{1.120 \times 10^{-1} \text{ mol}} = 0.121;$$

$$X(C_3H_8) = \frac{4.54 \times 10^{-4} \text{ mol}}{1.120 \times 10^{-1} \text{ mol}} = 4.06 \times 10^{-3};$$

Multiply the mole fractions by the total pressure to obtain the partial pressure (there should be two sig. figs. for ethane and propane since their masses are only known to two sig. figs.):
$p(CH_4) = (0.874)(2.35 \text{ atm}) = 2.05 \text{ atm};$
$p(C_2H_6) = (0.121)(2.35 \text{ atm}) = 0.28 \text{ atm};$
$p(C_3H_8) = (4.06 \times 10^{-3})(2.35 \text{ atm}) = 9.5 \times 10^{-3} \text{ atm}.$

5.39 Stoichiometric calculations require moles and a balanced chemical equation. Balance the equation, determine moles of CO_2 produced; and apply the ideal gas equation to calculate V.
The unbalanced chemical equation is:
$$C_6H_{12}O_6 + O_2 \rightarrow CO_2 + H_2O;$$
$$6C + 12H + 8O \rightarrow 1C + 2H + 3O$$

Give both CO_2 and H_2O a coefficient of 6 to balance C and H:
$$C_6H_{12}O_6 + O_2 \rightarrow 6CO_2 + 6H_2O;$$
$$6C + 12H + 8O \rightarrow 6C + 12H + 18O$$

This leaves 18 O on the product side. 6O are from glucose, so O_2 needs a coefficient of 6 to balance O and obtain the balanced chemical equation:

$$C_6H_{12}O_{6(g)} + 6\ O_{2(g)} \rightarrow 6\ CO_{2(g)} + 6\ H_2O_{(l)};$$

$$MM_{glucose} = 6(12.01\ \text{g/mol}) + 12(1.01\ \text{g/mol}) + 6(16.00\ \text{g/mol}) = 180.18\ \text{g/mol};$$

Use mass-mole conversions to determine the moles of CO_2 formed:

$$n(CO_2) = 4.65\ \text{g}\ C_6H_{12}O_6 \left(\frac{1\ \text{mol}}{180.18\ \text{g}} \right) \left(\frac{6\ \text{mol}\ CO_2}{1\ \text{mol}\ C_6H_{12}O_6} \right) = 1.548 \times 10^{-1}\ \text{mol};$$

$$V = \frac{nRT}{P} = \frac{(1.548 \times 10^{-1}\ \text{mol})(8.206 \times 10^{-2}\ \frac{\text{L atm}}{\text{mol K}})(310\ \text{K})}{1.00\ \text{atm}} = 3.94\ \text{L}.$$

5.41 This is a stoichiometry problem that involves gases. We are asked to determine the final pressure of the container (which is the pressure of chlorine gas). Begin by analyzing the chemistry. The starting materials are Na metal and Cl_2, a gas. The product of the reaction is NaCl.

The balanced chemical reaction is: $2\ Na_{(s)} + Cl_{2(g)} \rightarrow 2\ NaCl_{(s)}$

The problem gives information about the amounts of both starting materials, so this is a limiting reactant situation. We must calculate the number of moles of each species, construct a table of amounts, and use the results to determine the final pressure. Calculations of initial amounts:

$$n(Na) = 6.90\ \text{g} \left(\frac{1\ \text{mol}}{22.99\ \text{g}} \right) = 0.300\ \text{mol};$$

For chlorine gas, use the ideal gas equation to do pressure-mole conversions. The data must have the same units as those for R.

$$P(Cl_2) = 1.25 \times 10^3\ \text{torr} \left(\frac{1\ \text{atm}}{760\ \text{torr}} \right) = 1.64\ \text{atm};$$

$$n(Cl_2) = \frac{PV}{RT} = \frac{(1.64\ \text{atm})(3.00\ \text{L})}{(8.206 \times 10^{-2}\ \frac{\text{L atm}}{\text{mol K}})(300\ \text{K})} = 0.200\ \text{mol}$$

Divide each initial amount by the stoichiometric coefficient to determine the limiting reactant:

Na: $\dfrac{0.300\ \text{mol}}{2} = 0.150\ \text{mol}$ (LR)

Cl_2: 0.200 mol

Use the initial amounts and the balanced equation to construct the table of amounts:

Reaction:	2 Na(s) +	Cl$_2$(g) $\rightarrow$	2 NaCl(s)
Start (mol)	0.300	0.200	0.000
Change (mol)	–0.300	–0.150	+0.300
Final (mol)	0.000	0.050	0.300

Using the final amount of Cl$_2$ and the new temperature (47°C) obtain the pressure:

$$P = \frac{nRT}{V} = \frac{(5.0 \times 10^{-2}\ \text{mol})(8.206 \times 10^{-2}\ \frac{\text{L atm}}{\text{mol K}})(320\ \text{K})}{3.00\ \text{L}} = 0.438\ \text{atm}.$$

5.43 This is a stoichiometry problem that involves gases. We are asked to determine the mass of product produced. Begin by analyzing the chemistry. The starting materials are N$_2$ and H$_2$, both gases. The product of the reaction is NH$_3$.

The balanced chemical reaction is: N$_2$ + 3H$_2$ $\rightarrow$ 2NH$_3$

The problem gives information about the amounts of both starting materials, so this is a limiting reactant situation. We must calculate the number of moles of each species, construct a table of amounts, and use the results to determine the partial pressure.

Calculations of initial amounts:
Use the ideal gas equation to determine the initial amounts of each gas. The reactor initially contains the gases in 1:1 mole ratio, so

$$p_i = \frac{275\ \text{atm}}{2} = 137.5\ \text{atm for each gas}.$$

$$n_i = \frac{p_i V}{RT} = \frac{(137.5\ \text{atm})(8.75 \times 10^3\ \text{L})}{(8.206 \times 10^{-2}\ \frac{\text{L atm}}{\text{mol K}})(455\ \text{K})} = 2.01 \times 10^4\ \text{mol};$$

Since both gases have the same initial amount, the limiting reactant will be the one with the higher stoichiometric coefficient: H$_2$ is limiting. Here is the complete amounts table:

Reaction:	N$_2$(g) +	3 H$_2$(g) $\rightarrow$	2 NH$_3$ (g)
Start (10^4 mol)	2.01	2.01	0.00
Change(10^4 mol)	–(1/3)(2.01)	–2.01	+(2/3)(2.01)
Final(10^4 mol)	1.34	0.00	1.34

Obtain the mass of product formed from the final amount in the table:

$$m(NH_3) = n\, MM = 1.34 \times 10^4 \text{ mol} \left(\frac{17.04 \text{ g}}{1 \text{ mol}} \right) = 2.28 \times 10^5 \text{ g.}$$

5.45 The yield in a reaction is the ratio of the actual amount produced to the theoretical amount. The calculation in Problem 5.43 gives the theoretical amount:

$$\text{actual amount} = \text{ theoretical amount} \times \frac{\% \text{ yield}}{100\%}$$

$$\text{actual amount} = 2.28 \times 10^5 \text{ g} \left(\frac{13\%}{100\%} \right) = 3.0 \times 10^4 \text{ g.}$$

5.47 Table 5-3 gives the mole fractions of dry air (0% humidity), from which $X(O_2) = 0.2095$. Table 5-4 gives the vapor pressure of water at 20 °C, 17.535 torr; this is the partial pressure at 100% humidity for that temperature. To find the mole fraction of O_2 at 100% humidity, first determine the partial pressure of O_2.

$$p(H_2O) = 17.535 \text{ torr} \left(\frac{1 \text{ atm}}{760 \text{ torr}} \right) = 2.307 \times 10^{-2} \text{ atm;}$$

$$p(N_2 + O_2) = 1.000 \text{ atm} - 2.307 \times 10^{-2} \text{ atm} = 0.9769 \text{ atm;}$$

$$p(O_2) = (0.2095)(0.9769 \text{ atm}) = 0.2047 \text{ atm;}$$

$$X(O_2) = \frac{p(O_2)}{P_{total}} = \frac{0.2047 \text{ atm}}{1.000 \text{ atm}} = 0.2047;$$

$$\text{change} = 0.2095 - 0.2047 = +0.0048$$

5.49 The dew point is defined as the temperature at which the partial pressure of water vapor in the atmosphere matches the vapor pressure of water. Refer to Table 5-4 for vapor pressures: at 25 °C, $vp = 23.756$ torr.

$$p(H_2O) = 23.756 \text{ torr} \left(\frac{78\%}{100\%} \right) = 18.53 \text{ torr;}$$

This vapor pressure corresponds to a dew point between 20 and 25 °C. Assume a linear variation in vapor pressure between these two temperatures:

$$\text{dew point } = 20.0 \text{ °C } + \left[\frac{(18.53 - 17.535)}{(23.756 - 17.535)} \right] (5 \text{ °C}) = 20.0 + 0.8 = 20.8 \text{ °C}$$

5.51 We must calculate the volume of dry air that contains 10 mg of neon. Use moles, mole fractions from Table 5-3, and the ideal gas equation:

$$n_{Ne} = \frac{m}{MM} = 10 \text{ mg} \left(\frac{10^{-3} \text{ g}}{1 \text{ mg}} \right) \left(\frac{1 \text{ mol}}{20.180 \text{ g}} \right) = 5.0 \times 10^{-4} \text{ mol;}$$

From Table 5-3, $X_{Ne} = 1.82 \times 10^{-5} = \dfrac{n_{Ne}}{n_{air}}$;

Chapter 5

Thus, the number of moles of air containing 10 mg of neon is:

$$n_{air} = \frac{n_{Ne}}{X_{Ne}} = \frac{5.0 \times 10^{-4}\,\text{mol}}{1.82 \times 10^{-5}} = 27.5 \text{ mol};$$

The volume occupied by this amount of air can be calculated from the ideal gas equation:

$$V = \frac{nRT}{P} = \frac{(27.5\,\text{mol})(8.206 \times 10^{-2}\,\frac{\text{L atm}}{\text{mol K}})(298\,\text{K})}{1.00\,\text{atm}} = 6.7 \times 10^{2}\,\text{L}$$

(two sig. figs. because the moles of air should only have two sig. figs.)

5.53   Use the ideal gas equation to find molar density and the Avogadro constant to convert to molecular density.

$$\frac{n}{V} = \frac{P}{RT} = \frac{10^{-10}\,\text{atm}}{(8.206 \times 10^{-2}\,\frac{\text{L atm}}{\text{mol K}})(310\,\text{K})} = 4 \times 10^{-12}\,\text{mol/L};$$

$$\#/V = N_A\left(\frac{n}{V}\right) = \left(\frac{6.022 \times 10^{23}\,\text{molecules}}{1\,\text{mol}}\right)\left(\frac{4 \times 10^{-12}\,\text{mol}}{1\,\text{L}}\right) = 2 \times 10^{12}\,\text{molecules/L}.$$

5.55   This is an application of the ideal gas equation to changing gas conditions. The only constant quantity is n, so $\dfrac{PV}{T} = nR = $ constant. Therefore, $\dfrac{P_iV_i}{T_i} = \dfrac{P_fV_f}{T_f}$

Convert all values into units of R:
$P_i = 0.969$ atm,
$V_i = 0.963$ L,
$T_i = 22 + 273.15 = 295$ K;
$P_f = 1.00$ atm,
$T_f = 15 + 273.15 = 288$ K;

$$V_f = \left(\frac{P_iV_i}{T_i}\right)\left(\frac{T_f}{P_f}\right) = \frac{(0.969\,\text{atm})(0.963\,\text{L})(288\,\text{K})}{(295\,\text{K})(1.00\,\text{atm})} = 0.911\,\text{L}.$$

5.57   Concentrations in parts per billion can be converted to mole fractions, and the partial pressure can then be calculated using $p_i = X_iP_{total}$. To determine molecular concentration, first use the ideal gas equation to calculate mol, then multiply by N_A/V.
$X(SF_6) = (1\,\text{ppb})(10^{-9}/\text{ppb}) = 10^{-9}$;
$p(SF_6) = (10^{-9})(1\,\text{atm}) = 10^{-9}$ atm;

$$V = 1.0\,\text{cm}^3\left(\frac{10^{-3}\,\text{L}}{1\,\text{cm}^3}\right) = 1.0 \times 10^{-3}\,\text{L}$$

$$n = \frac{pV}{RT} = \frac{(10^{-9}\,\text{atm})(1.0 \times 10^{-3}\,\text{L})}{(8.206 \times 10^{-2}\,\frac{\text{L atm}}{\text{mol K}})(21 + 273\,\text{K})} = 4.1 \times 10^{-14}\,\text{mol};$$

$$\#/V = 4.1 \times 10^{-14} \, \text{mol} \left(\frac{6.022 \times 10^{23} \, \text{molecules}}{1 \, \text{mol}} \right) = 3 \times 10^{10} \, \text{molecules/cm}^3.$$

5.59 Use the ideal gas equation to determine moles, then do mole-mass-number conversions.

$$n = \frac{PV}{RT} = \left(\frac{(10.0 \, \text{atm})(725 \, \text{mL})}{(8.206 \times 10^{-2} \, \frac{\text{L atm}}{\text{mol K}})(925 + 273 \, \text{K})} \right) \left(\frac{10^{-3} \, \text{L}}{1 \, \text{mL}} \right) = 7.37 \times 10^{-2} \, \text{mol};$$

$$m = n \, MM = 7.37 \times 10^{-2} \, \text{mol} \left(\frac{83.80 \, \text{g}}{1 \, \text{mol}} \right) = 6.18 \, \text{g};$$

$$\text{atoms} = n \, N_A = 7.37 \times 10^{-2} \, \text{mol} \left(\frac{6.022 \times 10^{23} \, \text{atoms}}{1 \, \text{mol}} \right) = 4.44 \times 10^{22} \, \text{atoms}.$$

5.61 Molecular pictures show the relative numbers of molecules of various substances, which also represents the relative numbers of moles of those substances. Chamber A contains 6 atoms, B contains 12 atoms, and C contains 9 atoms, for a total of 27 atoms, and the chambers have equal volumes and are at the same temperature.
(a) Chamber B contains the most atoms, so it exhibits the highest pressure;
(b) The pressures in the chambers are proportional to the number of atoms:

$$P_A = P_B \left(\frac{\#A}{\#B} \right) = 1.0 \, \text{atm} \left(\frac{6}{12} \right) = 0.50 \, \text{atm};$$

(c) The pressure will increase in proportion to the number of atoms:

$$P(\text{new}) = P(\text{old}) \left(\frac{\#\text{new}}{\#\text{old}} \right) = 1.0 \, \text{atm} \left(\frac{27}{6} \right) = 4.5 \, \text{atm};$$

(d) When the valves are opened, the number of atoms in each chamber equalize: $27/3 = 9$.

$$P(\text{new}) = P(\text{old}) \left(\frac{\#\text{new}}{\#\text{old}} \right) = 0.50 \, \text{atm} \left(\frac{9}{12} \right) = 0.38 \, \text{atm}.$$

5.63 Much of the information in this problem is not needed, because molecular speed can be calculated from temperature and molar mass:

$$\overline{E}_{\text{kinetic}} = \frac{3RT}{2N_A}; = \tfrac{1}{2}mu^2, \text{ from which } u^2 = \frac{3RT}{mN_A} = \frac{3RT}{MM}$$

The mass must be converted to kg/mol for mass units to cancel (1 J = 1 kg m²/s²):

$$MM = 3 \left(\frac{16.00 \, \text{g}}{1 \, \text{mol}} \right) \left(\frac{1 \, \text{kg}}{1000 \, \text{g}} \right) = 48.00 \times 10^{-3} \, \text{kg/mol}$$

$$u_{\text{avg}} = \left(\frac{3RT}{MM} \right)^{1/2} = \left(\frac{3(8.314 \, \text{J mol}^{-1} \, \text{K}^{-1})(-25 + 273 \, \text{K})}{48.00 \times 10^{-3} \, \text{kg mol}^{-1}} \right)^{1/2} = 359 \, \text{m/s}.$$

5.65 Density can be calculated from the ideal gas equation and mole-mass conversions:

$$n = \frac{m}{MM} = \frac{PV}{RT}, \text{ from which } \frac{m}{V} = \frac{P \, MM}{RT}$$

$$P = 755 \text{ torr} \left(\frac{1 \text{ atm}}{760 \text{ torr}} \right) = 0.993 \text{ atm}$$

$$\frac{m}{V} = \frac{P\,MM}{RT} = \frac{(0.993 \text{ atm})(146.05 \text{ g/mol})}{(8.206 \times 10^{-2}\, \frac{\text{L atm}}{\text{mol K}})(27 + 273 \text{ K})} = 5.89 \text{ g/L.}$$

5.67 The vapor pressure of water is the partial pressure of gaseous water in the atmosphere under 100% humidity conditions. To determine the dew point, first calculate the partial pressure of water vapor at the existing temperature and then estimate the temperature at which this partial pressure equals the vapor pressure of water, assuming a linear variation of vapor pressure with temperature. Use data in Table 5-4:
$p(H_2O) = (vP)(\text{rel. hum.})/(100\%)$.

(a) $p(H_2O) = 42.175 \text{ torr} \left(\frac{80\%}{100\%} \right) = 33.74 \text{ torr;}$

$$\text{dew point} = 30.0 \text{ }^\circ\text{C} + \left[\frac{(33.74 - 31.824)}{(42.175 - 31.824)} \right] (5 \text{ }^\circ\text{C}) = 30.0 + 0.9 = 30.9 \text{ }^\circ\text{C}$$

(b) $p(H_2O) = 12.788 \text{ torr} \left(\frac{50\%}{100\%} \right) = 6.394 \text{ torr;}$

$$\text{dew point} = 0 \text{ }^\circ\text{C} + \left[\frac{(6.394 - 4.579)}{(6.543 - 4.579)} \right] (5 \text{ }^\circ\text{C}) = 0.0 + 4.6 = 4.6 \text{ }^\circ\text{C}$$

(c) $p(H_2O) = 23.756 \text{ torr} \left(\frac{30\%}{100\%} \right) = 7.1268 \text{ torr;}$

$$\text{dew point} = 5 \text{ }^\circ\text{C} + \left[\frac{(7.1268 - 6.543)}{(9.209 - 6.543)} \right] (5 \text{ }^\circ\text{C}) = 5.0 + 1.1 = 6.1 \text{ }^\circ\text{C}$$

5.69 The molar mass of gaseous oxygen can be determined using the ideal gas law:
$$MM = \frac{mRT}{PV}.$$
Pump out a bulb of known volume, weigh it, fill with oxygen at a measured pressure, and weigh again. If gaseous oxygen were monatomic, this experiment would give $MM = 16.00$ g/mol, while the diatomic gas gives $MM = 32.00$ g/mol.

5.71 Use Dalton's law of partial pressures to determine partial pressures, then apply the ideal gas equation to calculate masses.

$$X(C_3H_6) = \frac{1.00}{5.00} = 0.200, \; p(C_3H_6) = 0.200 \text{ atm;}$$
$$X(O_2) = 1.000 - 0.200 = 0.800, \; p(O_2) = 0.800 \text{ atm;}$$

$$m(C_3H_6) = \frac{pV(MM)}{RT} = \frac{(0.200 \text{ atm})(2.00 \text{ L})(42.078 \text{ g/mol})}{(8.206 \times 10^{-2}\, \frac{\text{L atm}}{\text{mol K}})(23.5 + 273.1 \text{ K})} = 0.692 \text{ g;}$$

$$m(O_2) = \frac{pV(MM)}{RT} = \frac{(0.800\,\text{atm})(2.00\,\text{L})(32.00\,\text{g/mol})}{(8.206 \times 10^{-2}\,\frac{\text{L atm}}{\text{mol K}})(23.5 + 273.1\ \text{K})} = 2.10\,\text{g}.$$

5.73   In this problem, two samples of gas are compared. All conditions except the volumes are different for the two samples.

(a) The bulb that contains the larger number of moles will contain the larger number of molecules:

$$n(H_2) = \frac{PV}{RT} = \frac{(2\,\text{atm})V}{(273\,\text{K})R} = 7 \times 10^{-3}\left(\frac{V}{R}\right)\text{mol};$$

$$n(O_2) = \frac{PV}{RT} = \frac{(1\,\text{atm})V}{(298\,\text{K})R} = 3 \times 10^{-3}\left(\frac{V}{R}\right)\text{mol};$$

The bulb of hydrogen contains more molecules.

(b) $m(H_2) = (7 \times 10^{-3})(V/R)\,\text{mol}\left(\dfrac{2.01\,\text{g}}{1\,\text{mol}}\right) = 1 \times 10^{-2}(V/R)\,\text{g};$

$m(O_2) = (3 \times 10^{-3})(V/R)\,\text{mol}\left(\dfrac{32.00\,\text{g}}{1\,\text{mol}}\right) = 1 \times 10^{-1}(V/R)\,\text{g};$

The bulb of oxygen contains more mass;

(c) The average kinetic energy of molecules depends only on the temperature of the gas, so the oxygen molecules, being at higher temperature, have greater average kinetic energy;

(d) The average molecular speed depends on $\sqrt{\dfrac{T}{m}}$: $u_{avg} = \sqrt{\dfrac{3RT}{mN_A}}$. The temperature

ratio of H_2:O_2 is $\left(\dfrac{273}{298}\right) = 0.916$, while the mass ratio is $\left(\dfrac{2}{32}\right) = 0.0625$. Thus, the

$\dfrac{T}{m}$ ratio is $\left(\dfrac{0.916}{0.0625}\right) = 14.6$, so the hydrogen molecules have greater average speed.

5.75   (a) The most probable kinetic energy at 300 K is 4×10^{-21} J

(b) $u_{\text{most probable}} = \sqrt{\dfrac{2(4 \times 10^{-21}\,\text{J})(6.022 \times 10^{23}\,\text{molecules mole}^{-1})}{32.00 \times 10^{-3}\,\text{kg mol}^{-1}}} = 4 \times 10^{2}$ m/s

5.77   Your explanation should include the fact that sulfur in coal combines with oxygen from the air to form SO_2, and that when SO_2 reacts with water, sulfuric acid is an end product. Chemical equations:

$S_{(s)} + O_{2(g)} \rightarrow SO_{2(g)};$

$2\,SO_{2(g)} + O_{2(g)} \rightarrow 2\,SO_{3(g)};$

$SO_{3(g)} + H_2O_{(l)} \rightarrow H_2SO_{4(aq)}.$

5.79　Each part of this problem represents a change of one or more conditions, for which a rearranged version of the ideal gas law can be used.

(a) *T* and *n* are fixed, so $PV = nRT =$ constant and $V_f = \dfrac{P_i V_i}{P_f}$;

$$V_f = \frac{(525 \text{ torr})(3.00 \text{ L})}{755 \text{ torr}} = 2.09 \text{ L};$$

(b) *T* and *n* are fixed, so $PV = nRT =$ constant and $P_f = \dfrac{P_i V_i}{V_f}$;

$$P_f = \frac{(525 \text{ torr})(3.00 \text{ L})}{2.00 \text{ L}} = 788 \text{ torr};$$

(c) *V* and *n* are fixed, so $\dfrac{P}{T} = \dfrac{nR}{V} =$ constant and $P_f = \dfrac{P_i T_f}{T_i}$;

$$P_f = \frac{(525 \text{ torr})(50.0 + 273.15 \text{ K})}{273.15 \text{ K}} = 621 \text{ torr};$$

5.81　Figure 5-8 illustrates atomic density and atomic motion of a sample of helium gas. Here is the initial picture which we will modify:

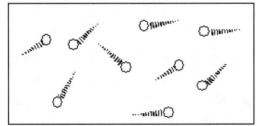

(a) Your drawing should show the same atomic density as Figure 5-8 but with longer "tails" on the atoms, indicating that they are moving at higher speeds.

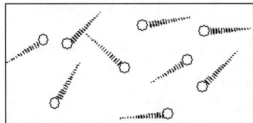

(a) Your drawing should show the same lengths of "tails" as Figure 5-8, but there should be 1/3 fewer atoms.

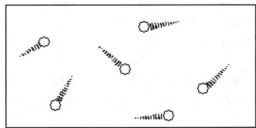

(c) Your drawing should be the same as Figure 5-8 except that half the atoms should be replaced with diatomic molecules, as shown below for a portion of the sample:

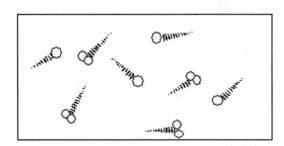

5.83 Use the ideal gas equation, $PV = nRT$, to determine the correct statements.

(a) At constant T and V, $\dfrac{P}{n} = \dfrac{RT}{V} = $ constant. Thus, P is directly proportional to number of moles of gas. The statement is false.

(b) At constant V and n, $\dfrac{P}{T} = \dfrac{nR}{V} = $ constant. The statement is true.

(c) At fixed V and n, $\dfrac{P}{T} = \dfrac{nR}{V} = $ constant. $\dfrac{P}{T}$ is a constant; PT is not. The statement is false.

5.85 Because Table 5-3 refers to dry air, the first step is to take into account the presence of H_2O at 50% relative humidity. The partial pressure of water must be subtracted from the total pressure to obtain the pressure due to dry air. Then $p_i = X(\text{dry air})_i\, P(\text{dry air})$. From Table 5-4, $vP(25\ ^oC) = 23.756$ torr.

$$p(H_2O) = 23.756\ \text{torr}\left(\frac{1\,\text{atm}}{760\,\text{torr}}\right)\left(\frac{50\%}{100\%}\right) = 1.6 \times 10^{-2}\ \text{atm};$$

$$p(\text{dry air}) = 765\ \text{torr}\left(\frac{1\,\text{atm}}{760\,\text{torr}}\right) - 1.6 \times 10^{-2}\ \text{atm} = 0.991\ \text{atm};$$

$p(N_2) = (0.7808)(0.991\ \text{atm}) = 0.774\ \text{atm};$

$p(O_2) = (0.2095)(0.991\ \text{atm}) = 0.208\ \text{atm};$

$p(Ar) = (9.34 \times 10^{-3})(0.991\ \text{atm}) = 9.26 \times 10^{-3}\ \text{atm};$

$p(CO_2) = (3.25 \times 10^{-4})(0.991\ \text{atm}) = 3.22 \times 10^{-4}\ \text{atm};$

$p(Ne) = (1.82 \times 10^{-5})(0.991\ \text{atm}) = 1.80 \times 10^{-5}\ \text{atm};$

$p(He) = (5.24 \times 10^{-6})(0.991\ \text{atm}) = 5.19 \times 10^{-6}\ \text{atm};$

$p(CH_4) = (1.4 \times 10^{-6})(0.991\ \text{atm}) = 1.4 \times 10^{-6}\ \text{atm}.$

5.87 This is an empirical formula problem, with P-V-T data added to allow calculation of the molar mass. First, determine the empirical formula from mass percentages. Consider 100 g of the compound.

C: $\left(\dfrac{71.22\,\text{g}}{100\,\text{g}}\right)\left(\dfrac{1\,\text{mol}}{12.011\,\text{g}}\right) = 5.930$ mol C/100 g;

H: $\left(\dfrac{14.94\,\text{g}}{100\,\text{g}}\right)\left(\dfrac{1\,\text{mol}}{1.0079\,\text{g}}\right) = 14.82$ mol H/100 g;

N: $\left(\dfrac{13.84\,\text{g}}{100\,\text{g}}\right)\left(\dfrac{1\,\text{mol}}{14.007\,\text{g}}\right) = 0.9881$ mol N/100 g;

Divide each by the smallest among them, 0.9881 mol N/100 g:

C: $\left(\dfrac{5.930\,\text{mol C}}{100\,\text{g}}\right)\left(\dfrac{100\,\text{g}}{0.9881\,\text{mol N}}\right) = 6.001$ mol C/mol N, round to 6 C/N;

H: $\left(\dfrac{14.82\,\text{mol H}}{100\,\text{g}}\right)\left(\dfrac{100\,\text{g}}{0.9881\,\text{mol N}}\right) = 15.00$ mol H/mol N;

Empirical formula: $C_6H_{15}N$, empirical $MM = 101$ g/mol;

Next use the *P-V-T* data to determine the actual molar mass:

$m = 250\,\text{mg}\left(\dfrac{10^{-3}\,\text{g}}{1\,\text{mg}}\right) = 0.250$ g

$P = 435\,\text{torr}\left(\dfrac{1\,\text{atm}}{760\,\text{torr}}\right) = 0.572$ atm;

$V = 150\,\text{mL}\left(\dfrac{10^{-3}\,\text{L}}{1\,\text{mL}}\right) = 0.150$ L

$T = 150 + 273\,\text{K} = 423$ K

$MM = \dfrac{mRT}{PV} = \dfrac{(0.250\,\text{g})(8.206 \times 10^{-2}\,\frac{\text{L atm}}{\text{mol K}})(423\,\text{K})}{(0.572\,\text{atm})(0.150\,\text{L})} = 101$ g/mol;

The molar mass is the same as the empirical molar mass, so the molecular formula is the same as the empirical formula.

5.89   This is a *P-V-T* problem. First calculate moles of CO_2, then use the ideal gas equation to calculate the final pressure.

$n = \dfrac{m}{MM} = 15.00\,\text{g}\left(\dfrac{1\,\text{mol}}{44.01\,\text{g}}\right) = 0.3408$ mol;

$p(CO_2) = \dfrac{(0.3408\,\text{mol})(8.206 \times 10^{-2}\,\frac{\text{L atm}}{\text{mol K}})(273.15\,\text{K})}{0.750\,\text{L}} = 10.2$ atm;

5.91   Density can be calculated from the ideal gas equation and mole-mass conversions:

$n = \dfrac{m}{MM} = \dfrac{PV}{RT}$, from which $\dfrac{m}{V} = \dfrac{P\,MM}{RT}$.

First, calculate the molar mass of dry air and of moist air. $MM(\text{air}) = \sum_i (X_i)(MM_i)$

MM(dry air) = (0.7808)(28.01 g/mol) + (0.2095)(32.00 g/mol) + (9.34 x 10^{-3})(39.948 g/mol) = 28.95 g/mol;

For moist air at 298 K, p(H$_2$O) = 23.756 torr$\left(\dfrac{1\,atm}{760\,torr}\right)$ = 3.13 x 10^{-2} atm,

and X(H$_2$O) = 0.0313. The total mole fraction of the dry air components is reduced from 1 to (1 − .0313) = 0.9687. Each dry air mole fraction must be multiplied by this factor to obtain the mole fractions in moist air.
Thus,

MM(moist air) = (3.13 x 10^{-2})(18.01 g/mol) + (0.9687)(28.95 g/mol) = 28.61 g/mol;

$$\frac{m}{V}(\text{dry air}) = \frac{(1\,atm)(28.95\,g/mol)}{(8.206 \times 10^{-2}\,\frac{L\,atm}{mol\,K})(298\,K)} = 1.18\,g/L.$$

$$\frac{m}{V}(\text{moist air}) = \frac{(1\,atm)(28.61\,g/mol)}{(8.206 \times 10^{-2}\,\frac{L\,atm}{mol\,K})(298\,K)} = 1.17\,g/L.$$

5.93   First determine moles of liquid oxygen in 150 L. Then calculate how many moles of dry air contain this amount of oxygen. Finally, use the ideal gas equation to calculate the volume.

$$n = \frac{\rho V}{MM} = 150\,L\left(\frac{10^3\,mL}{1\,L}\right)\left(\frac{1.14\,g}{1\,mL}\right)\left(\frac{1\,mol}{32.00\,g}\right) = 5.34 \times 10^3\,mol;$$

$$n(\text{air}) = \frac{n(O_2)}{X(O_2)} = \frac{5.34 \times 10^3\,mol}{0.2095} = 2.55 \times 10^4\,mol;$$

$$P = 750\,torr\left(\frac{1\,atm}{760\,torr}\right) = 0.987\,atm$$

$$V = \frac{nRT}{P} = \frac{(2.55 \times 10^4\,mol)(8.206 \times 10^{-2}\,\frac{L\,atm}{mol\,K})(298\,K)}{0.987\,atm} = 6.32 \times 10^5\,L.$$

5.95   In addition to a yield problem, information is given about both starting materials, so this is a limiting reactant problem that involves gases as well. We must calculate the number of moles of each species, construct a table of amounts, and use the results to determine the theoretical yield.

Water can be ignored, as CH$_4$ is the product of interest, and pressures can be used as the measures of amounts:

Reaction:	3 H$_{2(g)}$ +	CO$_{(g)}$ →	CH$_{4(g)}$
Start (atm)	20.0	10.0	0.0
Change (atm)	-20.0	-6.67	+6.67
Final (atm)	0.0	3.33	6.67

Use the ideal gas equation and mole-mass conversion to obtain the theoretical yield in g:

$$n = \frac{PV}{RT} = \frac{(6.67 \text{ atm})(100 \text{ L})}{(8.206 \times 10^{-2} \text{ L atm/ mol K})(575 \text{ K})} = 14.1 \text{ mol};$$

Theoretical amount $= n \, MM = 14.1 \text{ mol}\left(\dfrac{16.04 \text{ g}}{1 \text{ mol}}\right) = 226.2 \text{ g};$

Obtain the percent yield by dividing the actual yield by the theoretical yield:

% yield $= \dfrac{145 \text{ g}}{226.2 \text{ g}} \times 100\% = 64.1\%.$

5.97   The question looks like an empirical formula problem, but the only data provided are *P-V-T* data for the gaseous substance, which suggests that an ideal gas law calculation can be done. Use the gas data to calculate *n*, then use $n = \dfrac{m}{MM}$ to determine *MM* and subsequently determine the formula of the compound.

$$P = 552 \text{ torr}\left(\frac{1 \text{ atm}}{760 \text{ torr}}\right) = 0.726 \text{ atm}$$

$$n = \frac{PV}{RT} = \frac{(0.726 \text{ atm})(0.100 \text{ L})}{(8.206 \times 10^{-2} \text{ L atm/mol K})(30 + 273 \text{ K})} = 2.92 \times 10^{-3} \text{ mol};$$

$$MM = \frac{m}{n} = \frac{0.500 \text{ g}}{2.92 \times 10^{-3} \text{ mol}} = 171 \text{ g/mol};$$

The compound contains one nickel atom (*MM* = 58.69 g/mol) and *x* CO molecules (*MM* = 28.01 g/mol), so
171 g/mol = 58.69 g/mol + (*x*)(28.01 g/mol);

$$x = \frac{171 \text{ g/mol} - 58.69 \text{ g/mol}}{28.01 \text{ g/mol}} = 4; \text{ the formula is Ni(CO)}_4.$$

5.99   The chemical reaction in the Haber synthesis is $N_2 + 3 H_2 \rightarrow 2 NH_3$. The molecular picture shows 4 mol of N_2 and 12 mol of H_2, which represents stoichiometric proportions. If the reaction goes to completion, all molecules will react to form 8 mol of NH_3:

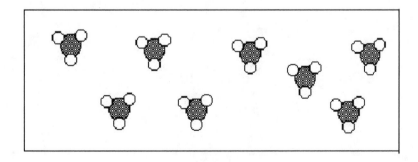

5.101 At higher elevations, Mount Everest, the atmospheric pressure, 250 mm, is much lower than at ground level, 760 mm. A lower pressure corresponds to a smaller molecular density. Thus when people say "the air is thin" they are actually referring to the fact that there is less air at higher elevations.

6.1   Calculate the molar volume of each metal from its density and molar mass, then convert to atomic volume using the Avogadro constant. Then estimate atomic diameter from the atomic volume by assuming that each atom occupies a cubical volume:

(a) $V_{atom} = \dfrac{[MM\,(\text{g/mol})][10^{-3}\,\text{kg/g}]}{[p(\text{kg/m}^3)][N_A]}$;

(b) $d \cong (V)^{1/3}$; and

(c) Layer thickness = (# of atoms)(atomic diameter).

Ag:

$V_{atom} = \left(\dfrac{107.868\,\text{g}}{1\,\text{mol}}\right)\left(\dfrac{10^{-3}\,\text{kg}}{1\,\text{g}}\right)\left(\dfrac{1\,\text{m}^3}{1.050 \times 10^4\,\text{kg}}\right)\left(\dfrac{1\,\text{mol}}{6.022 \times 10^{23}\,\text{atom}}\right) = 1.706 \times 10^{-29}\,\text{m}^3/\text{atom};$

$d \cong (V)^{1/3} = (1.706 \times 10^{-29}\,\text{m}^3)^{1/3} = 2.574 \times 10^{-10}\,\text{m};$

Thickness = $(6.5 \times 10^6\ \text{atoms})(2.574 \times 10^{-10}\,\text{m}) = 1.7 \times 10^{-3}\,\text{m};$

Pb:

$V_{atom} = \left(\dfrac{207.2\,\text{g}}{1\,\text{mol}}\right)\left(\dfrac{10^{-3}\,\text{kg}}{1\,\text{g}}\right)\left(\dfrac{1\,\text{m}^3}{1.134 \times 10^4\,\text{kg}}\right)\left(\dfrac{1\,\text{mol}}{6.022 \times 10^{23}\,\text{atom}}\right) = 3.034 \times 10^{-29}\,\text{m}^3/\text{atom};$

$d \cong (V)^{1/3} = (3.034 \times 10^{-29}\,\text{m}^3)^{1/3} = 3.119 \times 10^{-10}\,\text{m};$

Thickness = $(6.5 \times 10^6\ \text{atoms})(3.119 \times 10^{-10}\,\text{m}) = 2.0 \times 10^{-3}\,\text{m}.$

6.3   Gas pressure results from transfer of momentum between atoms and container walls, indicating that atoms have momentum and mass; mass spectrometers sort atomic and molecular ions according to their mass-to-charge ratios; all matter is made up of atoms and has mass.

6.5   At the atomic level, a layer of metal atoms appears as spheres nestled closely together:

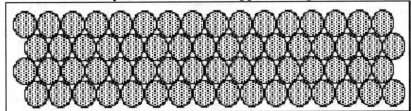

6.7   Use the eqn. $v = \dfrac{c}{\lambda}$ to solve these problems.

(a) $v = \left(\dfrac{2.998 \times 10^8\,\text{m/s}}{4.33\,\text{nm}}\right)\left(\dfrac{1\,\text{nm}}{10^{-9}\,\text{m}}\right) = 6.92 \times 10^{16}\,\text{Hz}$

(b) $v = \dfrac{2.998 \times 10^8\,\text{m/s}}{2.35 \times 10^{-10}\,\text{m}} = 1.28 \times 10^{18}\,\text{Hz}$

(c) $v = \left(\dfrac{2.998 \times 10^8\,\text{m/s}}{735\,\text{mm}}\right)\left(\dfrac{10^3\,\text{mm}}{1\,\text{m}}\right) = 4.08 \times 10^8\,\text{Hz}$

(d) $v = \left(\dfrac{2.998 \times 10^8 \, \text{m/s}}{4.57 \, \mu\text{m}} \right) \left(\dfrac{10^6 \, \mu\text{m}}{1 \, \text{m}} \right) = 6.56 \times 10^{13} \, \text{Hz}$

6.9 Use the eqn. $\lambda = \dfrac{c}{v}$ to solve these problems.

(a) $\lambda = \left(\dfrac{2.998 \times 10^8 \, \text{m/s}}{4.77 \, \text{GHz}} \right) \left(\dfrac{1 \, \text{GHz}}{10^9 \, \text{Hz}} \right) = 6.29 \times 10^{-2} \, \text{m}$

(b) $\lambda = \left(\dfrac{2.998 \times 10^8 \, \text{m/s}}{28.9 \, \text{kHz}} \right) \left(\dfrac{1 \, \text{kHz}}{10^3 \, \text{Hz}} \right) \left(\dfrac{10^2 \, \text{cm}}{1 \, \text{m}} \right) = 1.04 \times 10^6 \, \text{cm}$

(c) $\lambda = \left(\dfrac{2.998 \times 10^8 \, \text{m/s}}{60 \, \text{Hz}} \right) \left(\dfrac{10^3 \, \text{mm}}{1 \, \text{m}} \right) = 5 \times 10^9 \, \text{mm}$

(d) $\lambda = \left(\dfrac{2.998 \times 10^8 \, \text{m/s}}{2.88 \, \text{MHz}} \right) \left(\dfrac{10^6 \, \mu\text{m}}{1 \, \text{m}} \right) \left(\dfrac{1 \, \text{MHz}}{10^6 \, \text{Hz}} \right) = 1.04 \times 10^8 \, \mu\text{m}$

6.11 The energy of one photon of light is described by the equation, $E = hv = \dfrac{hc}{\lambda}$. To obtain energy per mol of photons, multiply by the Avogadro constant. Begin by converting the wavelength into meters and then determine the energy.

(a) $\lambda = 490.6 \, \text{nm} \left(\dfrac{10^{-9} \, \text{m}}{1 \, \text{nm}} \right) = 4.906 \times 10^{-7} \, \text{m};$

$E_{\text{photon}} = \dfrac{(6.626 \times 10^{-34} \, \text{J s})(2.998 \times 10^8 \, \text{m/s})}{4.906 \times 10^{-7} \, \text{m}} = 4.049 \times 10^{-19} \, \text{J/photon};$

$E_{\text{molar}} = 4.049 \times 10^{-19} \, \text{J} \left(\dfrac{6.022 \times 10^{23} \, \text{photons}}{1 \, \text{mol}} \right) \left(\dfrac{10^{-3} \, \text{kJ}}{1 \, \text{J}} \right) = 2.438 \times 10^2 \, \text{kJ/mol};$

(b) $\lambda = 25.5 \, \text{nm} \left(\dfrac{10^{-9} \, \text{m}}{1 \, \text{nm}} \right) = 2.55 \times 10^{-8} \, \text{m};$

$E_{\text{photon}} = \dfrac{(6.626 \times 10^{-34} \, \text{J s})(2.998 \times 10^8 \, \text{m/s})}{2.55 \times 10^{-8} \, \text{m}} = 7.79 \times 10^{-18} \, \text{J/photon};$

$E_{\text{molar}} = 7.79 \times 10^{-18} \, \text{J} \left(\dfrac{6.022 \times 10^{23} \, \text{photons}}{1 \, \text{mol}} \right) \left(\dfrac{10^{-3} \, \text{kJ}}{1 \, \text{J}} \right) = 4.69 \times 10^3 \, \text{kJ/mol};$

(c) $E_{\text{photon}} = (6.62608 \times 10^{-34} \, \text{J s})(2.5437 \times 10^{10} \, \text{s}^{-1}) = 1.6855 \times 10^{-23} \, \text{J/photon};$

$E_{\text{molar}} = 1.6855 \times 10^{-23} \, \text{J} \left(\dfrac{6.022 \times 10^{23} \, \text{photons}}{1 \, \text{mol}} \right) \left(\dfrac{10^{-3} \, \text{kJ}}{1 \, \text{J}} \right) = 1.0150 \times 10^{-2} \, \text{kJ/mol}.$

6.13 First, calculate the energy of one photon; then use $E_{total} = n\,E_{photon}$ to determine ***n*** (the number of photons):

$$\lambda = 337.1\text{ nm}\left(\frac{10^{-9}\text{ m}}{1\text{ nm}}\right) = 3.371 \text{ x } 10^{-7}\text{ m};$$

$$E_{photon} = \frac{(6.626 \text{ x } 10^{-34}\text{ J s})(2.998 \text{ x } 10^{8}\text{ m/s})}{3.371 \text{ x } 10^{-7}\text{ m}} = 5.893 \text{ x } 10^{-19}\text{ J/photon};$$

$$n = 10\text{ mJ}\left(\frac{10^{-3}\text{ J}}{1\text{ mJ}}\right)\left(\frac{1\text{ photon}}{5.893 \text{ x } 10^{-19}\text{ J}}\right) = 2 \text{ x } 10^{16}\text{ photons.}$$

6.15 (a) Convert from molar energy to energy per photon by dividing by the Avogadro constant, then use $E = \frac{hc}{\lambda}$ to find $\lambda = \frac{hc}{E}$.

$$E_{photon} = \left(\frac{745\text{ kJ}}{1\text{ mol}}\right)\left(\frac{10^{3}\text{ J}}{1\text{ kJ}}\right)\left(\frac{1\text{ mol}}{6.022 \text{ x } 10^{23}\text{ photons}}\right) = 1.237 \text{ x } 10^{-18}\text{ J};$$

$$\lambda = \frac{(6.626 \text{ x } 10^{-34}\text{ J s})(2.998 \text{ x } 10^{8}\text{ m/s})}{1.237 \text{ x } 10^{-18}\text{ J}} = 1.61 \text{ x } 10^{-7}\text{ m or 161 nm;}$$

$$\nu = \frac{c}{\lambda} = \frac{2.998\text{x}10^{8}\text{ m/s}}{1.61 \text{ x } 10^{-7}\text{ m}} = 1.86 \text{ x } 10^{15}\text{ Hz}$$

(b) $E_{photon} = 3.55 \text{ x } 10^{-19}$ J;

$$\lambda = \frac{(6.626 \text{ x } 10^{-34}\text{ J s})(2.998 \text{ x } 10^{8}\text{ m/s})}{3.55 \text{ x } 10^{-19}\text{ J}} = 5.60 \text{ x } 10^{-7}\text{ m or 560 nm}$$

$$\nu = \frac{c}{\lambda} = \frac{2.998 \text{ x } 10^{8}\text{ m/s}}{5.60 \text{ x } 10^{-7}\text{ m}} = 5.35 \text{ x } 10^{14}\text{ Hz}$$

6.17 (a) $\lambda = \frac{c}{\nu} = \frac{2.998 \text{ x } 10^{8}\text{ m/s}}{1.30 \text{ x } 10^{15}\text{ s}^{-1}} = 2.31 \text{ x } 10^{-7}$ m or 231 nm;

(b) First determine the energy of the photon, then use $bE = E_{photon} - kE_{electron}$

$E_{photon} = h\nu = (6.626 \text{ x } 10^{-34}\text{ J s})(1.30 \text{ x } 10^{15}\text{ s}^{-1}) = 8.61 \text{ x } 10^{-19}$ J;

$bE = 8.61 \text{ x } 10^{-19}\text{ J} - 5.2 \text{ x } 10^{-19}\text{ J} = 3.4 \text{ x } 10^{-19}$ J;

(c) The longest wavelength that will eject electrons corresponds to a photon with energy equal to the binding energy: $\lambda = \frac{hc}{bE}$.

$$\lambda = \frac{(6.626 \text{ x } 10^{-34}\text{ J s})(2.998 \text{ x } 10^{8}\text{ m/s})}{3.4 \text{ x } 10^{-19}\text{ J}} = 5.8 \text{ x } 10^{-7}\text{ m or 580 nm.}$$

6.19   Refer to Figure 6-8 for an example of an energy level diagram. The binding energy for cesium (Prob. 6.17) is 3.4 x 10^{-19} J, and the binding energy for chromium (Prob. 6.18) is 7.21 x 10^{-19} J. Your energy level diagram should show that chromium has a substantially greater binding energy than cesium. Note, if both metals are exposed to photons of the same wavelength the photoelectrons of Cs will have substantially greater kinetic energy than those emitted by Cr:

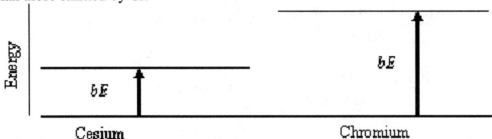

6.21   Use Figure 6-4 to answer these questions:
(a) Radio waves have wavelengths from 1 to 1000 m;
(b) Light with a wavelength of 5.8 x 10^{-7} m is yellow;
(c) A frequency of 4.5 x 10^8 Hz lies in the TV, FM region.

6.23   Figure 6-15 does not show an emission wavelength directly to the ground state from the state that emits 404 nm light, but the combination of emissions at 436 nm and 254 nm connects this state to the ground state. Use $E = \dfrac{hc}{\lambda}$ for each photon:

$$\Delta E(\text{kJ/mol}) = \frac{hcN_A}{\lambda_1} + \frac{hcN_A}{\lambda_2} = hcN_A\left(\frac{1}{\lambda_1} + \frac{1}{\lambda_2}\right)$$

$$\Delta E = 6.626 \times 10^{-34}\,\text{J s}\left(\frac{2.998 \times 10^8\,\text{m}}{1\,\text{s}}\right)\left(\frac{6.022 \times 10^{23}\,\text{atoms}}{1\,\text{mol}}\right)\left(\frac{1}{436\,\text{nm}} + \frac{1}{254\,\text{nm}}\right)\left(\frac{10^9\,\text{nm}}{1\,\text{m}}\right)$$

$$\Delta E = 745 \times 10^3\,\text{J/mol}$$

Convert to kJ/mol:

$$\Delta E = = 745 \times 10^3\,\text{J/mol}\left(\frac{10^{-3}\,\text{kJ}}{1\,\text{J}}\right) = 745\,\text{kJ/mol}$$

6.25   Energies and wavelengths for transitions in hydrogen atoms can be calculated from the equation for hydrogen atom energy levels:

$$E_n = \frac{-2.18 \times 10^{-18}\,\text{J}}{n^2}$$

$$\Delta E_{8-1} = (E_8 - E_1) = \frac{-2.18 \times 10^{-18} \text{ J}}{8^2} - \frac{-2.18 \times 10^{-18} \text{ J}}{1^2} = 2.146 \times 10^{-18} \text{ J};$$

$$\lambda = \frac{hc}{E} = \frac{(6.626 \times 10^{-34} \text{ Js})(2.998 \times 10^8 \text{ m/s})}{2.146 \times 10^{-18} \text{ J}} = 9.257 \times 10^{-8} \text{ m};$$

Convert to nm:

$$\lambda = 9.257 \times 10^{-8} \text{ m} \left(\frac{10^9 \text{ nm}}{1 \text{ m}} \right) = 92.57 \text{ nm};$$

$$\Delta E_{9-1} = (E_9 - E_1) = \frac{-2.18 \times 10^{-18} \text{ J}}{9^2} - \frac{-2.18 \times 10^{-18} \text{ J}}{1^2} = 2.153 \times 10^{-18} \text{ J};$$

$$\lambda = \frac{hc}{E} = \frac{(6.626 \times 10^{-34} \text{ Js})(2.998 \times 10^8 \text{ m/s})}{2.153 \times 10^{-18} \text{ J}} = 9.227 \times 10^{-8} \text{ m};$$

Convert to nm:

$$\lambda = 9.227 \times 10^{-8} \text{ m} \left(\frac{10^9 \text{ nm}}{1 \text{ m}} \right) = 92.27 \text{ nm};$$

These photons lie in the ultraviolet region of the spectrum (see Figure 6-4).

6.27   Similar to the ball and staircase of Figure 6-13, the ball can be raised from the ground to any step. When the ball falls down to the ground it can hit several steps along the way or none at all. Similarly, an electron can be raised to a specific energy level and then fall along several different paths to the ground state(each path corresponding to the electron emitting photons of different energies and therefore different frequencies). For example, let's assume the ball (or an electron) can be raised up to, but not beyond, the 5th step. In this case the ball can only absorb 5 different amounts of energy (the energies corresponding to raising the ball from the ground level to each step up to the 5th step). When the ball (electron) falls, it can fall from steps 5 to 4 (emitting a photon of one frequency) then from 4 to 2 (emitting a photon of a different frequency) and then to the ground state (emitting a photon of a third frequency). The ball/electron could also fall directly to the ground (emitting a photon of a completely different frequency than the three above). There are, of course, several other possibilities of how the ball/electron can fall from the 5th step to the ground. Hence there will be more lines in the emission spectra than the absorption spectra.

6.29   Use the mass of the electron and the Avogadro constant to calculate the molar mass:

$$m_{\text{molar}} = 9.1093897 \times 10^{-31} \text{ kg} \left(\frac{10^3 \text{ g}}{1 \text{ kg}} \right) \left(\frac{6.0221367 \times 10^{23} \text{ e}^-}{1 \text{ mol}} \right) = 5.4857990 \times 10^{-4} \text{ g/mol}.$$

6.31   To determine the wavelength of an electron from the kinetic energy, first determine the speed of the electron and then convert to wavelength. The speed of an electron is determined using $u = \sqrt{\dfrac{2E_{\text{kinetic}}}{m}}$. Remember, for the units to work out, mass must be in kg and the energy in J such that $\dfrac{J}{kg} = \dfrac{kg\ m^2/s^2}{kg} = m^2/s^2$. For wavelength calculations use

$\lambda = \dfrac{h}{mu}$:

(a) $u = \sqrt{\dfrac{2(1.15 \times 10^{-19}\,J)}{9.109 \times 10^{-31}\,kg}} = 5.02 \times 10^5$ m/s;

$\lambda = \dfrac{6.626 \times 10^{-34}\,J\,s}{(9.109 \times 10^{-31}\,kg)(5.02 \times 10^5\,m/s)} = 1.45 \times 10^{-9}$ m or 1.45 nm.

(b) $E_{electron} = \left(\dfrac{3.55\,kJ}{1\,mol} \right)\left(\dfrac{10^3\,J}{1\,kJ} \right)\left(\dfrac{1\,mol}{6.022 \times 10^{23}\,photons} \right) = 5.90 \times 10^{-21}$ J

$u = \sqrt{\dfrac{2(5.90 \times 10^{-21}\,J)}{9.109 \times 10^{-31}\,kg}} = 1.14 \times 10^5$ m/s;

$\lambda = \dfrac{6.626 \times 10^{-34}\,J\,s}{(9.109 \times 10^{-31}\,kg)(1.14 \times 10^5\,m/s)} = 6.38 \times 10^{-9}$ m or 6.38 nm.

(c) $E_{electron} = \left(\dfrac{7.45 \times 10^{-3}\,J}{1\,mol} \right)\left(\dfrac{1\,mol}{6.022 \times 10^{23}\,electrons} \right) = 1.24 \times 10^{-26}$ J;

$u = \sqrt{\dfrac{2(1.24 \times 10^{-26}\,J)}{9.109 \times 10^{-31}\,kg}} = 1.65 \times 10^2$ m/s;

$\lambda = \dfrac{6.626 \times 10^{-34}\,J\,s}{(9.109 \times 10^{-31}\,kg)(1.65 \times 10^2\,m/s)} = 4.41 \times 10^{-6}$ m or 4.41 μm.

6.33 First determine the speed of the electron from its wavelength and then use the speed to determine the kinetic energy. Kinetic energy and wavelength of an electron are related through $\lambda = \dfrac{h}{mu}$ and $E_{kinetic} = \dfrac{mu^2}{2}$.

(a) $\lambda = 3.75$ nm$\left(\dfrac{10^{-9}\,m}{1\,nm} \right) = 3.75 \times 10^{-9}$ m

$u = \dfrac{6.626 \times 10^{-34}\,J\,s}{(9.109 \times 10^{-31}\,kg)(3.75 \times 10^{-9}\,m)} = 1.94 \times 10^5$ m/s

$E_{kinetic} = \dfrac{(9.109 \times 10^{-31}\,kg)(1.94 \times 10^5\,m/s)^2}{2} = 1.71 \times 10^{-20}$ J.

(b) $\quad u = \dfrac{6.626 \times 10^{-34}\,\text{J s}}{(9.109 \times 10^{-31}\,\text{kg})(4.66\,\text{m})} = 1.56 \times 10^{-4}\,\text{m/s}$

$E_{\text{kinetic}} = \dfrac{(9.109 \times 10^{-31}\,\text{kg})(1.56 \times 10^{-4}\,\text{m/s})^2}{2} = 1.11 \times 10^{-38}\,\text{J.}$

(c) $\quad \lambda = 8.85\,\text{mm}\left(\dfrac{10^{-3}\,\text{m}}{1\,\text{mm}}\right) = 8.85 \times 10^{-3}\,\text{m}$

$u = \dfrac{6.626 \times 10^{-34}\,\text{J s}}{(9.109 \times 10^{-31}\,\text{kg})(8.85 \times 10^{-3}\,\text{m})} = 8.22 \times 10^{-2}\,\text{m/s}$

$E_{\text{kinetic}} = \dfrac{(9.109 \times 10^{-31}\,\text{kg})(8.22 \times 10^{-2}\,\text{m/s})^2}{2} = 3.08 \times 10^{-33}\,\text{J.}$

6.35 The designation $6p$ specifies that $n = 6$ and $l = 1$; the other two quantum numbers m_l and m_s, can take any of their possible values. Remember that m_l must be an integer with value between $+l$ and $-l$, while m_s is either $+1/2$ or $-1/2$. There are 6 valid sets:

n	l	m_l	m_s
6	1	+1	+1/2
6	1	+1	−1/2
6	1	0	+1/2
6	1	0	−1/2
6	1	−1	+1/2
6	1	−1	−1/2

6.37 Remember that l is restricted to positive integers less than n, m_l is restricted to integers between $-l$ and $+l$, and $m_s = +1/2$ or $-1/2$. For $n = 3$, l can be 0 with $m_l = 0$, l can be 1 with $m_l = +1, 0,$ or -1; and l can be 2 with $m_l = +2, +1, 0, -1,$ or -2.

6.39 Remember that n must be a positive integer, l is restricted to zero and positive integers less than n, m_l is restricted to integers between $-l$ and $+l$, and $m_s = +1/2$ or $-1/2$.
(a) non-existent: m_s must be $+1/2$ or $-1/2$; (b) actual; (c) non-existent: l must be less than n; and (d) actual.

6.41 The designation $3d$ specifies that $n = 3$ and $l = 2$; the largest possible value for m_l is $+l$, in this case $+2$; and "spin up" means $m_s = +1/2$. Thus, the values are:
3, 2, 2, +1/2.

6.43 The designation $n = 2$, $l = 1$ refers to a $2p$ orbital. This orbital has no nodes, so its electron density rises and then falls. Figure 6-23 shows a contour diagram. For this orbital, an electron density drawing looks very similar to a contour drawing.

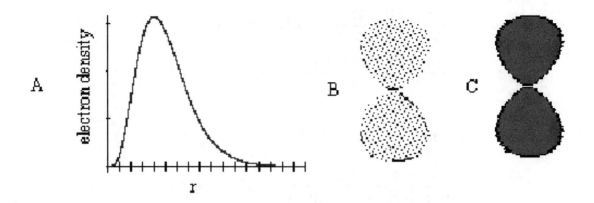

6.45   The p_y orbital runs lengthwise along the y-axis. It has the same view along the x and z-axis. Along the y-axis you will only see one lobe of the orbital:

x- and z- axis y-axis

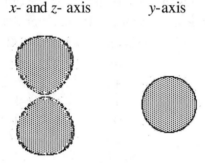

6.47   Orbital (a) has the shape characteristic of a p orbital, while orbitals (b) and (c) have shapes characteristic of d orbitals. Because orbital (a) appears smaller than the $3s$ orbital, it has n < 3 and is a $2p$ orbital; because orbitals (b) and (c) have similar sizes as the $3s$ orbital, they have n = 3 and are $3d$ orbitals.
(a) = $2p$, n = 2, l = 1; (b), (c) = $3d$, n = 3, l = 2.

6.49   Electron density plots accurately show how the wave function varies along one axis, but they fail to show whether or not an orbital has spherical symmetry, and they do not directly show electron densities. Thus, they fail to give a good picture of how an orbital appears in three dimensions.

6.51   A contour drawing of an s orbital is a spherical surface. The orbitals should grow larger as n increases:

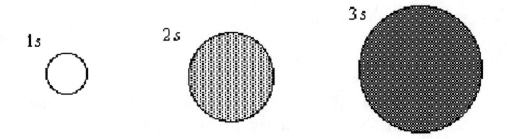

6.53 Each picture should show a molecule absorbing a photon and fragmenting appropriately:
N_2:

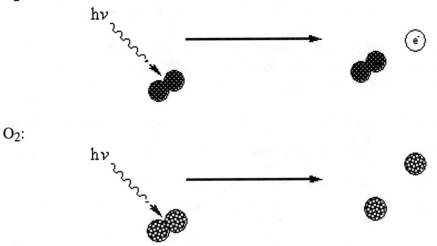

O_2:

6.55 Calculate the fragmentation energy using $E = \dfrac{hcN_A}{\lambda}$:

$$\lambda = 340 \text{ nm}\left(\frac{10^{-9}\,\text{m}}{1\,\text{nm}}\right) = 3.40 \times 10^{-7}\,\text{m};$$

$$hcN_A = 6.626 \times 10^{-34}\text{ J s}\left(\frac{2.998 \times 10^8\,\text{m}}{1\,\text{s}}\right)\left(\frac{10^{-3}\,\text{kJ}}{1\,\text{J}}\right)\left(\frac{6.022 \times 10^{23}\,\text{photons}}{1\,\text{mol}}\right)$$

$$hcN_A = 1.196 \times 10^{-4}\text{ kJ m/mol}$$

The minimum energy to fragment an ozone molecule is:

$$\Delta E = \frac{1.196 \times 10^{-4}\,\text{kJ m / mol}}{3.40 \times 10^{-7}\,\text{m}} = 352 \text{ kJ/mol};$$

The kinetic energy of the fragments is the difference between the energy of the photon causing the fragmentation and the minimum energy required for fragmentation:

$$\lambda = 250 \text{ nm} \left(\frac{10^{-9} \text{ m}}{1 \text{ nm}} \right) = 2.50 \times 10^{-7} \text{ m};$$

$$E_{\text{kinetic}} = E_{\text{photon}} - \Delta E$$

$$= \frac{(6.626 \times 10^{-34} \text{ J s})(2.998 \times 10^{8} \text{ m/s})}{2.50 \times 10^{-7} \text{ m}} - \left(\frac{352 \times 10^{3} \text{ J}}{1 \text{ mol}} \right) \left(\frac{1 \text{ mol}}{6.022 \times 10^{23} \text{ photons}} \right)$$

$$E_{\text{kinetic}} = (7.95 \times 10^{-19} \text{ J}) - (5.85 \times 10^{-19} \text{ J}) = 2.10 \times 10^{-19} \text{ J}.$$

6.57 Use the figure in the Chemistry and the Environment Box to determine which region of the atmosphere absorbs the given wavelengths:
(a) Below 200 nm, O_2, N_2, and O_3 all absorb effectively (thermosphere);
(b) From 240-310 nm, only O_3 absorbs effectively (stratosphere).

6.59 (a) According to Figure 6-25, a pressure of 10^{-3} atm occurs at ~50 km.
(b) the balloon is on the border of the stratosphere and mesosphere.
(c) the chemistry of this region is the formation of ozone.

6.61 To calculate photon properties, use $\lambda v = c$ and the equations for photons in Table 6-1:
(a) $\lambda = \dfrac{c}{v} = \dfrac{2.998 \times 10^{8} \text{ m/s}}{4.5 \times 10^{13} \text{ s}^{-1}} = 6.7 \times 10^{-6} \text{ m};$
(b) $E = hv = (6.626 \times 10^{-34} \text{ J s})(4.5 \times 10^{13} \text{ s}^{-1}) = 3.0 \times 10^{-20} \text{ J};$
(c) $t = \dfrac{\text{distance}}{\text{speed}} = 2.86 \times 10^{5} \text{ mi} \left(\dfrac{1.609 \text{ km}}{1 \text{ mi}} \right) \left(\dfrac{10^{3} \text{ m}}{1 \text{ km}} \right) \left(\dfrac{1 \text{ s}}{2.998 \times 10^{8} \text{ m}} \right) = 1.53 \text{ s}.$

6.63 Calculate the wavelength using $\lambda = \dfrac{hcN_A}{E}$:

$$hcN_A = 6.626 \times 10^{-34} \text{ J s} \left(\frac{2.998 \times 10^{8} \text{ m}}{1 \text{ s}} \right) \left(\frac{10^{-3} \text{ kJ}}{1 \text{ J}} \right) \left(\frac{6.022 \times 10^{23} \text{ photons}}{1 \text{ mol}} \right)$$

$$hcN_A = 1.196 \times 10^{-4} \text{ kJ m/mol}$$

$$\lambda = \left(\frac{1.196 \times 10^{-4} \text{ kJ m}}{1 \text{ mol}} \right) \left(\frac{1 \text{ mol}}{243 \text{ kJ}} \right) = 4.92 \times 10^{-7} \text{ m or 492 nm};$$

This wavelength is in the visible region of the spectrum, which reaches the surface of the Earth. Thus, Cl_2 in the troposphere will be dissociated by sunlight.

6.65 A 4p electron must have $n = 4$, $l = 1$, but m_l has three possible values and m_s has two possible values. There are six possible sets of values:

n	l	m_l	m_s
4	1	1	1/2
4	1	1	−1/2
4	1	0	1/2

4	1	0	$-1/2$
4	1	-1	$1/2$
4	1	-1	$-1/2$

6.67 When the frequency is doubled and the amplitude is kept the same the peaks are thinner and occur twice as often:

Original   2x frequency

6.69 The longest wavelength that can eject an electron from the metal surface is the wavelength that just overcomes the binding energy. Calculate the wavelength using $\lambda = \dfrac{hcN_A}{E}$:

$$hcN_A = 6.626 \times 10^{-34} \text{ J s}\left(\frac{2.998 \times 10^{8} \text{ m}}{1 \text{ s}}\right)\left(\frac{10^{-3} \text{ kJ}}{1 \text{ J}}\right)\left(\frac{6.022 \times 10^{23} \text{ photons}}{1 \text{ mol}}\right)$$

$$hcN_A = 1.196 \times 10^{-4} \text{ kJ m/mol}$$

$$\lambda = \left(\frac{1.196 \times 10^{-4} \text{ kJ m}}{1 \text{ mol}}\right)\left(\frac{1 \text{ mol}}{216.4 \text{ kJ}}\right) = 5.527 \times 10^{-7} \text{ m or } 552.7 \text{ nm.}$$

6.71 The photoelectric effect is described by an energy balance equation:
$E_{\text{kinetic, electron}} = E_{\text{photon}} - bE$.
(a) $bE = E_{\text{photon}} - E_{\text{electron}} = 6.00 \times 10^{-19} \text{ J} - 2.70 \times 10^{-19} \text{ J} = 3.30 \times 10^{-19} \text{ J}$;

(b) $\lambda = \dfrac{hc}{E} = \dfrac{(6.626 \times 10^{-34} \text{ J s})(2.998 \times 10^{8} \text{ m/s})}{6.00 \times 10^{-19} \text{ J}} = 3.31 \times 10^{-7}$ m;

(c) For electrons, $\lambda = \dfrac{h}{mu}$;

$$u = \sqrt{\frac{2E_{\text{kinetic}}}{m}} = \sqrt{\frac{2(2.70 \times 10^{-19} \text{ J})}{9.109 \times 10^{-31} \text{ kg}}} = 7.70 \times 10^{5} \text{ m/s} \; ;$$

$$\lambda = \frac{(6.626 \times 10^{-34} \text{ kg m}^2 / \text{s})}{(9.109 \times 10^{-31} \text{ kg})(7.70 \times 10^{5} \text{ m/s})} = 9.45 \times 10^{-10} \text{ m.}$$

6.73 For electromagnetic radiation, $E = h\nu$ and $\lambda = \dfrac{c}{\nu}$.

$$\lambda = \left(\frac{2.998 \times 10^8 \, \text{m/s}}{27.3 \, \text{MHz}}\right)\left(\frac{1 \, \text{MHz}}{10^6 \, \text{Hz}}\right)\left(\frac{1 \, \text{Hz}}{1 \, \text{s}^{-1}}\right) = 11.0 \, \text{m};$$

$$E = (6.626 \times 10^{-34} \, \text{J s})(27.3 \, \text{MHz})\left(\frac{10^6 \, \text{Hz}}{1 \, \text{MHz}}\right)\left(\frac{1 \, \text{s}^{-1}}{1 \, \text{Hz}}\right) = 1.81 \times 10^{-26} \, \text{J}.$$

6.75 Make use of the figure legends to determine the answers to the questions.
(a) At an altitude of 60 km, $P \sim 2 \times 10^{-4}$ atm;
(b) examples are: N_2, O_2, and O_3
(c) At a pressure of 8 torr (or 0.0105 atm) the altitude should be ~35 km
(d) stratosphere

6.77 Calculate the frequencies using $v = \frac{c}{\lambda}$ and the energies using $E = \frac{hcN_A}{\lambda}$.

$hcN_A = 1.196 \times 10^{-4}$ kJ m/mol. The calculations for 487 nm are shown:

$$\lambda = 487 \, \text{nm}\left(\frac{1 \, \text{m}}{10^9 \, \text{nm}}\right) = 4.87 \times 10^{-7} \, \text{m};$$

$$v = \left(\frac{2.998 \times 10^8 \, \text{m s}^{-1}}{4.87 \times 10^{-7} \, \text{m}}\right) = 6.16 \times 10^{14} \, \text{s}^{-1};$$

$$E = \frac{hcN_A}{\lambda} = \frac{1.196 \times 10^{-4} \, \text{kJ m mol}^{-1}}{4.87 \times 10^{-7} \, \text{m}} = 246 \, \text{kJ/mol};$$

λ (nm)	487	514	543	553	578
v (10^{14} s^{-1})	6.16	5.83	5.52	5.42	5.19
E (kJ/mol)	246	233	220	216	207

6.79 Energies and frequencies for transitions in hydrogen atoms can be calculated from the equation for hydrogen atom energy levels:

$$E_n = \frac{-2.18 \times 10^{-18} \, \text{J}}{n^2}$$

$$\Delta E_{5-1} = (E_5 - E_1) = \frac{-2.18 \times 10^{-18} \, \text{J}}{5^2} - \frac{-2.18 \times 10^{-18} \, \text{J}}{1^2} = 2.09 \times 10^{-18} \, \text{J}$$

$$\lambda = \frac{hc}{E} = \frac{(6.626 \times 10^{-34} \, \text{Js})(2.998 \times 10^8 \, \text{m/s})}{2.09 \times 10^{-18} \, \text{J}} = 9.50 \times 10^{-8} \, \text{m}$$

$$\lambda = 9.50 \times 10^{-8} \, \text{m}\left(\frac{10^9 \, \text{nm}}{1 \, \text{m}}\right) = 95.0 \, \text{nm}$$

These photons lie in the ultraviolet, just above the X-ray region.

6.81 According to Figure 6-4, visible light has wavelengths in the 4 - 7.2 x 10^{-7} m range. The highest frequency of visible light is $\nu = \dfrac{2.998\text{x}10^{8}\,\text{m/s}}{4\text{x}10^{-7}\,\text{m}}$ ~ 7.5 x 10^{14} s^{-1}. This is smaller than the threshold frequency, so Mg cannot be used in photoelectric devices..

6.83 Volume is related to radius through the equation, $V = \dfrac{4}{3}\pi r^3$.

$$V_{\text{atom}} = \frac{4\pi(10^{-10}\,\text{m})^3}{3} = 4 \text{ x } 10^{-30}\ \text{m}^3;$$

$$V_{\text{nucleus}} = \frac{4\pi(10^{-15}\,\text{m})^3}{3} = 4 \text{ x } 10^{-45}\ \text{m}^3$$

$\dfrac{4 \text{ x } 10^{-45}\,\text{m}^3}{4 \text{ x } 10^{-30}\ \text{m}^3} = \dfrac{1}{1 \text{ x } 10^{15}}$ is the fraction of the volume that the nucleus takes up.

6.85 First calculate the energy of one photon of wavelength 510 nm. Then divide the detection limit by this energy to find the number of photons.

$$\lambda = 510\ \text{nm}\left(\frac{10^{-9}\,\text{m}}{1\,\text{nm}}\right) = 5.10 \text{ x } 10^{-7}\ \text{m};$$

$$E_{\text{photon}} = \frac{hc}{\lambda} = \frac{(6.626 \text{ x } 10^{-34}\ \text{Js})(2.998 \text{ x } 10^{8}\ \text{m/s})}{5.10 \text{ x } 10^{-7}\ \text{m}} = 3.90 \text{ x } 10^{-19}\ \text{J}$$

$$n = \frac{2.35 \text{ x } 10^{-18}\ \text{J}}{3.90 \text{ x } 10^{-19}\ \text{J}} = 6 \text{ photons.}$$

6.87 The wavelengths associated with particles are calculated from $\lambda = \dfrac{h}{mu}$, 5.000 % of the speed of light is $u = \left(\dfrac{5.000\ \%}{100\%}\right)(2.998 \text{ x } 10^{8}\ \text{m/s}) = 1.499 \text{ x } 10^{7}\ \text{m/s}$.

$$\text{electron: } \lambda = \frac{h}{mu} = \frac{(6.626 \text{ x } 10^{-34}\ \text{Js})}{(9.109 \text{ x } 10^{-31}\ \text{kg})(1.499 \text{ x } 10^{7}\ \text{m/s})} = 4.853 \text{ x } 10^{-11}\ \text{m};$$

$$\text{proton: } \lambda = \frac{h}{mu} = \frac{(6.626 \text{ x } 10^{-34}\ \text{Js})}{(1.673 \text{ x } 10^{-27}\ \text{kg})(1.499 \text{ x } 10^{7}\ \text{m/s})} = 2.642 \text{ x } 10^{-14}\ \text{m}.$$

6.89 The energy level diagram shows the energy differences among the various levels. By examining it, we can see that the smallest energy difference is associated with Photon$_1$ and the largest energy difference with Photon$_2$. The energy of a photon is inversely proportional to its wavelength, so Photon$_1$ has the longest wavelength and Photon$_2$ the shortest wavelength. Thus the assignments are as follows:

$$\text{Photon}_1 = 565 \text{ nm } (d \text{ to } c)$$
$$\text{Photon}_2 = 121 \text{ nm } (c \text{ to } b)$$
$$\text{Photon}_3 = 152 \text{ nm } (b \text{ to } a)$$

Level b:

$$\lambda = 152 \text{ nm} \left(\frac{10^{-9} \text{ m}}{1 \text{ nm}} \right) = 1.52 \times 10^{-7} \text{ m};$$

$$E_{photon} = \frac{hc}{\lambda} = \frac{(6.626 \times 10^{-34} \text{ Js})(2.998 \times 10^8 \text{ m/s})}{1.52 \times 10^{-7} \text{ m}} = 1.31 \times 10^{-18} \text{ J}$$

Level c:

$$\lambda = 121 \text{ nm} \left(\frac{10^{-9} \text{ m}}{1 \text{ nm}} \right) = 1.21 \times 10^{-7} \text{ m};$$

$$E = \text{level b} + \frac{hc}{\lambda}$$

$$= 1.31 \times 10^{-18} \text{ J} + \frac{(6.626 \times 10^{-34} \text{ J s})(2.998 \times 10^8 \text{ m/s})}{1.21 \times 10^{-7} \text{ m}} = 2.95 \times 10^{-18} \text{ J}$$

Level d:

$$\lambda = 565 \text{ nm} \left(\frac{10^{-9} \text{ m}}{1 \text{ nm}} \right) = 5.65 \times 10^{-7} \text{ m};$$

$$E = \text{level c} + \frac{hc}{\lambda}$$

$$= 2.95 \times 10^{-18} \text{ J} + \frac{(6.626 \times 10^{-34} \text{ J s})(2.998 \times 10^8 \text{ m/s})}{5.65 \times 10^{-7} \text{ m}} = 3.30 \times 10^{-18} \text{ J}$$

6.91 (a) Calculate the energies of the photons using $E = \dfrac{hc}{\lambda}$:

$$E_{488} = \frac{(6.626 \times 10^{-34} \text{ J s})(2.998 \times 10^8 \text{ m/s})}{4.88 \times 10^{-7} \text{ m}} = 4.07 \times 10^{-19} \text{ J};$$

$$E_{514} = \frac{(6.626 \times 10^{-34} \text{ J s})(2.998 \times 10^8 \text{ m/s})}{5.14 \times 10^{-7} \text{ m}} = 3.86 \times 10^{-19} \text{ J};$$

(b) Each transition leaves the ion in the same state, so the state from which 488 nm emission occurs must be slightly higher in energy than that from which 514 nm emission occurs:

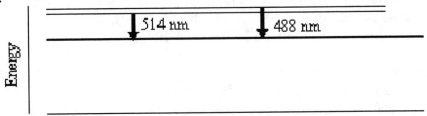

(c) Calculate frequency and wavelength using $v = \dfrac{E}{h}$ and $\lambda = \dfrac{c}{v}$:

$$v = \frac{E}{h} = \frac{2.76 \times 10^{-18}\,J}{6.626 \times 10^{-34}\,J\,s} = 4.17 \times 10^{15}\,s^{-1};$$

$$\lambda = \frac{c}{v} = \frac{2.998 \times 10^{8}\,m/s}{4.17 \times 10^{15}\,s^{-1}} = 7.19 \times 10^{-8}\,m \text{ or } 71.9 \text{ nm.}$$

6.93   (a) Calculate the energies of the photons using $E = \dfrac{hcN_A}{\lambda}$:

$$hcN_A = 6.626 \times 10^{-34}\,J\,s \left(\frac{2.998 \times 10^{8}\,m}{1\,s}\right)\left(\frac{10^{-3}\,kJ}{1\,J}\right)\left(\frac{6.022 \times 10^{23}}{1\,mol}\right) =$$

$$hcN_A = 1.196 \times 10^{-4}\,kJ\,m/mol$$

$$E_{589.6} = \frac{1.196 \times 10^{-4}\,kJ\,m/mol}{5.896 \times 10^{-7}\,m} = 202.8 \text{ kJ/mol;}$$

$$E_{590.0} = \frac{1.196 \times 10^{-4}\,kJ\,m/mol}{5.900 \times 10^{-7}\,m} = 202.7 \text{ kJ/mol;}$$

(b) The two levels are very close together at about 203 kJ/mol above the ground state:

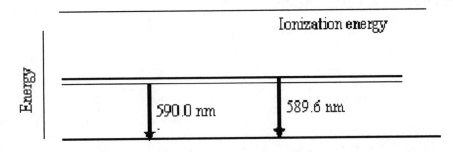

(c) The energy required to ionize the atom from this state is the difference in energies:

$E = 486 - 202.7 = 283$ kJ/mol. Calculate the wavelength using $\lambda = \dfrac{hcN_A}{E}$:

$$\lambda = \frac{hcN_A}{E} = \left(\frac{1.196 \times 10^{-4}\,kJ\,m}{1\,mol}\right)\left(\frac{1\,mol}{283\,kJ}\right) = 4.23 \times 10^{-7}\,m \text{ or } 423 \text{ nm.}$$

6.95 (a) Binding energy is the energy required to eject an electron from a metal.
For metal A, electrons are not ejected until $4.0 \times 10^{14}\,s^{-1}$ is reached, therefore the binding energy will be the energy corresponding to this frequency:
$E = h\nu = (6.626 \times 10^{-34}\,Js)(4.0 \times 10^{14}\,s^{-1}) = 2.7 \times 10^{-19}\,J$

Similarly for metal B at $6.5 \times 10^{14}\,s^{-1}$:
$E = h\nu = (6.626 \times 10^{-34}\,Js)(6.5 \times 10^{14}\,s^{-1}) = 4.3 \times 10^{-19}\,J$;
Therefore metal B has the higher binding energy since higher frequency photons are required to generate photoelectrons.

(b) kinetic energy = photon energy – binding energy;
$$\lambda = 125\ nm\left(\frac{10^{-9}\,m}{1\,nm}\right) = 1.25 \times 10^{-7}\,m;$$

$$E_{photon} = \frac{hc}{\lambda} = \frac{(6.626 \times 10^{-34}\,Js)(2.998 \times 10^{8}\,m/s)}{1.25 \times 10^{-7}\,m} = 1.59 \times 10^{-18}\,J$$

Hence, using the photon energy above and the binding energy calculated in part (a), the kinetic energies are:
Metal A: $1.59 \times 10^{-18}\,J - 0.27 \times 10^{-18}\,J = 1.32 \times 10^{-18}\,J$;
Metal B: $1.59 \times 10^{-18}\,J - 0.43 \times 10^{-18}\,J = 1.16 \times 10^{-18}\,J$.

(c)   The wavelength range that electrons can be ejected from one electron but not the other corresponds to the frequency range of $4.0 \times 10^{14}\,s^{-1}$ to $6.5 \times 10^{14}\,s^{-1}$.

At $4.0 \times 10^{14}\,s^{-1}$, wavelength, $\lambda = \dfrac{c}{\nu} = \dfrac{2.998 \times 10^{8}\,m/s}{4.0 \times 10^{14}\,s^{-1}} = 7.5 \times 10^{-7}\,m$ or 750 nm;

At $6.5 \times 10^{14}\,s^{-1}$, wavelength, $\lambda = \dfrac{c}{\nu} = \dfrac{2.998 \times 10^{8}\,m/s}{6.5 \times 10^{14}\,s^{-1}} = 4.6 \times 10^{-7}\,m$ or 460 nm;

Thus, the wavelength range over which photons can eject electrons from one metal but not the other is 460 nm to 750 nm.

7.1   Orbital stability in multi-electron atoms depends on the value of *n* (stability decreases with increasing value), *Z* (stability increases with increasing value), *l* (stability decreases with increasing value), and amount of screening (stability increases as other electrons are removed).
(a) He 1*s* is more stable than He 2*s* because of its lower value of *n*.
(b) Kr 5*s* is more stable than Kr 5*p* because of its lower value of *l*.
(c) He$^+$ 2*s* is more stable than He 2*s* because it is less screened by 1*s* electrons.

7.3   A hydrogen atom contains just one electron, so there is no screening effect. In the absence of screening, orbital energy depends only on *n* and *Z*, so all *n* = 3 orbitals have identical energy. In a helium atom, on the other hand, an electron in an *n* = 3 orbital is screened from the nucleus by the second electron, and the amount of screening decreases as *l* increases.

7.5   (a) The ionization energy (I.E.) of the He 2*p* orbital (0.585×10^{-18} J) is not much larger than that of the H 2*p* orbital (0.545×10^{-18} J). The similar values indicate nearly equal effective nuclear charges due to nearly complete screening; in the absence of screening, the He 2*p* orbital would have four times the ionization energy as the H 2*p* orbital.
(b) The I.E. of the He$^+$ 2*p* orbital (2.18×10^{-18} J) is four times larger than that of the H 2*p* orbital (0.545×10^{-18} J). A four-fold increase when *Z* doubles indicates a Z^2 dependence.

7.7

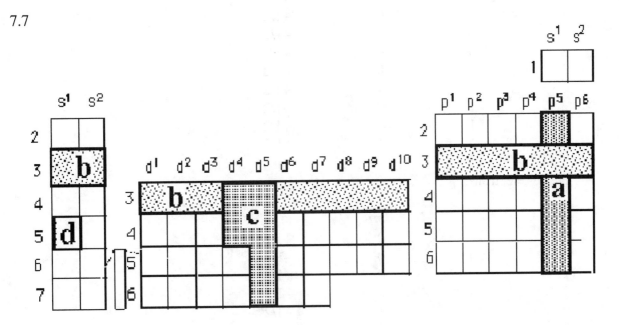

7.9   The element below lead in the periodic table would have atomic number 114 (count from Mt, element 109: 3 elements needed to complete the *d* block, 2 more to reach column 14 in the *p* block). The block that is filling would match that of lead, but with one higher *n*-value: 7*p*.

7.11   Element 111 would fall directly below gold in Group 11, in row 7 (*n* = 6 for the *d* block).

7.13 The column location of an element is the indicator of how many valence electrons it has. Remember that *s* electrons as well as those in the filling block count as valence electrons.
O, fourth column of *p* block, 4 *p* + 2 *s* = 6 valence electrons;
V, third column of *d* block, 3 *d* + 2 *s* = 5 valence electrons;
Rb, 1 valence electron;
Sn, second column of *p* block, 2 *s* + 2 *p* = 4 valence electrons; and
Cd, end of *d* block, 2 valence electrons.

7.15 Use the aufbau principle to determine the orbital occupancies and quantum numbers for the electrons in a configuration. When there are equivalent configurations, unpair as many electrons as possible in accordance with Hund's rule.
Be, 4 electrons:

n	l	m_l	m_s
1	0	0	+1/2
1	0	0	−1/2
2	0	0	+1/2
2	0	0	−1/2

O, 8 electrons:

n	l	m_l	m_s
1	0	0	+1/2
1	0	0	−1/2
2	0	0	+1/2
2	0	0	−1/2
2	1	1	+1/2
2	1	0	+1/2
2	1	−1	+1/2
2	1	1	−1/2

Ne, 10 electrons:

n	l	m_l	m_s
1	0	0	+1/2
1	0	0	−1/2
2	0	0	+1/2
2	0	0	−1/2
2	1	1	+1/2
2	1	1	−1/2
2	1	0	+1/2
2	1	0	−1/2
2	1	−1	+1/2
2	1	−1	−1/2

P, 15 electrons: first 10 electrons are the same as for Ne; remaining 5 electrons are:

n	l	m_l	m_s
3	0	0	+1/2
3	0	0	−1/2
3	1	1	+1/2
3	1	0	+1/2
3	1	−1	+1/2

7.17 Atoms with unpaired electrons in their configurations are paramagnetic. Of the atoms in Problem 7.15, O and P have unpaired electrons. Here are the orbital occupancy diagrams for the least stable occupied orbitals:

7.19 (a) This is Pauli-forbidden: *s* orbitals can hold no more than two electrons.
(b) This is Pauli-forbidden: *s* orbitals can hold no more than two electrons.
(c) This is an excited-state configuration: the 1*s* and 2*s* orbitals are not full.
(d) This is an excited-state configuration: the 2*s* orbital is not full.
(e) This is the ground-state configuration.
(f) This configuration uses a non-existent orbital: there is no 1*p* orbital.
(g) This configuration uses a non-existent orbital: there is no 2*d* orbital.

7.21 The element with the most unpaired electrons will have the higher spin. Mo has ground-state configuration $[Kr]\,5s^1\,4d^5$; that of Tc is $[Kr]\,5s^2\,4d^5$:

7.23 (a) When orbitals are near-degenerate, exceptions to the normal filling order exist. Normally, the 4*s* orbital fills immediately after 3*p*, starting with element 19. Exceptions that indicate near-degeneracy of 3*d* and 4*s* are Cr, $4s^1\,3d^5$; and Cu, $4s^1\,3d^{10}$.

(b) The first two elements in column 6, Cr and Mo, have $s^1\,d^5$ configurations, while the second two elements in this column, W and Sg, have $s^2\,d^4$ configurations.

(c) The 6*d* and 5*f* orbitals fill in the second *f*-block, the actinides. In this block, four elements have both these orbitals partly filled, indicating near-degeneracy: Pa, U, Np, and Cm.

7.25 The ground state for N is $1s^2\,2s^2\,2p^3$. Excited states with no electron having $n > 2$ can be formed by moving electrons out of the 1*s* and/or 2*s* orbital and placing them in the 2*p* orbital.

There are seven ways to do this:

$1s^1 2s^2 2p^4$; $\qquad$ $2s^2 2p^5$;

$1s^1 2s^1 2p^5$; $\qquad$ $1s^2 2s^1 2p^4$;

$1s^2 2p^5$; $\qquad$ $2s^1 2p^6$; and

$1s^1 2p^6$.

7.27 Ionization energy (*IE*) decreases with increasing **n** and increases with increasing *Z*. Cs, which has the largest **n**-value, has the smallest *IE*; K, which has the next larger **n**-value, has the next smallest *IE*; because of its larger *Z* value, Ar has a larger *IE* than Cl:

Ar > Cl > K > Cs.

7.29 The value of IE_2 is almost ten times that of IE_1, indicating that the second electron removed is a core electron rather than a valence electron. The elements in column 1 contain only one valence electron, so this is a column 1 element. An electron affinity around –50 kJ/mol matches this assignment (see Figure 7-18). (The element is Cs).

7.31 The valence configurations are N, $2s^2 2p^3$; Mg, $3s^2$; and Zn, $4s^2 3d^{10}$.

The added electron adds to an already-occupied orbital. Electron-electron repulsion makes this an unfavorable process.

The added electron adds to the next higher orbital, which is significantly higher in energy.

The added electron adds to the next higher orbital, which is significantly higher in energy.

7.33 Br⁻ has 36 electrons. Isoelectronic ions with less than 4 units of net charge are: As³⁻, Se²⁻, Rb⁺, Sr²⁺, and Y³⁺. For isoelectronic ions, size decreases with increasing *Z*:

Y³⁺ < Sr²⁺ < Rb⁺ < Br⁻ < Se²⁻ < As³⁻.

7.35 Stable atomic anions form for those elements that are one electron short of filled *p* orbitals; these elements have high electron affinities: F, Cl, Br, I, (At).

7.37 To form [K⁺I⁻] from neutral gaseous atoms we must remove an electron from potassium, add an electron to iodine and then form the coulombic interaction. The sequence of reactions for doing this calculation is (all gas phase):
K → K⁺ + e⁻, $\Delta E = IE_1 = 418.8$ kJ/mol;

$I + e^- \rightarrow I^-, \Delta E = EA = -295.3$ kJ/mol;

$K^+ + I^- \rightarrow KI, \Delta E = E_{coulomb}$; (use Equation 7-1).

$$E_{coulomb} = \frac{k(q^+)(q^-)}{r} = \frac{(1.389 \times 10^5 \text{ kJ pm/mol})(+1)(-1)}{(133 \text{ pm} + 220 \text{ pm})} = -393 \text{ kJ/mol}$$

The total energy is the sum of the above energies:

$\Delta E_{total} = (418.8$ kJ/mol$) + (-295.3$ kJ/mol$) + (-393$ kJ/mol$) = -270.$ kJ/mol.

7.39 Stable anions form from elements in columns 16 (dianions) and 17 (monoanions). Stable cations form from various metals.
Ca, Cu, Cs, and Cr, all metals, may be found in ionic compounds as cations.
Cl may be found in ionic compounds as a –1 anion.
C is not found in ionic compounds.

7.41 First identify the chemical eqn for the formation of 1 LiF(s):

$$Li(s) + \frac{1}{2}F_2(g) \rightarrow LiF(s)$$

Now determine the steps needed for this reaction. Remember that in this salt, Li will form a +1 cation and F will form a –1 anion.

(1) $Li(s) \rightarrow Li(g)$ ΔE = energy of vaporization = 159 kJ/mol
(2) $Li(g) \rightarrow Li^+(g) + e^-$ $\Delta E = IE$ = 520.2 kJ/mol
(3) $\frac{1}{2}F_2(g) \rightarrow F(g)$ ΔE = ½ bond energy = 155/2 kJ/mol = 77.5 kJ/mol
(4) $F(g) + e^- \rightarrow F^-(g)$ $\Delta E = EA$ = -328.0 kJ/mol
(5) $F^-(g) + Li^+(g) \rightarrow LiF(s)$ ΔE = lattice energy = -1036 kJ/mol

Summing up all of these energies gives the overall energy change for the formation of lithium fluoride:

(159 + 520.2 + 77.5 + -328.0 + -1036) kJ/mol = -607 kJ/mol

7.43 For a Born-Haber diagram you need to show energy of vaporization of the metal, ionization energy of the metal, energy to break the molecular bonds, electron affinity of anions, and the lattice energy that brings the ions of the salt together.

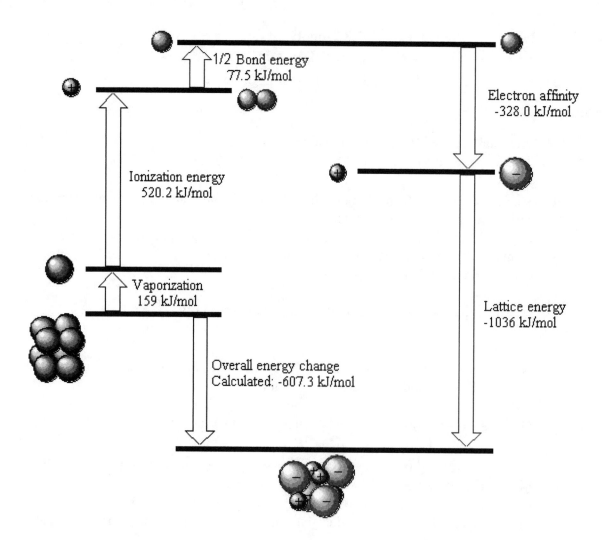

7.45   Non-metals are found in the upper right portion of the periodic table, metalloids along a
diagonal running through the *p* block, and all other elements are metals. In Column 16,
Te is classified as a metalloid. The elements above it, O, S, and Se, are non-metals, and
the element below it, Po, is a metal.

7.47   Non-metals are found in the upper right portion of the periodic table, metalloids along a
diagonal running through the *p* block, and all other elements are metals. Thus, C and Cl
are non-metals, and Ca, Cu, Cs, and Cr are metals.

7.49   The metalloids occupy a diagonal strip across the *p* block of the periodic table. The
metals immediately to the left of this strip are likely to be metalloid-like: Al, Ga, Sn, Bi.

7.51   Use the filling rules to determine the correct configuration, being aware of exceptions.
C (6 electrons): $1s^2\,2s^2\,2p^2$;
Cr (24 electrons, exception): $1s^2\,2s^2\,2p^6\,3s^2\,3p^6\,4s^1\,3d^5$;
Sb (51 electrons): $1s^2\,2s^2\,2p^6\,3s^2\,3p^6\,4s^2\,3d^{10}\,4p^6\,5s^2\,4d^{10}\,5p^3$;
Br (35 electrons): $1s^2\,2s^2\,2p^6\,3s^2\,3p^6\,4s^2\,3d^{10}\,4p^5$.

7.53 To determine the correct configuration of a cation, first determine the configuration of the neutral atom, then remove the appropriate number of electrons, removing valence s electrons before valence d electrons:

Mn (25 electrons): $1s^2\,2s^2\,2p^6\,3s^2\,3p^6\,4s^2\,3d^5$;

Mn^{2+} (remove 2 $4s$ electrons from Mn): $1s^2\,2s^2\,2p^6\,3s^2\,3p^6\,3d^5$;

The least stable occupied orbital is one of the five degenerate $3d$ orbitals whose electrons are distributed to produce maximum spin:

n	l	m_l	m_s
3	2	2	+1/2
3	2	1	+1/2
3	2	0	+1/2
3	2	-1	+1/2
3	2	-2	+1/2

7.55 Total electron spin depends on the number of unpaired electrons. In partially filled shells, each unpaired electron contributes 1/2 to the total spin.

P: [Ne] $3s^2\,3p^3$; each $3p$ orbital is half filled, so net spin = $3(1/2) = 3/2$;

Br^-: [Kr], all orbitals are filled, so net spin = 0;

Cu^+: [Ar] $3d^{10}$, all orbitals are filled, so net spin = 0.

7.57 Use the periodic table to determine the correct configurations.

$Z = 9$ is F: [He] $2s^2\,2p^5$;

$Z = 20$ is Ca: [Ar] $4s^2$; and

$Z = 33$ is As: [Ar] $4s^2\,3d^{10}\,4p^3$.

7.59 Ionization energy decreases with increasing value of n and, for the same value of n, increases with increasing Z (increasing Coulombic attraction). Na is the only species in this set with an $n = 3$ valence electron, so it has the smallest IE. Na^+ has the largest Z, so it has the largest IE. O has its last two electrons paired, reducing its IE below that of N despite its higher Z because of increased electron-electron repulsion:

$Na < O < N < Ne < Na^+$.

7.61 Size increases with increasing n value and, for the same value of n, decreases with increasing Z (increasing Coulombic attraction). Cl, Cl^-, and K^+ all have $n = 3$ for the valence electrons, while Br^- has $n = 4$, so Br^- is the largest of this set. Among the others, the extra electron in Cl^- makes it larger than Cl, and K^+ has higher Z than Cl:

$Br^- > Cl^- > Cl > K^+$.

7.63   (a) In all one-electron atoms, orbital energy depends only on n and Z, both of which are the same for the hydrogen atom $2s$ and $2p$ orbitals.
(b) In multi-electron atoms, n orbitals with different l values are screened to different extents. The orbital with the lower l value is less screened, hence more stable.

7.65   This problem provides an exercise in graph reading.
(a) The greatest drop within a row is from element 80 (Hg) to 81 (Tl);
(b) Cs, element 55, has the lowest value shown in the figure;
(c) The value changes least from $Z = 56$ to 71, $Z = 39$ to 46, and $Z = 20$ to 29;
(d) Elements with values between 925 and 1050 kJ/mol are 15 (P), 16 (S), 33 (As), 34 (Se), 53 (I), 80 (Hg), 85 (At), and 86 (Rn).

7.67   Remember that in the transition metal cations, the ns orbital is less stable than $(n-1)d$:
   Cu$^+$ is $4s^0 3d^{10}$ Mn^{2+} is $4s^0 3d^5$ Au^{3+} is $6s^0 5d^8$

(bottom orbitals are d, top orbitals are s)

7.69   Unpaired electrons occur only among the valence orbitals. Construct the configuration of neutrals, then remove one electron to form S$^+$ and add one electron to form S$^-$:
   S: [Ne] $3s^2 3p^4$;   S$^+$: [Ne] $3s^2 3p^3$;   S$^-$: [Ne] $3s^2 3p^5$;

S$^+$, with three unpaired electrons, has more unpaired than S or S$^-$.

7.71   Identify the orbitals using shapes and relative sizes, given that (a) has $n = 3$. (a) and (b) are both spherical, with (b) smaller than (a). A is $3s$, B is $2s$. C has the unique shape of the $3d_{z^2}$ orbital.

(a) (b) is most stable (smallest n value), and (a) is more stable than (c);
(b) $n = 3$, $l = 0$, $m_l = 0$, $m_s = +1/2$; $n = 3$, $l = 0$, $m_l = 0$, $m_s = -1/2$;
(c) Any element from magnesium to calcium has $3s^2$ but $3d^0$: $Z = 12$ to $Z = 20$.
(d) The $3d$ transition metal cations have partially occupied $3d$ orbitals, for example, Fe^{3+};
(e) There are eight other $n = 3$ orbitals: $3s$, three $3p$, and four additional $3d$;
(f) Whenever an electron is removed from a doubly-occupied orbital, electron-electron repulsion decreases for the remaining electron, so the orbital becomes smaller.

7.73   The filling pattern after $7p$ should mirror the pattern after $6p$: $8s$, then $5g$ (or perhaps $6f$).

7.75   The data in Appendix C indicate that noble gas elements have unfavorable electron affinities (> 0) and relatively large first ionization energies (> 1350 kJ/mol). Consequently, they are unlikely to either gain or lose electrons to form stable chemical compounds. Nevertheless, Xe and, to a lesser extent Kr, do form a few compounds.

7.77 In an excited state configuration, one electron is shifted to a less stable orbital. Start by constructing the ground state configuration, then move an electron from the most stable occupied to the least stable unoccupied orbital:

Be: [He] $2s^2$; the next orbital is $2p$:   [He] $2s^1\, 2p^1$;
O^{2-}: [He] $2s^2\, 2p^6$; the next orbital is $3s$:   [He] $2s^2\, 2p^5\, 3s^1$;
Br^-: [Ar] $4s^2\, 3d^{10}\, 4p^6$; the next orbital is $5s$: [Ar] $4s^2\, 3d^{10}\, 4p^5\, 5s^1$;
Ca^{2+}: [Ar]; the next orbital is $4s$:   [Ne] $3s^2\, 3p^5\, 4s^1$;
Sb^{3+}: [Kr] $4d^{10}\, 5s^2$; the next orbital is $5p$:   [Kr] $4d^{10}\, 5s^1\, 5p^1$.

7.79 Examine the configurations of the atoms and ions to determine the reason for the ionization energy differences:

Li: $1s^2\, 2s^1$ Be: $1s^2\, 2s^2$;
Li^+: $1s^2$ Be^+: $1s^2\, 2s^1$;

The first ionization energies involve the $2s$ orbital for both atoms, so Be, with larger Z, has a larger *IE*. The second electron removed from Li is a core electron, so the *IE* is much greater than the second *IE* for Be.

7.81 Consult Figures 7-6, 7-7, and 7-16 for examples of this kind of drawing.

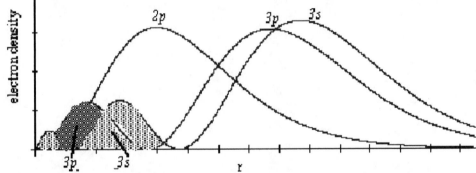

The shaded portions of the $3s$ and $3p$ indicate the region that is ineffectively screened by the $2p$ orbital. $3s$ and $3p$ electrons have greater probability of being in the larger areas mostly outside the $2p$ orbital. Thus, the $2p$ orbital effectively screens electrons in both the $3s$ and $3p$ orbitals.

8.1   Determine a configuration from the position of an element in the periodic table. The electrons with the highest principal quantum number will be involved in bond formation. Be has the configuration $1s^2\,2s^2$. Its two $n = 2$ electrons will be involved in bond formation.

8.3   Atoms with one valence electron form bonds by overlap of their valence orbitals. The configuration of Na is [Ne] $3s^1$. Two sodium atoms form a bond by overlap of their $3s$ orbitals to produce an orbital with high electron density between the nuclei:

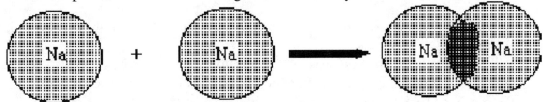

8.5   The orbitals that participate in bond formation are the valence orbitals; that is, the occupied orbitals of largest principal quantum number and any partially-filled orbitals of lower principal quantum number. Consult the configuration of the atom to determine what these orbitals are:

Al, [Ne] $3s^2\,3p^1$: the $3s$ and $3p$ orbitals participate in bonding;

As, [Ar] $3d^{10}\,4s^2\,4p^3$: the $4s$ and $4p$ orbitals participate in bonding;

F, [He] $2s^2\,2p^5$: the $2s$ and $2p$ orbitals participate in bonding;

Sn, [Kr] $4d^{10}\,5s^2\,5p^2$: the $5s$ and $5p$ orbitals participate in bonding.

8.7   Atoms with one valence electron form bonds by overlap of their valence orbitals. The configuration of Li is [He] $2s^1$. The $2s$ orbital of a lithium atom can overlap with the $1s$ orbital of a hydrogen atom to produce an orbital with high electron density between the nuclei:

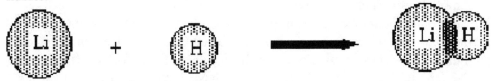

8.9   The periodic trends predict that electronegativity increases across a row (because of increasing Z which binds electrons more tightly) and decreases down a column (because of increasing n, which means less tightly bound electrons). Consult Figure 8-7 in your textbook to find exceptions to these trends:

Exceptions across the rows: Cr (1.6) and Mn (1.5); Cu (1.9) and Zn (1.6); Pd (2.2), Ag (1.9), and Cd (1.7); Au (2.4) and Hg (1.9);

Exceptions down the columns: Cr (1.6) and Mo (1.8); Mn (1.5) and Tc (1.9).

8.11   Ionic substances have $\Delta\chi > 2.0$, non-polar substances have $\Delta\chi = 0$, and polar substances have intermediate values:

Non-polar: F_2; ionic: NaF and CaO; polar: HF and NaH.

8.13   Electronegativities describe the tendency of each element to attract bonding electrons from another. In any pair, the element with higher electronegativity attracts bonding electrons more strongly. Use figure 8-7 to obtain the electronegativity of each element:

(a) N (3.0) will attract electrons more than C (2.5);
(b) S (2.5) will attract electrons more than H (2.1);
(c) I (2.5) will attract electrons more than Zn (1.6); and
(d) S (2.5) will attract electrons more than As (2.0).

8.15   Bond polarity increases with the difference in electronegativity of the bond-forming elements.
In these compounds, H is the least electronegative element, so bond polarity increases with the electronegativity of the other element. The electronegativity order is P < S < N < O, so the order of bond polarity is $PH_3 < H_2S < NH_3 < H_2O$.

8.17   Use the procedure in your text for determining the Lewis structures.
KBr is an ionic compound with K^+ and Br^- ions. K^+ has no valence electrons, Br^- has 8 valence electrons.

$$K^+ \quad :\!\ddot{Br}\!:^-$$

Both HBr and CBr_4 are covalent compounds.
1. Count the valence electrons.
    HBr: Hydrogen has 1 valence electron, Br (Group 17) has 7 for a total of 8 valence electrons.
    CBr_4: C (Group 14) has four valance electrons and each bromine atom contributes seven electrons for a total of 32 electrons.
2. Assemble the bonding framework.
    HBr is a diatomic molecule so simply connect the two atoms together. CBr_4, the carbon must be an inner atom with 4 outer bromine atoms. Remember to determine the number of electrons remaining (each bond uses 2 electrons). HBr now has 8 - 2 = 6 remaining valence electrons, CBr_4 has 32-2(4bonds) = 24 electrons

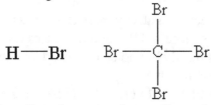

3. Add 3 nonbonding electron pairs to all non-hydrogen outer atoms. Note that this uses up the remaining valence electrons and gives us the final Lewis structures.

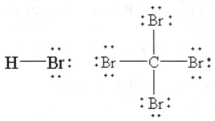

8.19 Determine the Lewis structures using standard procedures.
Na_2SO_3 is an ionic compound and should be treated as Na^+ and SO_3^{2-}.
Na^+ has no valence electrons and the Lewis structure is Na^+
SO_3^{2-}:
1) There are $6 + 3(6) + 2 = 26$ valence electrons.
2) Three electron pairs are used in forming the bonding framework, leaving
   $26 - 3(2) = 20$ electrons.
3) Place three pairs of electrons around each outer O atom leaving
   $20 - 3(6) = 2$ electrons.
4) Assign the remaining 2 electrons to the inner S atom
5) The resulting structure has $FC_S = 6 - 5 = +1$. Use a lone pair from one of the outer O
   atoms to form a double bond and reduce the formal charge to 0 (there are three
   resonance structures):

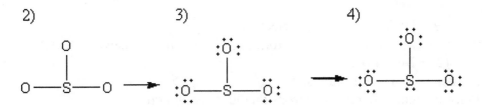

5) Resonance structures

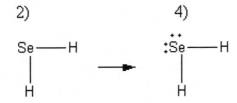

H_2Se:
1) There are $2 + 6 = 8$ valence electrons.
2) Two electron pairs are used in forming the bonding framework, leaving
   $8 - 2(2) = 4$ electrons.
3) All outer atoms are H atoms.
4) Assign the remaining 4 electrons to the inner Se atom.
5) The resulting structure has $FC_{Se} = 6 - 4 - 2 = 0$, so this is the Lewis structure:

2)   4)

Se —— H   :Se —— H

  | |

  H   H

$AlCl_3$:
1) There are $3 + 3(7) = 24$ valence electrons.
2) Three electron pairs are used in forming the bonding framework, leaving
   $24 - 3(2) = 18$ electrons.

3) Place three pairs of electrons around each outer Cl atom, leaving
   $18 - 3(6) = 0$ electrons.
4) No remaining electrons.
5) The resulting structure has 6 electrons surrounding the inner Al atom. Our rules call
   for making a double bond to Al, but this would give Al a negative formal charge,
   which is unlikely, so we leave the structure as it is; compare with triethylaluminum in
   your textbook:

   2)   3)

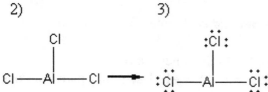

8.21 Determine the Lewis structures using standard procedures.
     NH_3:
     1) There are $5 + 3(1) = 8$ valence electrons.
     2) Three electron pairs are used in forming the bonding framework, leaving
        $8 - 3(2) = 2$ electrons.
     3) All outer atoms are H atoms.
     4) Place the remaining two electrons on the inner N atom.
     5) There is an octet around the inner N atom, so this is the correct Lewis structure:

        2)   4)

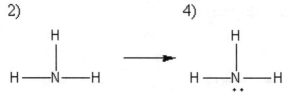

NH_4NO_3 is an ionic compound and should be treated as NH_4^+ and NO_3^-.
NO_3^-:
1) There are $5 + 3(6) + 1 = 24$ valence electrons.
2) Three electron pairs are used in forming the bonding framework, leaving
   $24 - 3(2) = 18$ electrons.
3) Use the remaining electrons to place three pairs of electrons around each outer O
   atom.
4) No remaining electrons.
5) The resulting structure has 6 electrons on the inner N atom. Use a lone pair from the
   outer O atoms to form a double bond and complete the octet on the N (there are three
   resonance structures):

2)   3)

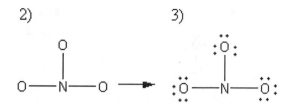

5) Resonance structures

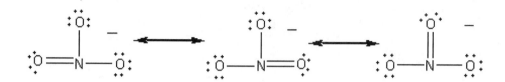

NH_4^+:
1) There are $5 + 4(1) - 1 = 8$ valence electrons.
2) Four electron pairs are used in forming the bonding framework, leaving no electrons left.
3) All outer atoms are H atoms.
4) No remaining electrons.
5) The resulting structure has 8 electrons on the inner N atom:

HNO_3:
1) There are $5 + 3(6) + 1 = 24$ valence electrons.
2) Four electron pairs are used in forming the bonding framework, leaving $24 - 4(2) = 16$ electrons.
3) Place three pairs of electrons around each outer O atom, leaving $16 - 2(6) = 4$ electrons.
4) Place the remaining 4 electrons on the inner O atom.
5) The resulting structure has 6 electrons on the inner N atom. Use a lone pair from the outer O atoms to form a double bond and complete the octet on the N (there are two resonance structures):

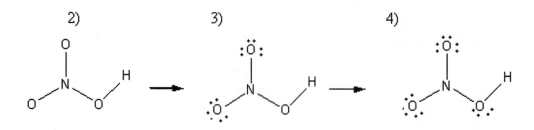

2)   3)   4)

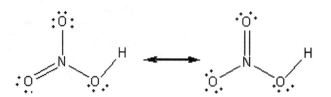

5) Resonance structures

8.23   Determine the Lewis structures using standard procedures.
LiOH is an ionic compound and should be treated as Li^+ and OH^-:
OH$^-$:
1) There are $6 + 1 + 1 = 8$ valence electrons.
2) One electron pair is used in forming the bonding framework, leaving $8 - 2 = 6$ electrons.
3) Place 3 pairs of electrons on the O atom:

$$H\!-\!\ddot{\underset{\cdot\cdot}{O}}:{}^{-}$$

KH$_2$PO$_4$ is an ionic compound and should be treated as K^+ and $H_2PO_4^-$
H$_2$PO$_4^-$:
1) There are $2(1) + 5 + 4(6) + 1 = 32$ valence electrons.
2) Six electron pairs are used in forming the bonding framework, leaving $32 - 6(2) = 20$ electrons.
3) Place 3 pairs of electrons on both outer O atoms, leaving $20 - 2(6) = 8$ electrons.
4) Place 2 pairs of electrons on each inner O atom.
5) The resulting structure has $FC_P = 5 - 4 = +1$, use a lone pair from an outer O atom to form a double bond and reduce the formal charge to 0:

2)   3)   4)

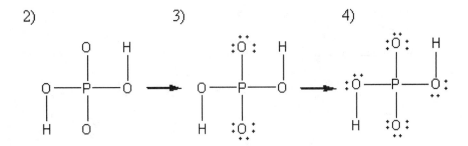

5) Resonance structures

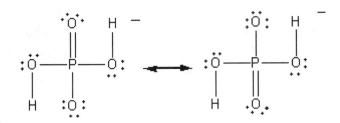

8.25   Determine the Lewis structures using standard procedures.
        NaCN: the polyatomic ion is CN^-
        CN^-:
        1) There are $4 + 5 + 1 = 10$ valence electrons.
        2) One electron pair is used in forming the bonding framework, leaving
            $10 - 2 = 8$ electrons.
        3) Place 3 pairs of electrons on the more electronegative N atom, leaving $8 - 6 = 2$
            electrons.
        4) Place the remaining 2 electrons on the C atom.
        5) Use two electron pairs from the N atom to form a triple bond, completing the octet on
            both atoms:

2)   3)   4)   5)

C—N → C—N̈: → :C—N̈: → :C≡N:⁻

(NH$_4$)$_2$CrO$_4$ has two polyatomic ions: NH_4^+ and CrO_4^{2-}
NH_4^+:
1) There are $5 + 4(1) - 1 = 8$ valence electrons.
2) Four electron pairs are used in forming the bonding framework, leaving no electrons.
3) All outer atoms are H atoms.
4) No remaining electrons.
5) The resulting structure has 8 electrons on the inner N atom.

$$\begin{array}{c} H \\ | \quad + \\ H-\!\!\!-N-\!\!\!-H \\ | \\ H \end{array}$$

$CrO_4{}^{2-}$:

1) There are $6 + 4(6) + 2 = 32$ valence electrons.
2) Four electron pairs are used in forming the bonding framework, leaving $32 - 4(2) = 24$ electrons.
3) Place 3 pairs of electrons on each outer O atom, leaving $24 - 4(6) = 0$ electrons.
4) No remaining electrons.
5) The resulting structure has $FC_{Cr} = 6 - 4 = +2$, use lone pairs from the outer atoms to form two double bonds to reduce the formal charge to 0. There are 6 resonance structures (not shown).

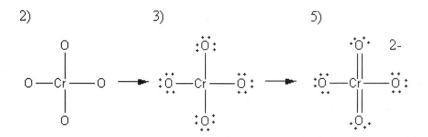

2)   3)   5)

8.27   Determine the Lewis structures using standard procedures.
Both compounds are ionic. The resulting polyatomic ions are $NO_3{}^-$, $NH_4{}^+$, and $CO_3{}^{2-}$ ions.
$NO_3{}^-$:

1) There are $5 + 3(6) + 1 = 24$ valence electrons.
2) Three electron pairs are used in forming the bonding framework, leaving $24 - 3(2) = 18$ electrons.
3) Use the remaining electrons to place three pairs of electrons around each outer O atom
4) No remaining electrons
5) The resulting structure has 6 electrons on the inner N atom. Use a lone pair from the outer O atoms to form a double bond and complete the octet on the N (there are three resonance structures):

2)   3)

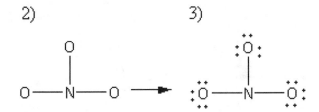

5) Resonance structures

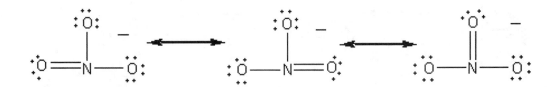

NH_4^+:
1) There are $5 + 4(1) - 1 = 8$ valence electrons.
2) Four electron pairs are used in forming the bonding framework, leaving no electrons left.
3) All outer atoms are H atoms.
4) No remaining electrons.
5) The resulting structure has 8 electrons on the inner N atom.

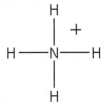

CO_3^{2-}:
1) There are $4 + 3(6) + 2 = 24$ valence electrons.
2) Three electron pairs are used in forming the bonding framework, leaving $24 - 2(3) = 18$ electrons.
3) Use the remaining electrons to place 3 electron pairs on each outer O atom.
4) No remaining electrons.
5) The resulting structure has 6 electrons on the inner C atom; use a lone pair from an O atom to form a double bond with the C atom.

2)   3)

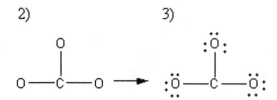

5) Resonance structures

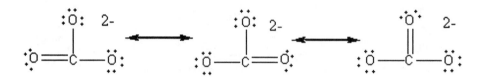

8.29 A description of bonding and geometry always starts with a Lewis structure and a count of bonds and non-bonding pairs around inner atoms.

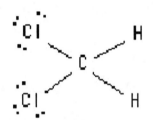

The steric number of the carbon atom is 4. The bonding about it can be described using sp^3 hybrid orbitals overlapping with $1s$ orbitals of H and $3p$ orbitals of Cl.

The geometry about the carbon atom is tetrahedral. The bonds can be depicted as follows:

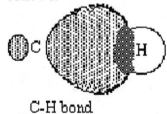

C-H bond

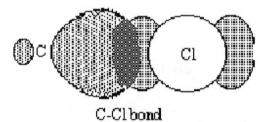

C-Cl bond

8.31 Construct the bonding framework from the formula, knowing that the more electronegative atoms (Cl) will be in outer positions. In 1,2-dichloroethane, each carbon atom is bonded to the other, and there is one Cl atom bonded to each carbon atom.

There are $2(7) + 2(4) + 4(1) = 26$ valence electrons. The bonding framework contains 7 bonds, and there are three lone pairs on each Cl atom, accounting for all the valence electrons. Thus, the bonding framework is the correct Lewis structure of 1,2-dichloroethane. Each carbon atom has SN = 4 and tetrahedral geometry. Your ball-and-stick model should reflect this.

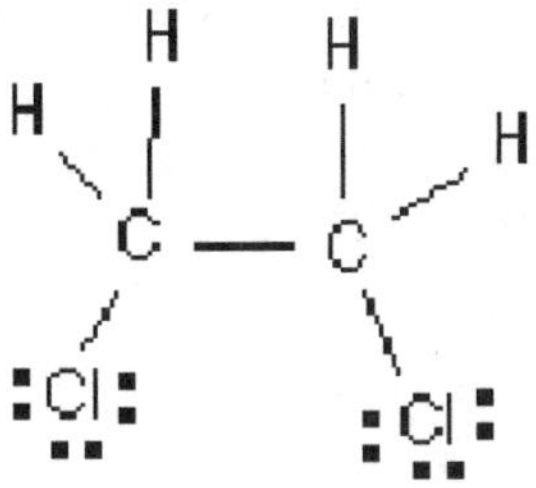

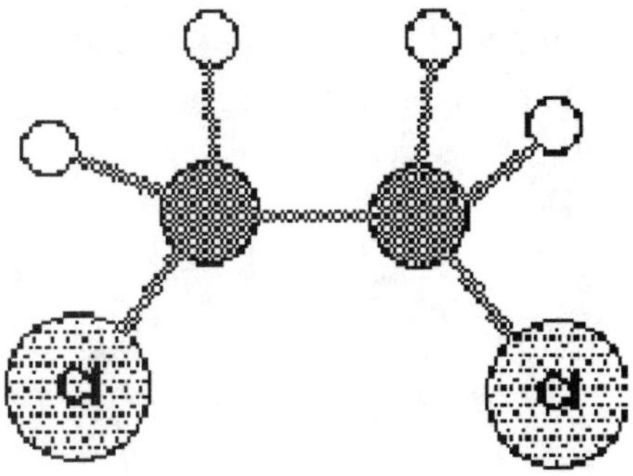

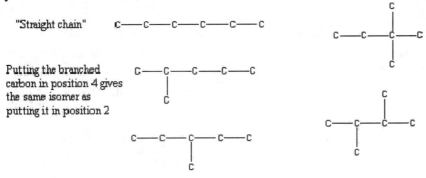

8.33 To draw structural isomers of an alkane, start with the "straight chain" compound, then rearrange the bonding arrangement to make other isomers. It is convenient to work with only the carbon skeleton:

"Straight chain"   C—C—C—C—C—C

Putting the branched   C—C—C—C—C
carbon in position 4 gives |
the same isomer as   C
putting it in position 2

C—C—C—C—C
        |
        C

            C
            |
C—C—C—C
            |
            C

        C
        |
C—C—C—C
        |
        C

8.35 Determine the Lewis structure following the standard procedures.

1) There are 2(4) + 6(1) + 6 = 20 valence electrons.
2) The chemical formula indicates that the bonding framework includes an –OH group attached to one of the carbon atoms. The bonding framework contains 8 bonds = 16 electrons, leaving 4 valence electrons:

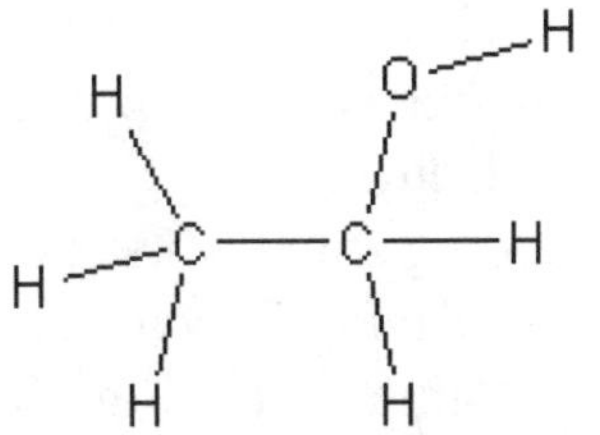

3) All the outer atoms are hydrogen, so no electrons can be placed on outer atoms.
4) Place the remaining two pairs on the oxygen atom. The resulting structure has an octet around each row 2 atom, so it is the correct Lewis structure.

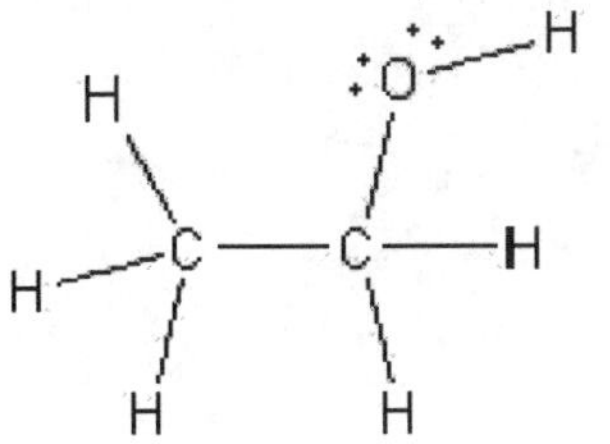

Methane molecule with left hydrogen replaced with methyl, top hydrogen replaced with -OH

131

Each inner atom (both C's and the O atom) has SN = 4, so the appropriate hybridization is sp^3 and the geometry about each inner atom is tetrahedral.

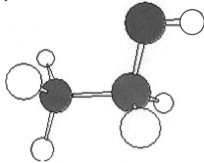

8.37   Determine the Lewis structure following the standard procedures.
1) There are $4(4) + 4 + 12(1) = 32$ valence electrons.
2) The chemical formula indicates that the bonding framework includes four –CH$_3$ groups attached to the silicon atom. The bonding framework contains 16 bonds = 32 electrons. All electrons are placed, so this structure is the correct Lewis structure. Each inner atom has SN = 4, sp^3 hybridization, and the geometry about each inner atom is tetrahedral.

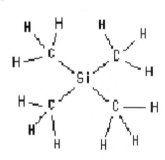 

8.39   Determine the Lewis structure following the standard procedures.
1) There are $8 + 4(7) + 6 = 42$ valence electrons.
2) Xe is the only inner atom, so the bonding framework has Xe surrounded by the other atoms. Five bonds require 10 electrons, leaving 32 valence electrons to place.
3) Place three pairs around each outer atom for a total of 30, leaving 2 valence electrons.
4) Place the remaining pair on the xenon atom. FC$_{Xe}$= $8 – 5 – 2 = +1$.
5) Make one double bond to Xe by shifting an electron pair from the least electronegative outer atom, oxygen:
Provisional: Complete:

   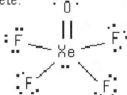

Xe has SN = 6, octahedral electron pair geometry, and the appropriate hybridization is d^2sp^3. One position is occupied by a lone pair, giving a square pyramid with bond angles near 90°.

8.41 Determine the Lewis structures following the standard procedures.
GeF$_4$:
1) There are 4 + 4(7) = 32 valence electrons.
2) Four electron pairs are used in forming the bonding framework, leaving 32 – 4(2) = 24 electrons.
3) Place three pairs of electrons around each outer F atom, leaving 24 – 4(6) = 0 electrons.
4) No remaining electrons.
5) The resulting structure has FC_{Ge} = 4 – 4 = 0. So the structure is correct.

SeF$_4$:
1) There are 6 + 4(7) = 34 valence electrons.
2) Four electron pairs are used in forming the bonding framework, leaving 34 – 4(2) = 26 electrons.
3) Place three pairs of electrons around each outer F atom, leaving 26 – 4(6) = 2 remaining electrons.
4) Place the remaining two electrons on the inner Se atom.
5) The resulting structure has FC_{Se} = 6 – 4 –2 = 0. So the structure is correct.

XeF$_4$:
1) There are 8 + 4(7) = 36 valence electrons.
2) Four electron pairs are used in forming the bonding framework, leaving 36 – 4(2) = 28 electrons.
3) Place three pairs of electrons around each outer F atom, leaving 28 – 4(6) = 4 remaining electrons.
4) Place the remaining four electrons on the inner Xe atom.
5) The resulting structure has FC_{Xe} = 8 – 4 – 4 = 0. So the structure is correct.

In GeF$_4$, SN$_{Ge}$ = 4, so this molecule is tetrahedral and the bonding about the germanium can be described using sp^3 hybridization. In SeF$_4$, SN$_{Se}$ = 5 with a lone pair so this molecule has seesaw geometry and the bonding about the Se atom can be described using dsp^3 hybridization. In XeF$_4$, SN$_{xe}$ = 6 with two lone pairs so this molecule has square planar geometry and the bonding about the Xe atom can be described using d^2sp^3 hybridization.

8.43   Determine Lewis structures using the standard procedures. Only asymmetric molecules have dipole moments. Use the Lewis structures to determine steric numbers and ascertain whether or not the molecules are asymmetric.

SiF_4: There are $4 + 4(7) = 32$ valence electrons. 4 pairs are used in the bonding framework, and 3 pairs are placed on each outer F atom. This leaves $FC_{Si} = 4-4 = 0$, so this is the correct Lewis structure.

H_2S: There are $6 + 2 = 8$ valence electrons. 2 pairs are used in the bonding framework, the remaining 4 electrons are placed on the S atom.

XeF_2: There are $8 + 2(7) = 22$ electrons. 2 pairs are used in the bonding framework, three pairs are placed on each outer F atom, the remaining 6 electrons are placed on the inner Xe atom. The resulting structure has $FC_{Xe} = 8 - 8 = 0$, so this is the correct Lewis structure

$GaCl_3$: There are $3 + 3(7) = 24$ valence electrons. 3 pairs are used in the bonding framework. Add 3 pairs to each outer atom (using the remaining electrons). The inner Ga atom has 6 electrons around it, but our rules call for making a double bond to Ga, but this would give Ga a negative formal charge, which is unlikely, so we leave the structure as it is; compare with triethylaluminum in your textbook.

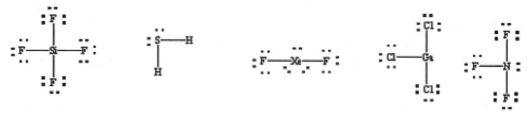

SiF_4 is tetrahedral and symmetric; it has no dipole moment; H_2S, like H_2O, is bent and has a dipole moment; XeF_2 has SN $=5$, with its 3 lone pairs in equatorial positions, so the molecule is linear without a dipole moment; $GaCl_3$ has SN$=3$, so the molecule is trigonal planar and has no dipole moment; NF_3, like NH_3, is pyramidal and has a dipole moment.

8.45   Bond angles can be predicted from the steric number of the inner atom, which determines the preferred electron pair geometry. The presence of lone pairs on the inner atom reduces bond angles slightly due to increased electron-electron repulsion for lone pairs.

(a) SiF_4, SN $= 4$, tetrahedral, $109.5°$;

(b) H_2S, SN$=4$ with 2 lone pairs, bent, $<109.5°$. (This angle is close to $90°$ because the sulfur inner atom is large and reduces repulsions between the two hydrogens);

(c) XeF_2, SN$=5$ with 3 lone pairs, linear, $180°$;

(d) $GaCl_3$, SN$=3$, trigonal planar, $120°$;

(e) NF_3, SN$=3$ with one lone pair, trigonal pyramidal, $<109.5°$.

8.47   Determine Lewis structures following the standard procedures. Draw ball-and-stick models that illustrate the molecular geometries determined by the steric numbers of the inner atoms.

(a) Cl_2O has $2(7) + 6 = 20$ valence electrons. Its Lewis structure has two bonding pairs, three lone pairs on each chlorine atom, and two lone pairs on the inner oxygen atom. $SN_O = 4$, leading to tetrahedral electron pair geometry and bent molecular shape:

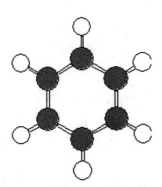

(b) C_6H_6 has $6(4) + 6(1) = 30$ valence electrons. After constructing the bonding framework, six electrons remain to be placed. Place these on alternate carbon atoms; then complete octets around the other carbon atoms by shifting lone pairs to make double bonds. There are two equivalent resonance structures:

Provisional structure

resonance structures

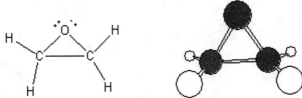

$SN_C = 3$, leading to trigonal planar geometry about each carbon atom and a flat, hexagonal molecular shape.

(c) C_2H_4O has $2(4) + 4(1) + 6 = 18$ electrons. When the bonding framework is complete, four electrons remain. These are placed on the oxygen atom. The triangular ring constrains this molecule to have 60° bond angles, even though the steric numbers of the atoms are four.

8.49   Determine the Lewis structures using standard procedures. All the species in this problem have four halogen atoms bonded to an inner atom. Both CI_4 and $SiCl_4$ have 32 electrons and SeF_4 has 34 valence electrons. All the 32-electron species have the same Lewis structure and the same geometry.

CI_4 and $SiCl_4$:
1) There are $6 + 4(7) = 32$ valence electrons.
2) Four electron pairs are used in forming the bonding framework, leaving $32 - 4(2) = 24$ electrons.
3) Place three pairs of electrons around each outer atom, leaving $24 - 4(6) = 0$ electrons
4) No remaining electrons.
5) The resulting structure has $FC_{Si} = 4 - 4 = 0$ and C has an octet. So the structures are correct.

SeF_4:
1) There are $6 + 4(7) = 34$ valence electrons.
2) Four electron pairs are used in forming the bonding framework, leaving $34 - 4(2) = 26$ electrons.
3) Place three pairs of electrons around each outer F atom, leaving $26 - 4(6) = 2$ electrons.
4) Place the remaining two electrons on the inner Se atom.
5) The resulting structure has $FC_{Se} = 6 - 4 - 2 = 0$. So the structure is correct.

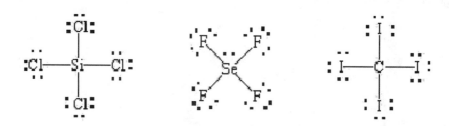

The 32-electron species have $SN = 4$ and are tetrahedral; SeF_4 is a seesaw.

8.51   Determine the number of different isomers by constructing the different possible geometric possibilities. Two X atoms can be placed at opposite ends of one Cartesian axis; all structures with this arrangement are equivalent, because the molecule can be rotated about an axis to superimpose these positions. The three X atoms can all be placed at right angles to one another, giving a different isomer. Thus, there are two isomers:

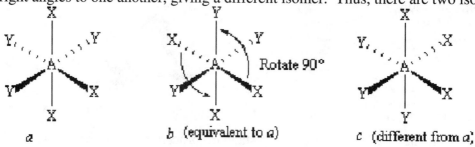

8.53 A description of bonding in a molecule starts with its Lewis structure, from which steric number, geometry, and hybridization can be deduced. $AlCl_3$ has $3 + 3(7) = 24$ valence electrons. The bonding framework requires three pairs, and the remaining 18 electrons are placed around the three chlorine atoms. $FC_{Al} = 3 - 3 = 0$, so this is the correct structure. The structure has $SN_{Al} = 3$, indicating that the molecule has trigonal planar structure with Al-Cl bonds that can be represented using sp^2 hybrids on Al and $3p$ orbitals on Cl.

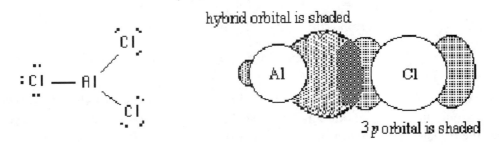

8.55 The empirical chemical formula of the silicon-oxygen network of zircon, an orthosilicate, is SiO_4^{4-}. The Si-O bonds are formed by the overlap of the oxygen's $2p$ orbital with one of the silicon's sp^3 orbitals. The molecular geometry of the SiO_4^{4-} anions are tetrahedral. Orthosilicates networks are ionic, containing discrete (not connected) SiO_4^{4-} anions and metal cations.

8.57 Determine Lewis structures following the standard procedures.
(a) Bromate, BrO_3^-, contains $7 + 3(6) + 1 = 26$ valence electrons. Three pairs are required for the bonding framework, nine pairs go on the oxygen atoms, and the remaining pair is placed on the bromine atom. The resulting structure has $FC_{Br} = +2$, shift two electron pairs to make two double bonds and reduce the formal charge on bromine to 0:

$FC_{Br} = 7-5 = +2$
Shift 2 electron pairs to make 2 double bonds

There are two additional resonance structures

(b) Nitrite, NO_2^-, contains $5 + 2(6) + 1 = 18$ valence electrons. Two pairs are required for the bonding framework, six pairs go on the O atoms, and the remaining pair is placed on N. Shift one electron pair to make a double bond and complete the octet on nitrogen. There are two equivalent structures. $FC_N = 5-5 = 0$:

2 equivalent structures

(c) Phosphate, PO_4^{3-}, contains $5 + 4(6) + 3 = 32$ valence electrons. Four pairs are required for the bonding framework and 12 pairs go on the oxygen atoms, giving a provisional structure with $FC_P = 5 - 4 = +1$. Shift one electron pair to make a double bond and reduce the formal charge to zero. There are four equivalent structures.

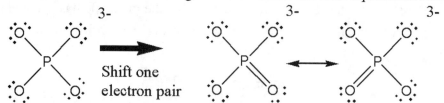

There are two additional equivalent structures

(d) Hydrogen carbonate, HCO_3^-, contains $1 + 4 + 3(6) + 1 = 24$ valence electrons. Four pairs are required for the bonding framework, three pairs go on each outer oxygen atom, and the remaining two pairs go on the inner oxygen atom. Shift one electron pair from an outer O atom to form a double bond to C and complete its octet. There are two equivalent structures:

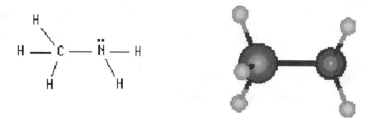

8.59 Determine the Lewis structures following the standard procedures.
H_3CNH_2 has $5(1) + 4 + 5 = 14$ valence electrons. Six pairs are required to complete the bonding framework and the remaining pair goes on the nitrogen atom, giving a structure in which both carbon and nitrogen have complete octets.

The steric number is 4 for both C and N, so the bond angles about carbon are 109.5° and those about nitrogen are < 109.5° because of the extra repulsion created by the lone pair on the nitrogen atom.

8.61 Determine Lewis structures following the standard procedures.
SF_2 has $6 + 2(7) = 20$ valence electrons. Two electron pairs are required for the bonding framework, three pairs are placed around each outer F atom, and the remaining four electrons are put on the S atom. $FC_S = 6 - 6 = 0$, so this is the correct structure.

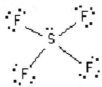

SF_4 has $6 + 4(7) = 34$ valence electrons. Four pairs are used in forming the bonding framework, three pairs are placed around each outer F atom, and the remaining two electrons are given to the S atom. $FC_S = 6 - 6 = 0$, so this is the correct structure.

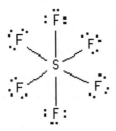

SF_6 has $6 + 6(7) = 48$ valence electrons. Use 6 electron pairs for the bonding framework and 3 pairs are placed around each outer F atom. $FC_S = 6 - 6 = 0$, so this is the correct structure.

8.63 Determine molecular geometries and hybridizations from steric numbers, which can be found using the Lewis structures.

SF_2 has SN = 4 with two lone pairs, bent geometry and sp^3 hybridization;

SF_4 has SN = 5 with one lone pair, seesaw geometry and dsp^3 hybridization;

SF_6 has SN = 6, octahedral geometry and d^2sp^3 hybridization.

8.65 A description of bonding and geometry starts with determination of the Lewis structure. $(CH_3)_3C^+$ has $4(4) + 9(1) - 1 = 24$ electrons. All the electrons are required to complete the bonding framework, so there are no lone pairs to shift and the provisional structure is correct even though the central carbon atom has only six electrons associated with it (as this suggests, carbocations are highly reactive species).

The central carbon has SN = 3, trigonal planar geometry, and its bonding orbitals can be described using sp^2 hybrid orbitals. The methyl carbons have SN = 4, tetrahedral geometry, and their bonding orbitals can be described using sp^3 hybrid orbitals.

8.67 A bond angle of 92.2° indicates that the bonding can be well described using p orbitals rather than hybrid orbitals, whereas an angle of 104.5° indicates significant distortion from pure p orbital bonding. Space-filling models of the two molecules shows that the smaller oxygen atom cannot accommodate two H atoms at right angles, but the larger S atom can.

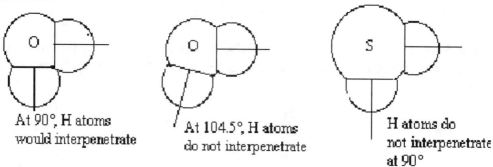

At 90°, H atoms would interpenetrate

At 104.5°, H atoms do not interpenetrate

H atoms do not interpenetrate at 90°

8.69 Determine Lewis structures using the standard procedures.
HNO₃:
1) There are 1 + 5 + 3(6) = 24 valence electrons.
2) The bonding framework requires 4 pairs of electrons, leaving 24-4(2) = 16 electrons.
3) Place 3 pairs of electrons on each outer O atom.
4) Place the remaining 4 electrons on the inner O atom.
5) Shift an electron pair from an outer O atom to complete the octet on nitrogen (one resonance structure also exists).

HClO₂:
1) There are 1 + 7 + 2(6) = 20 valence electrons.
2) The bonding framework requires 3 pairs of electrons, leaving 24-3(2) = 18 electrons.
3) Place 3 pairs of electrons on the outer O atom.
4) Place 2 pairs of electrons on the inner O atom and the remaining 4 electrons on the inner Cl atom.
5) The resulting structure has FC_{Cl} = 7 – 4 –2 = 1. Shift an electron pair from the outer O atom to reduce the formal charge to zero.

H₂SO₄:
1) There are 2(1) + 6 + 4(6) = 32 valence electrons.
2) The bonding framework requires 6 pairs of electrons, leaving 32-6(2) = 20 electrons.
3) Place 3 pairs of electrons on each outer O atom, leaving 8 electrons.
4) Place 2 pairs of electrons on each inner O atom.
5) The resulting structure has FC_S = 6 – 4 = 2. Shift two electron pairs from the outer O atoms to reduce the formal charge to zero.

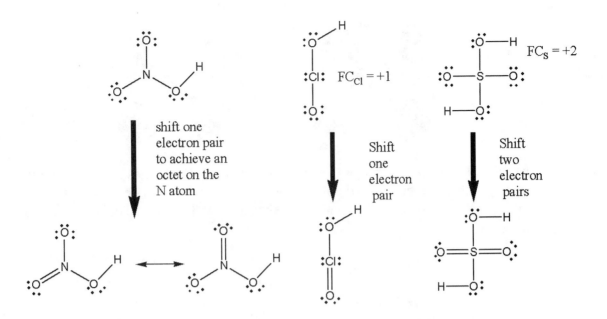

8.71   The electronegativity of hydrogen (2.1) is the smallest among the elements involved in these chemical bonds, so the ranking of bond polarity is from smallest to largest electronegativity of the bonding partner. Remember that electronegativity decreases down a column and increases across a row:

Si-H < C-H < N-H < O-H < F-H.

8.73   All positions around a tetrahedral center are equivalent, so there is only one isomer of CH_2Br_2. The two structures show different views of the same compound.

8.75   The Lewis structures of these molecules show octets around the inner atom and $FC_X = 0$ for compounds with formula XF_3, making them stable. Compounds with formula XF_5 also have $FC_X = 0$ but have five electron pairs associated with the inner atom. This is possible for phosphorus, a third row element that has *d* orbitals available for bonding. It is not possible for nitrogen, a second row element which lacks valence *d* orbitals.

8.77   (a) Square planar XY_2Z_2 molecules may have like atoms adjacent to or opposite each other:

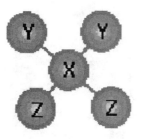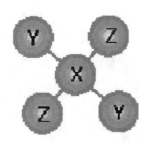

(b) When like atoms are opposite each other, their bond polarities oppose each other and cancel, so this isomer has no dipole moment. When like atoms occupy adjacent positions, the bond polarities do not entirely cancel and the molecule has a net dipole moment.

8.79   The shape of a molecule indicates the steric number of its inner atom and the number of atoms bonded to the inner atom. Consult Table 8-6 of your textbook for details. Tellurium, in column 16 of the periodic table, has six valence electrons, and each fluorine contributes seven valence electrons. Species with an even number of fluorine atoms are neutral, but those with an odd number must have one unit of charge (+1 or −1) in order to contain an even number of electrons.

(a) Bent indicates TeF_2;

(b) T-shape indicates SN = 5 on Te, three F atoms, and two lone pairs on Te, for a total of $4 + 3(8) = 28$ valence electrons. TeF_3 would have $6 + 3(7) = 27$ valence electrons, so the formula of this species must be TeF_3^-;

(c) square pyramid indicates SN = 6 on Te, five F atoms, and one lone pair on Te, for a total of $2 + 5(8) = 42$ valence electrons. TeF_5 would have $6 + 5(7) = 41$ valence electrons, so the formula of this species must be TeF_5^-;

(d) trigonal bipyramid indicates SN = 5 on Te and five F atoms, for a total of $5(8) = 40$ valence electrons. TeF_5 would have $6 + 5(7) = 41$ valence electrons, so the formula of this species must be TeF_5^+;

(e) octahedron indicates SN = 6 and six F atoms, formula TeF_6;

(f) seesaw indicates SN = 5 on Te, four F atoms, and one lone pair on Te, formula TeF_4.

8.81   A compound XY_7 is possible if X has d orbitals available for bonding, but it is sterically crowded because of the large number of Y atoms around the central X atom. For this compound to exist, Y must be as small as possible and X must be as large as possible. Thus, Y is the halogen with the smallest size, fluorine; and X is iodine, the halogen with the largest size (apart from astatine, which is not involved). The compound is IF_7.

8.83   Determine Lewis structures following the standard procedures.

(a) $OPCl_3$ contains $6 + 5 + 3(7) = 32$ valence electrons. Four pairs are required for the bonding framework and 12 pairs go on the outer atoms, giving a structure in which the inner phosphorus atom has $FC_P = 5 - 4 = +1$. Shift one electron pair from oxygen to form a double bond:

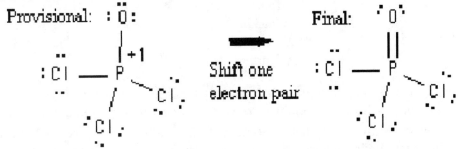

(b) HN_3 has $1 + 3(5) = 16$ valence electrons. Three pairs are required for the bonding framework and three pairs go on the outer nitrogen atom, leaving two pairs which should be placed on the nitrogen bonded to hydrogen. Then shift two electron pairs to form multiple bonds to the "central" nitrogen atom and complete its octet. There are two ways to do this, giving two equivalent structures:

(c) $SeCl_6$ has $6 + 6(7) = 48$ valence electrons. Six pairs are required for the bonding framework and $3(6) = 18$ pairs go on the outer chlorine atoms. The resulting structure has $FC_{Se} = 0$ and is correct.

(d) We are told that this compound has an S-S bond. Remember that symmetry is preferred, therefore give each S atom three O atoms and 1 H atom (which should be bonded to an O). $H_2S_2O_6$ has $2(1) + 2(6) + 6(6) = 50$ valence electrons. Nine pairs are required for the bonding framework, $3(4) = 12$ pairs go on the outer oxygen atoms, and two pairs go on each inner oxygen atom. In the provisional structure, $FC_S = 6 - 4 = +2$ for each sulfur atom. Shift electron pairs from outer oxygen atoms to make double bonds and reduce these formal charges to zero:

Provisional: Final:

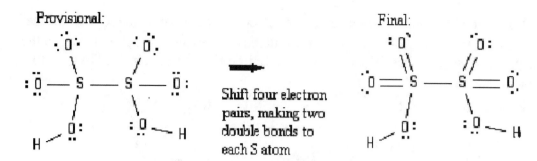

Shift four electron
pairs, making two
double bonds to
each S atom

8.85   The line structure of spiroheptane indicates that it is made up entirely of C and H atoms,
connected entirely by single bonds. Convert the line structure into a bond framework.
The framework shows that each C atom has four bonds, SN = 4, uses sp^3 hybrids, and
should have tetrahedral geometry. The four-carbon rings are distorted squares, however,
so the two rings are strained with bond angles around 90° rather than the tetrahedral
109.5°

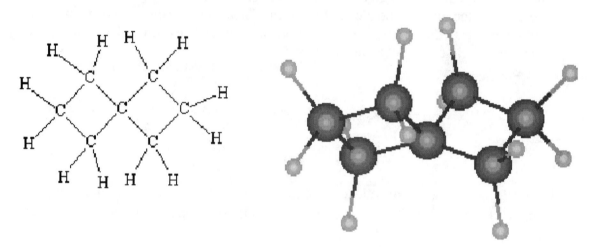

Problems 9.1-9.6 all deal with organic molecules whose line structures are provided. To determine the bonding pattern from a line structure, first convert the line structure into a molecular structure by adding C and H atoms. Molecular structures contain enough information to deduce steric numbers of C, N, and O inner atoms without determining the complete Lewis structure. An inner C, N, or O atom with no multiple bonds has SN = 4 and bonding that can be described using sp^3 hybrid orbitals. One with one π bond has SN = 3 and bonding that can be described using sp^2 hybrid orbitals and one p orbital. One with two π bonds has SN = 2 and bonding that can be described using sp hybrid orbitals and two p orbitals.

9.1   Acetone has three inner C atoms. Two have only single bonds (one C–C and three C–H); these have SN = 4. The atom bonded to O has one π bond and SN = 3. The σ bonds are:

LIST OF	Bond	Number	Orbitals
σ BONDS:	C–H	6	sp^3 - $1s$
	C–C	2	sp^3 - sp^2
	C–O	1	sp^2 - $2p$

Between C and O there is also a π bond formed by side-by-side overlap of $2p$ orbitals on each atom. Here are sketches of the bonding orbitals:

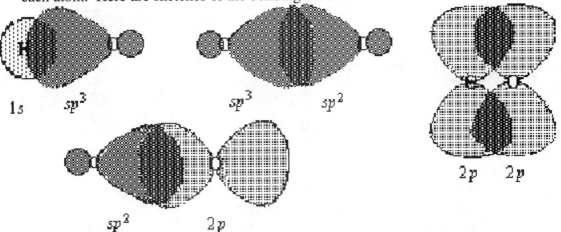

9.3   The line structure of neocembrene in the Chemistry and Life Box shows that it contains only inner carbon atoms and outer hydrogen atoms. Some of the inner atoms have all single bonds, SN = 4, sp^3 hybridization, and tetrahedral geometry; others have one double bond, SN = 3, sp^2 hybridization, and trigonal planar geometry. The SN values for the 20 carbon atoms are shown in the line structure below.

9.5 From the line structures of the compounds, we determine the steric number of each carbon atom. These are shown on the line structures below:

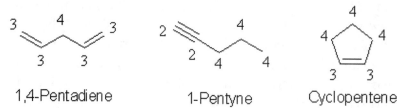

1,4-Pentadiene   1-Pentyne   Cyclopentene

C atoms designated "4" have only single bonds, SN = 4, sp^3 hybridization, and tetrahedral geometry. Those with a double bond, designated "3," have SN = 3, sp^2 hybridization, and trigonal planar geometry. Two C atoms, designated "2," have two double bonds, sp hybridization, and linear geometry. The σ bonds are described by overlap of hybrid orbitals, with each C–H bond described as a hybrid overlapping with a hydrogen $1s$ orbital. The π bonds are described by side-by-side overlap of $2p$ orbitals. In 1–pentyne, the two C atoms designated "2" have three bonds: a σ bond formed from sp hybrids and two π bonds, at right angles to each other, as in acetylene. Here are sketches of the different types of bonds:

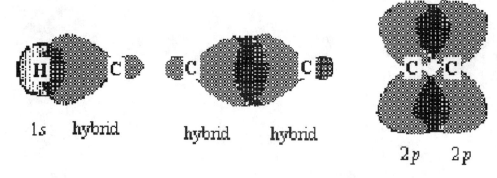

1s   hybrid   hybrid   hybrid

2p   2p

9.7 The determinants of bond length, in order of importance, are principal quantum number of the valence orbitals, bond order, and bond polarity. Among these bonds, H–N is shortest because it involves an $n = 1$ orbital and Cl–N is longest because it involves an $n = 3$ orbital. Of the three bonds involving $n = 2$ orbitals, the N–N single bond is longest and the N≡N triple bond is shortest: H–N < N≡N < C=O < N–N < Cl–N

9.9 Bond strengths, like bond lengths, depend on principal quantum number, bond order, and bond polarity, but the correlation is not as strong as for bond lengths, so tabulated values must be consulted. Here are the values from Table 9-2 of your textbook, listed in increasing order:

Bond	C–C	<	H–N	<	C=C	<	C=O	<	N≡N
Energy (kJ/mol)	345		390		615		750		945

Strong bonding and bond polarity of the $n = 1$ orbital makes H–N > C–C; Multiple bonding makes C=C > H–N; Bond polarity makes C=O > C=C; and Multiple bonding makes N≡N > C=O.

9.11 To estimate the energy change in a reaction, determine Lewis structures to obtain types of bonds, list the number of bonds of each type in reactants and products, and subtract the sum of average bond energies for products from the sum of average bond energies for reactants, using values from Table 9-2 of your textbook.

The reaction is $N_2 + 2\,O_2 \rightarrow N_2O_4$

Here are the Lewis structures (resonance forms not shown):

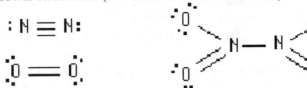

Reactants			Products		
Bond	No.	Energy(kJ/mol)	Bond	No.	Energy(kJ/mol)
N≡N	1	945	N–N	1	160
O=O	2	495	N–O	2	200
			N=O	2	605

$\Delta E_{reaction} \cong [1\ mol(945\ kJ/mol) + 2\ mol(495\ kJ/mol)]$
$- [1\ mol(160\ kJ/mol) + 2\ mol(200\ kJ/mol) + 2\ mol(605\ kJ/mol)]$
$= 1935\ kJ - 1770\ kJ \cong 165\ kJ$

9.13 To estimate the energy change in a reaction, determine Lewis structures to obtain types of bonds, list the number of bonds of each type in reactants and products, and subtract the sum of average bond energies for products from the sum of average bond energies for reactants, using values from Table 9-2 of your textbook.

$$2\,H_2 + O_2 \rightarrow 2\,H_2O$$

Reactants			Products		
Bond	No.	Energy(kJ/mol)	Bond	No.	Energy (kJ/mol)
H–H	2	435	O–H	4	460
O=O	1	495			

$\Delta E_{reaction} \cong [2\ mol(435\ kJ/mol) + 1\ mol(495\ kJ/mol)] - [4\ mol(460\ kJ/mol)] = -475\ kJ$

$$3\,H_2 + N_2 \rightarrow 2\,NH_3$$

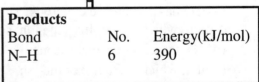

Reactants			Products		
Bond	No.	Energy(kJ/mol)	Bond	No.	Energy(kJ/mol)
H–H	3	435	N–H	6	390
N≡N	1	945			

$\Delta E_{reaction} \cong [3\ mol(435\ kJ/mol) + 1\ mol(945\ kJ/mol)] - [6\ mol(390\ kJ/mol)] = -90\ kJ$

$$3\,H_2 + CO \rightarrow CH_4 + H_2O$$

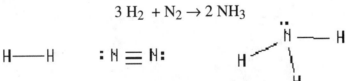

Reactants				Products		
Bond	No.	Energy(kJ/mol)		Bond	No.	Energy(kJ/mol)
H–H	3	435		C–H	4	415
C≡O	1	1070		O–H	2	460

$\Delta E_{reaction} \cong [3\ mol(435\ kJ/mol) + 1\ mol(1070\ kJ/mol)]$
$- [4\ mol(415\ kJ/mol) + 2\ mol(460\ kJ/mol)] = 2375\ kJ - 2580\ kJ = -205\ kJ$

9.15 Orbital energy ladders for diatomic molecules have the form shown in Figure 9-17 of your textbook. Count valence electrons, then place them in available orbitals using the Pauli and aufbau principles. Stability is determined by the number of bonding and antibonding electrons. Na_2 has two valence electrons, one from each atom:

The configuration has two bonding and no antibonding electrons, indicating a stable molecule with a single bond.

9.17 Stability of a diatomic molecule is determined by the number of bonding and antibonding electrons. To compare species, determine how many valence electrons each species possesses, then place them in the available orbitals following the Pauli and aufbau principles. Here are the configurations for nine and ten valence electrons:

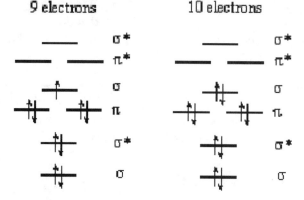

(a) CO has 10 valence electrons. To make CO^+, an electron is removed from a bonding orbital, which destabilizes the ion, so CO has a stronger bond.
(b) N_2 has 10 valence electrons. To make N_2^+, an electron is removed from a bonding orbital, which destabilizes the ion, so N_2 has a stronger bond.
(c) CN^- has 10 valence electrons. To make CN, an electron is removed from a bonding orbital, which destabilizes the species, so CN^- has a stronger bond.

9.19 Determine how many valence electrons the ion possesses, then place them in the available orbitals following the Pauli and aufbau principles. Stability is determined by the number of bonding and antibonding electrons.
ClO^- has 14 valence electrons, six from O, seven from Cl, plus one due to its charge:

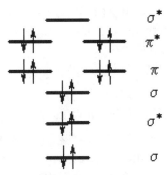

There are eight bonding electrons and six antibonding electrons, for a net of two bonding electrons and a single bond for this ionic species.

9.21 Orbital sketches for diatomic molecular orbitals are constructed by overlapping the appropriate atomic orbitals. See the figures in Section 9.3 of your textbook. The configuration of CO (10 valence electrons) is $(\sigma_s)^2 (\sigma_s^*)^2 (\pi_x)^2 (\pi_y)^2 (\sigma_p)^2$, indicating that there are four bonding orbitals and one anti-bonding orbital to depict. The two π orbitals appear identical except for their orientations relative to each other, so four pictures are needed:

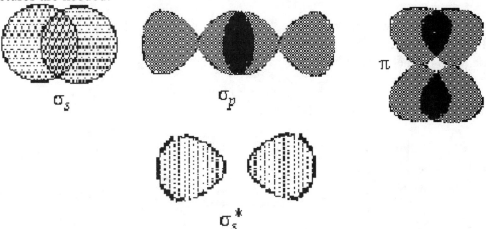

9.23 Determine the Lewis structure using the usual procedures. The molecule has $2(6) + 4 = 16$ valence electrons:

$$\ddot{\underset{..}{S}} = C = \ddot{\underset{..}{S}}$$

The bonding pattern for CS_2 is like that for CO_2, except that the S atoms have $n = 3$ valence orbitals rather than $n = 2$ orbitals. The inner atom can be described using sp hybrids which overlap with $3p$ orbitals from the S atoms to form two σ bonds. There are two delocalized π systems at right angles to each other, each made up of a $2p$ orbital on the C atoms overlapping side-by-side with a $3p$ orbital on each S atom. As in CO_2, eight electrons occupy the delocalized π orbitals, four in bonding and four in non-bonding orbitals.

9.25   Determine the Lewis structures using the usual procedures. The cation has
        $2(6) + 5 - 1 = 16$ valence electrons, the anion has $2(6) + 5 + 1 = 18$ valence electrons:

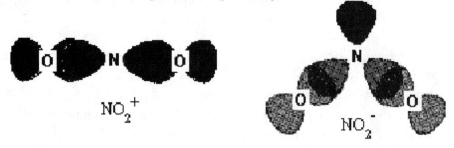

NO_2^+ is isoelectronic with CO_2. It has no lone pair on the inner N atom, giving SN = 2,
sp hybridization, and a linear geometry. NO_2^- is isoelectronic with O_3. Its 18 valence
electrons include a lone pair on the inner atom, SN = 3 and sp^2 hybridization to
accommodate three electron pairs (bent geometry):

9.27   Determine the Lewis structure using the usual procedures. The molecule has
        $3(5) + 1 = 16$ valence electrons. Begin by drawing the bonding framework and placing
        lone pairs on the outer atoms:

This uses up all the electrons. Form double and triple bonds to complete the octet on the
inner N atom:

The N atom bonded to H has SN = 3 and 4 in the two structures, so it has bent geometry
with an angle between 109° and 120° (the experimental value is 112° 39'). The best
hybridization is sp^2, leaving one *p* orbital free to form a π bond. The other inner N atom
has SN = 2, *sp* hybridization, and bond angles of 180°. There are two π networks, one
delocalized over all three N atoms and the other localized between the outer and adjacent
inner N atoms.

9.29   Conjugation stabilizes a molecule whenever there are more than two adjacent atoms
        involved in π bond formation, thereby leading to delocalized π orbitals. Of the molecules
        whose structures appear in the Chemistry and Life Box of your textbook, only xanthin
        has a delocalized π system in which all 13 π bonds contribute.

9.31   We can determine hybridizations directly from line structures: Each carbon atom with no
        double bonds, and each inner oxygen atom, has SN = 4 and sp^3 hybrids can describe the
        bonding. A C atom with one double bond has SN = 3 and sp^2 hybridization.

(a) Hybridization is as shown on the line drawing ($3 = sp^3$, $2 = sp^2$).

(b - c) The Lewis structure shows 2 π bonds, on adjacent atoms, so as in butadiene there are four electrons in the extended π system

(d) The four atoms screened in the line drawing lie in the same plane.

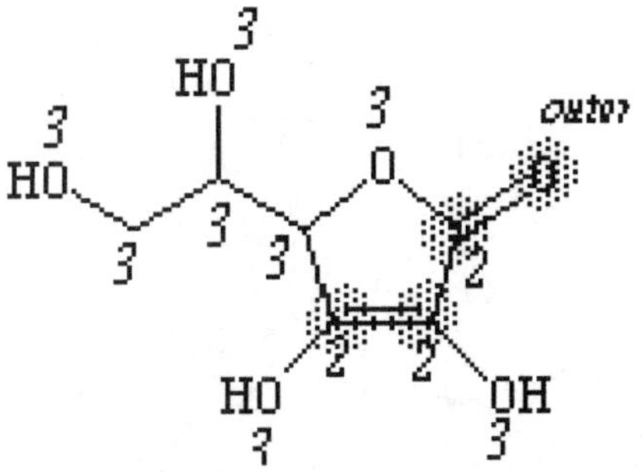

9.33   Determine the Lewis structure using the usual procedures. The molecule has $3(4) + 2(6) = 24$ valence electrons. Begin by drawing the bonding framework and adding in lone pairs of electrons to the outer atoms. Use the lone pairs to form double bonds to complete the octets on the inner carbons:

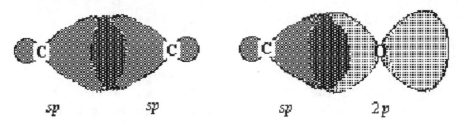

Each carbon atom has SN = 2, so the molecule is linear and the σ bond network can be described using sp hybrids. There are two sp-sp bonds between carbon atoms and two sp-$2p$ bonds between carbon and oxygen atoms:

If the molecular axis is the z-axis, a p_x and a p_y orbital on each atom overlaps side-by-side, giving two sets of delocalized π orbitals, each extending over all five atoms:

9.35   A description of bonding in a molecule starts with its Lewis structure, from which steric number, geometry, hybridization, and extension of π bonds can be deduced.

ClO_4^- has $7 + 4(6) + 1 = 32$ valence electrons.

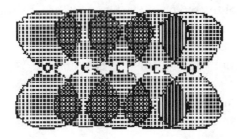

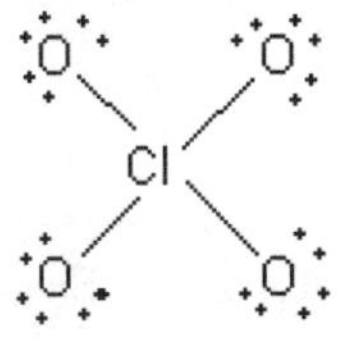

The provisional Lewis structure has $FC_{Cl} = 7 - 4 = 3$, so three electron pairs are shifted to make double bonds:

$SN_{Cl} = 4$, indicating that the molecule has tetrahedral geometry and the σ bond framework can be described using sp^3 hybrids on Cl overlapping with $2p$ orbitals from O. The resonance structures signal that the π system is extended over all five atoms between the p orbitals on the O atoms and the d orbitals of the Cl atom, and there are three bonding π orbitals occupied by six electrons.

9.37 Doped semiconductors are *p*-type if the dopant has fewer valence electrons than the bulk element and *n*-type if the dopant has more valence electrons than the bulk element:
(a and c) GaP and CdSe are stoichiometric compounds, hence are undoped.
(b) InSb (Groups 13 and 15) doped with Te (Group 16) is *n*-type.

9.39 The bonding of any metal can be described by constructing bands from the valence orbitals and then distributing the valence electrons into these bands. Iron and potassium are both row 4 elements with similar band structures, but whereas each potassium atom contributes just one valence electron to the valence (bonding) band of the solid, each iron atom contributes eight electrons. Thus iron has much greater bonding, making it harder and giving it a higher melting point than potassium.

9.41 To find exceptions to normal trends, consult the appropriate listings in tabulated values. Table 9-1 lists bond lengths by *n*-value. For *n* = 2 it lists four *X–X* bonds, C–C, N–N, O–O, and F–F; and three *X–Y* bonds, C–N, C–O, and C–F.
(a) C–N is shorter but weaker than C–C;
(b) C–N is also longer but stronger than N–N.

9.43 To predict stability from bond energies, we must tabulate the number of bonds of each type and add up their contributions. Start with Lewis structures in order to count bonds of each type:

Ethanol: Diethyl ether:

5 C–H bonds, 1 C–O bond   6 C–H bonds, 2 C–O bonds
1 O–H bond, 1 C–C bond
We can ignore the 5 C–H bonds and 1 C–O bond that the two compounds have in common, leaving O–H and C–C in ethanol vs. C–H and C–O in diethyl ether:

Ethanol: 460 kJ/mol (O–H) + 345 kJ/mol (C–C) = 805 kJ/mol
Diethyl ether: 415 kJ/mol (C–H) + 360 kJ/mol (C–O) = 775 kJ/mol

Based on average bond energies, ethanol is more stable by 30 kJ/mol.

9.45   Semiconductor type is determined by the number of valence electrons of the host element and its dopant. When the dopant contains more valence electrons than the host, the semiconductor is *n*-type; when it contains fewer valence electrons, the semiconductor is *p*-type.
(a) Ge (Group 14) doped with Sb (Group 15) is *n*-type;
(b) As (Group 15) doped with Si (Group 14) is *p*-type; and
(c) Si (Group 14) doped with In (Group 13) is *p*-type.

9.47   The determinants of bond length, in order of relative importance, are principal quantum number of the valence orbitals, bond order, and bond polarity.
The bonds found in these molecules are H–O, H–C, C=O, and C≡N. Of these, H–O and H–C are shorter than the others because they form from an *n* = 1 orbital; H–O is shorter than H–C because it is a more polar bond. C=O has bond order 2 and is longer than C≡N because bond order is a stronger determinant of length than bond polarity.

9.49   Bond strengths, like bond lengths, depend on principal quantum number, bond order, and bond polarity, but the correlation is not as strong as for bond lengths, so tabulated values must be consulted. Here are the values from Table 9.2 of your textbook, listed in increasing order:

Bond	H–C	<	H–O	<	C=O	<	C≡N
Energy (kJ/mol)	415		460		750		890

H–O is stronger than H–C because of bond polarity; C=O is stronger than H–O because of multiple bonding; and C≡N is stronger than C=O because of multiple bonding.

9.51   Describe bonding and geometry starting with a Lewis structure and a count of bonds and non-bonding pairs around inner atoms. The provisional structure of each compound has a positive formal charge on the row 3 sulfur atom, which is reduced to zero by making double bonds to each oxygen atom:

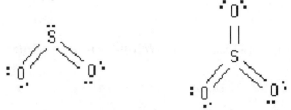

The sulfur atom in each compound has SN = 3, so *sp²* hybrids overlapping with oxygen 2*p* orbitals describe the σ bonds: two in SO_2 and three in SO_3. SO_2 is bent, like O_3, and SO_3 is trigonal planar, like NO_3^-. The Lewis structures indicate the presence of two π bonds in SO_2 and three π bonds in SO_3. All the π orbitals extend over the entire molecule. Because sulfur is a third-row element, its 3*d* orbitals contribute to the extended π bonding orbitals, which form from side-by-side overlap of oxygen $2p$ orbitals with sulfur $3p$ and $3d$ orbitals.

9.53 Follow the standard procedure to obtain the Lewis structure of a molecule, then use the Lewis structure to determine hybridization, geometry, and identify the extended π system.

(a) This molecule has 4 N (5 e⁻) + 2 C (4 e⁻) + 2 O (6 e⁻) + 4 H (1 e⁻) = 44 valence electrons, 22 of which are used for the 11 bonds in the framework. Add 6 electrons around each outer O atom, then place the remaining 5 pairs on inner N atoms, which are more electronegative than C atoms, giving a provisional structure in which the three second-row atoms marked with *3* have only three pairs of electrons. Complete their octets by shifting electron pairs to form double bonds in the final Lewis structure:

(b) The end N atoms have SN = 4, sp^3 hybridization, and tetrahedral geometry. The C atoms and the center N atoms have SN = 3, sp^2 hybridization, and trigonal planar geometry.

(c) The four atoms with SN = 3 plus the two outer O atoms form a 6-atom plane with an extended π system. There are three extended π orbitals containing 6 electrons.

9.55 A calculation using average bond energies requires a list of all bonds in the reactants and products. To obtain an estimate for reaction energy per mole of material, use the equation, $\Delta E \cong \Sigma B.E._{reactants} - \Sigma B.E._{products}$:

Ethane reaction:	H₃C–CH₃ +	7/2 O₂ →	2 CO₂ +	3 H₂O
Bonds:	1 C–C, 6 C–H	7/2 (1 O=O)	2 (2 C=O)	3 (2 O–H)
Energies (kJ/mol):	345, 415	495	800	460

$\Delta E \cong [1(345 \text{ kJ/mol}) + 6(415 \text{ kJ/mol}) + 7/2(495 \text{ kJ/mol})] - [4(800 \text{ kJ/mol}) + 6(460 \text{ kJ/mol})]$

$\Delta E \cong 4568 \text{ kJ/mol} - 5960 \text{ kJ/mol} = -1392 \text{ kJ/mol}$

$\Delta E \text{(per gram)} = \dfrac{\Delta E}{MM} = \left(\dfrac{-1392 \text{ kJ}}{1 \text{ mol}}\right)\left(\dfrac{1 \text{ mol}}{30.08 \text{ g}}\right) = -46.3 \text{ kJ/g}$

Ethylene reaction:	H₂C=CH₂ +	3 O₂ →	2 CO₂ +	2 H₂O

Bonds:	1 C=C, 4 C–H	3 (1 O=O)	2 (2 C=O)	2 (2 O–H)
Energies (kJ/mol):	615, 415	495	800	460

$\Delta E \cong [1(615 \text{ kJ/mol}) + 4(415 \text{ kJ/mol}) + 3(495 \text{ kJ/mol})] - [4(800 \text{ kJ/mol}) + 4(460 \text{ kJ/mol})]$

$\Delta E \cong 3760 \text{ kJ/mol} - 5040 \text{ kJ/mol} = -1280 \text{ kJ/mol}$

$\Delta E(\text{per gram}) = \dfrac{\Delta E}{MM} = \left(\dfrac{-1280 \text{ kJ}}{1 \text{ mol}}\right)\left(\dfrac{1 \text{ mol}}{28.06 \text{ g}}\right) = -45.6 \text{ kJ/g}$

Acetylene reaction:	HC≡CH +	5/2 O$_2$ →	2 CO$_2$ +	H$_2$O
Bonds:	1 C≡C, 2 C–H	5/2 (1 O=O)	2 (2 C=O)	2 O–H
Energies (kJ/mol):	835, 415	495	800	460

$\Delta E \cong [1(835 \text{ kJ/mol}) + 2(415 \text{ kJ/mol}) + 2.5(495 \text{ kJ/mol})]$
$\quad\quad - [4(800 \text{ kJ/mol}) + 2(460 \text{ kJ/mol})] = 2903 \text{ kJ/mol} - 4120 \text{ kJ/mol}$

$\Delta E = -1217 \text{ kJ/mol}$

$\Delta E(\text{per gram}) = \dfrac{\Delta E}{MM} = \left(\dfrac{-1217 \text{ kJ}}{1 \text{ mol}}\right)\left(\dfrac{1 \text{ mol}}{26.04 \text{ g}}\right) = -46.7 \text{ kJ/g}$

These estimates indicate that acetylene releases slightly more energy per unit mass than ethane or ethylene, but experimental values show that ethane releases the most. Remember that average bond energy calculations provide estimates, not exact values.

9.57   The appearance of a band gap diagram depends on the nature of the semiconductor.

Silicon (Group 14) doped with antimony (Group 15) is an *n*-type semiconductor, with the dopant providing an excess of electrons. Thus, its band gap diagram shows a full valence band and a small population of electrons in the conduction band:

9.59   To estimate the energy change in a reaction, determine Lewis structures to obtain types of bonds, list the number of bonds of each type in reactants and products, and subtract the sum of average bond energies for products from the sum of average bond energies for reactants, using values from Table 9.2 of your textbook.

$$H–Cl + H_2C=CH_2 \rightarrow H_3C–CH_2Cl$$

Reactants			Products		
Bond	No.	Energy (kJ/mol)	Bond	No.	Energy (kJ/mol)
H–Cl	1	430	C–Cl	1	330
C–H	4	415	C–H	5	415
C=C	1	615	C–C	1	345

$\Delta E_{\text{reaction}} \cong$ [1 mol(430 kJ/mol) + 4 mol(415 kJ/mol) + 1 mol(615 kJ/mol)]

— [1 mol(330 kJ/mol) + 5 mol(415 kJ/mol) + 1 mol(345 kJ/mol)]

$\Delta E_{\text{reaction}}$ = 2705 kJ – 2750 kJ = – 45 kJ

$$Cl_2 + H_2C=CH_2 \rightarrow ClH_2C–CH_2Cl$$

Reactants			Products		
Bond	No.	Energy (kJ/mol)	Bond	No.	Energy (kJ/mol)
Cl–Cl	1	240	C–Cl	2	330
C–H	4	415	C–H	4	415
C=C	1	615	C–C	1	345

$\Delta E_{\text{reaction}} \cong$ [1 mol(240 kJ/mol) + 4 mol(415 kJ/mol) + 1 mol(615 kJ/mol)]

— [2 mol(330 kJ/mol) + 4 mol(415 kJ/mol) + 1 mol(345 kJ/mol)]

$\Delta E_{\text{reaction}}$ = – 150 kJ

9.61   To determine which orbitals atoms use in a molecule, we must determine steric numbers from either the line structure or the Lewis structure. Any inner C, N, or O atom with no double bonds has SN = 4; with one double bond, an inner C or N atom has SN = 3. Remember that outer atoms (including H) always use atomic orbitals.

9.63   To determine the Lewis structure, first expand the line structure into a complete structural formula, then add lone pairs and look for resonance structures. The structure of acrolein is straightforward:

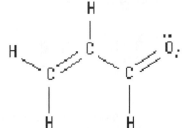

Each carbon atom has SN = 3, has trigonal planar geometry, and can be described using sp^2 hybridization. This leaves one p orbital on each atom, plus one from the oxygen atom, to form a delocalized π network extending over four atoms.

9.65   We can determine hybridizations directly from line structures (remember that C–H bonds and lone pairs are not explicitly shown in line structures): Each C or N atom with no double bonds, and each inner O atom, has SN = 4 and sp^3 hybrids can describe the bonding. C, N, and O with one double bond have SN = 3 and sp^2 hybridization.

(a) The Lewis structure shows three π bonds, all isolated.

(b) Atoms A and D have SN = 3, so sp^3 hybrids are appropriate. Atom C has SN = 3, so it is appropriately described using sp^2 hybrids. Atom B is an outer atom using atomic $2p$ orbitals.

(c) Atoms D and E have SN = 4, tetrahedral geometry, and 109.5° bond angles. Atom F has SN = 3, trigonal planar geometry, and 120° bond angle.

(d) The structure of saxitoxin in your text is correct except that it is missing the lone pairs of electrons. Since all atoms in this molecule are row 2 atoms (except the H's) simply add lone pairs to complete the octets. Remember that there are H atoms that are not shown for each carbon atom, so the only atoms in the structure that need lone pairs are the N and O atoms.

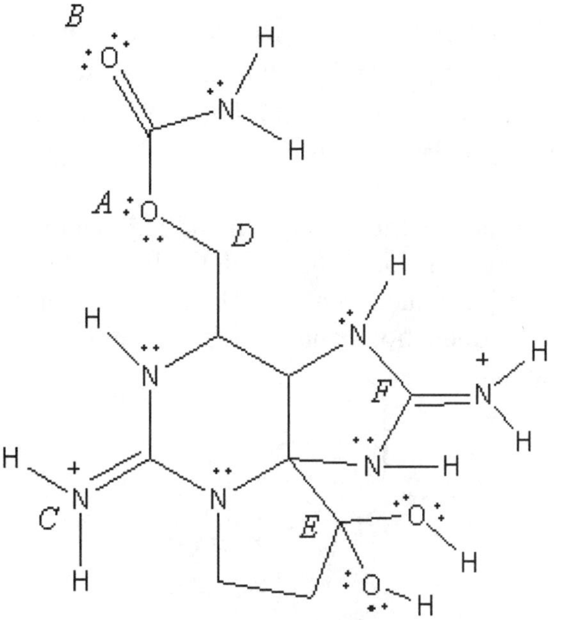

9.67   The standard procedure gives a Lewis structure from which hybridization, geometry, and π bonding are identified. Ketene has 2 C (4 e⁻) + 1 O (6 e⁻) + 2 H (1 e⁻) =16 valence electrons, 8 of which are used for the 4 bonds in the framework. Add 6 electrons around the outer O atom, then place the remaining pairs on one C atom. Complete the octet around the other C atom by shifting electron pairs to form 2 double bonds in the final Lewis structure:

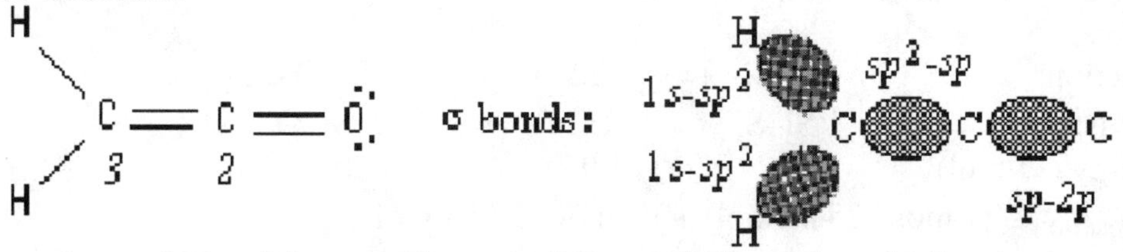

The steric numbers of the two C atoms differ, as indicated on the Lewis structure. All atoms lie in a common plane, with a σ bond framework described as shown in the sketch. The C atom with SN = 3 has trigonal planar geometry and 120 ° angles. The C atom with

SN = 2 has linear geometry and 180 ° angles. There is a two-center π bond in the plane of the molecule and an extended π bond perpendicular to the molecular plane:

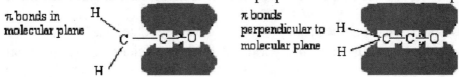

9.69 Orbital descriptions of second-row diatomic molecules make use of orbital configurations. Here are the configurations:

N_2: $(\sigma_s)^2 (\sigma_s*)^2 (\pi_x)^2 (\pi_y)^2 (\sigma_p)^2$

O_2: $(\sigma_s)^2 (\sigma_s*)^2 (\sigma_p)^2 (\pi_x)^2 (\pi_y)^2 (\pi*_x)^1 (\pi*_y)^1$

F_2: $(\sigma_s)^2 (\sigma_s*)^2 (\sigma_p)^2 (\pi_x)^2 (\pi_y)^2 (\pi*_x)^2 (\pi*_y)^2$

Each molecule has eight bonding electrons. N_2 has two antibonding electrons, for a net bond order of three; O_2 has four antibonding electrons, for a net bond order of two; F_2 has six antibonding electrons, for a net bond order of one. The bond lengths, which increase in the order $N_2 < O_2 < F_2$, reflect this trend in bond order.

9.71 As Figure 9-33 shows, an electrical potential applied to a material distorts the bands. As Figure 9-35 shows, a semiconductor has a relatively small band gap between its filled valence band and empty conduction band. The distortion of the bands allows electrons to "hop" from valence to conduction band without changing energy, leading to electron mobility and conduction:

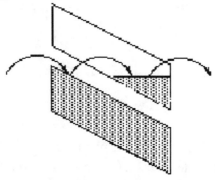

9.73 Before calculating estimated reaction energies, we must determine what the net reaction is and how many bonds are involved. When a $Y–Y$ chain is converted into a $Y–O–Y$ chain, each $Y–Y$ bond is replaced by two $Y–O$ bonds:

Reaction:	–Si–Si– +	1/2 O_2 $\rightarrow$	–Si–O–Si–
Bonds:	1 Si–Si	1/2 O=O	2 Si–O
Energy (kJ/mol):	225	495	450

$\Delta E_{reaction} \cong [1 \text{ mol}(225 \text{ kJ/mol}) + 0.5 \text{ mol}(495 \text{ kJ/mol})]$
$- [2 \text{ mol}(450 \text{ kJ/mol})] = 470 \text{ kJ} - 900 \text{ kJ} = -430 \text{ kJ}$

Reaction:	–C–C– +	1/2 O_2 $\rightarrow$	–C–O–C–
Bonds:	1 C–C	1/2 O=O	2 C–O
Energy (kJ/mol):	345	495	360

$$\Delta E_{reaction} \cong [1\ mol(345\ kJ/mol) + 0.5\ mol(495\ kJ/mol)]$$
$$- [2\ mol(360\ kJ/mol)] = 590\ kJ - 720\ kJ = -130\ kJ$$

As the results of these calculations imply, Si–O chains are considerably more stable than C–O chains.

9.75   We obtain the charges on the ions from the chemical formulas: KO_2 contains one K^+ per O_2 unit, so superoxide has one negative charge; K_2O_2 contains two K^+ per O_2 unit, so peroxide has two negative charges. These ions are second-row diatomic species whose electron configurations can be written using the orbital energy level diagrams in Figure 9-17 of your textbook ($Z \geq 8$). Superoxide has one more electron than O_2, and peroxide has two more:

O_2: $(\sigma_s)^2 (\sigma_s^*)^2 (\sigma_p)^2 (\pi_x)^2 (\pi_y)^2 (\pi_x^*)^1 (\pi_y^*)^1$ Bond order = $1/2(8-4) = 2$

O_2^-: $(\sigma_s)^2 (\sigma_s^*)^2 (\sigma_p)^2 (\pi_x)^2 (\pi_y)^2 (\pi_x^*)^2 (\pi_y^*)^1$ Bond order = $1/2(8-5) = 1.5$

O_2^{2-}: $(\sigma_s)^2 (\sigma_s^*)^2 (\sigma_p)^2 (\pi_x)^2 (\pi_y)^2 (\pi_x^*)^2 (\pi_y^*)^2$ Bond order = $1/2(8-6) = 1$

From the bond orders, we see that O_2 has the strongest, shortest bond and O_2^{2-} the longest, weakest bond.
From the number of unpaired electrons, we see that O_2 and O_2^- are both magnetic, with O_2 showing the largest magnetism.

9.77   The strength of a bond increases with the number of electrons shared by the two atoms, so a triple bond, with three electron pairs, is stronger than a double bond, with two electron pairs. However, the energy *per electron pair* is less for a triple bond than for a double bond: C=C, 615/2 = 310 kJ/mol electron pair; C≡C, 835/3 = 280 kJ/mol electron pair. Thus, the triple bond is weaker per electron pair, and this accounts for it being more reactive.

9.79   A description of bonding in a molecule starts with its Lewis structure, from which steric number, geometry, hybridization, and extension of π bonds can be deduced.
ClO_3^- has $7 + 3(6) + 1 = 26$ valence electrons. Draw the bonding framework, add 3 lone pairs to each outer oxygen atom and place the two remaining electrons on the inner Cl atom:

The provisional Lewis structure has $FC_{Cl} = 7 - 5 = 2$, so two electron pairs are shifted to make double bonds:

159

$SN_{Cl} = 4$, indicating that the molecule has tetrahedral geometry and the σ bond framework can be described using sp^3 hybrids on Cl overlapping with $2p$ orbitals from O. The resonance structures signal that the π system is extended over all four atoms, and there are two bonding π orbitals occupied by four electrons.

ClO_2^- has $7 + 2(6) + 1 = 20$ valence electrons. The provisional Lewis structure has two single bonds and two lone pairs on Cl, giving $FC_{Cl} = 7 - 2 - 2(2) = 1$, so one electron pair is shifted to make a double bond:

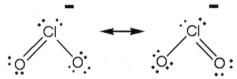

$SN_{Cl} = 4$, indicating that the electron pair geometry is tetrahedral, the molecule is bent, and the σ bond framework can be described using sp^3 hybrids on Cl overlapping with $2p$ orbitals from O. The resonance structures signal that the π system is extended over all three atoms, with one bonding π orbital occupied by two electrons.

9.81   Bond order measures the amount of electron sharing, based on Lewis structures or orbital configurations. It is based on models and cannot be directly measured by experiments. The more electrons shared between atoms, however, the stronger and shorter the bond generally will be, so both bond energy and bond length are correlated with bond order. These bond properties can be measured experimentally, so we obtain confirmation of the bond orders predicted by bonding models by seeing whether experimental bond strengths and lengths match predicted bond orders. Higher bond order should be accompanied by increased bond strength and decreased bond length.

9.83   The bonding in H–X molecules can be represented by overlap between the $1s$ orbital of the H atom and the $n\,p$ orbital of the X atom. As n increases, the p orbital becomes larger but more diffuse because the electron is spread out over a larger volume. Consequently, electron density decreases:

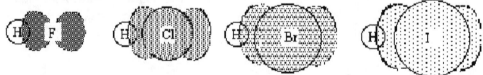

In these sketches, the shading represents the electron density within the orbital. Lighter shading represents lower electron density, hence a weaker bond.

9.85   The differences in behavior between $n = 2$ elements and their $n = 3$ counterparts from the same column of the periodic table can be explained by the difference in side-by-side π overlap for $n = 2$ and $n = 3$. Multiple bonds involving π overlap are strong for N and O, but π overlap is very slight for P and S. Consequently, P–P and S–S diatomic molecules are minimally stabilized by π bonding, and these elements are more stable in chains with single bonds.

9.87   The three bond energy trends are (1) Bonds become stronger as more electrons are shared between the atoms; (2) The greater the electronegativity difference ($\Delta\chi$) between bonded atoms, the stronger the bond; (3) Bonds become stronger as orbital overlap increases. Consult Table 9-2 in your textbook to find examples involving N (other than N–O, which is used in the text) that illustrate these features. There are many examples; here is one set:
(1) N=N (420 kJ) stronger than N–N (160 kJ)
(2) N–F (285 kJ) stronger than N–N (160 kJ)
(3) N–F (285 kJ) stronger than N–Cl (200 kJ)

9.89   A description of bonding always starts with the Lewis structure for the molecule. Here is the Lewis structure for acrylonitrile. The steric numbers of the inner atoms are shown:

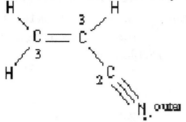

(a) The carbon atoms with SN = 3 have σ bonding that can be described using sp^2 hybrid orbitals, and the σ bonding of the carbon atom with SN = 2 can be described using sp hybrids.

(b) The entire molecule lies in the same plane. The σ bond network falls within this plane, and there is a localized π bond in this plane. Perpendicular to the molecular plane is an extended π network:

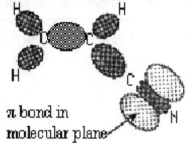
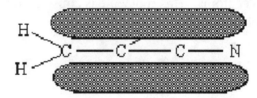

(c) There is an extended π network that includes the three C atoms and the N atom. It has four π electrons in two bonding orbitals that span the four atoms, in orbitals that are perpendicular to the molecular plane.

9.91   The bonding in second-row diatomic molecules can be described using the orbital energy level diagrams shown in Figure 9-18 of your textbook. CN has 9 valence electrons and the orbital energy level diagram for $Z < 8$:

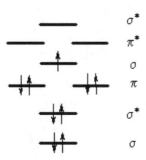

161

There are 7 bonding electrons and 2 antibonding electrons, for a net of 5, or bond order 2.5. The least stable occupied orbital is a π orbital:

9.93 A description of bonding and geometry starts with determination of the Lewis structure. ClO_2 has $7 + 2(6) = 19$ electrons. Thus, from a Lewis structure point of view, it is unusual in having an odd number of electrons. The bond framework requires two pairs, six additional electrons are placed on each O atom, leaving three lone electrons on the Cl atom. $FC_{Cl} = 7 - 2 - 3 = +2$, so shift two pairs to make double bonds:

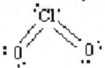

The Cl atom has SN = 4 (remember that a lone electron requires an orbital, just as an electron pair does), so its geometry is tetrahedral, the σ bonds can be described using sp^3 hybrids from Cl, and the molecule has a bond angle near 109°. There is an extended π system formed from d orbitals on Cl overlapping side-by-side with $2p$ orbitals from the two O atoms. This molecule is considered unusual because it has an odd number of electrons.

Chapter 10

10.1 The condensation temperature of a substance measures the strength of the interatomic attractions. Therefore, since Xe has the highest condensation temperature it also has the highest interatomic attractions; Ar with the lowest condensation temperature has the lowest interatomic attractions. At 140 K Xe is in a condensed phase because the kinetic energy of the atoms has not yet exceeded the interatomic attractions. Kr and Ar however, are both gases because the kinetic energy is greater than their interatomic attractions. At 100K only Argon's kinetic energy is larger than its interatomic attractions and is therefore the only gas (Xe and Kr are both in condensed phases).

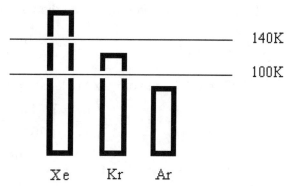

10.3 For any given substance, intermolecular attractions are constant. Their relative importance depends on two features: How close together molecules are, on average; and how much kinetic energy of motion molecules have, on average.
For any given substance, molecular volume is constant. Its relative importance depends on how close together molecules are, on average.
(a) Molecules are farther apart, making intermolecular attractions less significant. Also molecular size is less significant.
(b) Molecules are closer together, so intermolecular attractions and molecular sizes are more significant.

10.5 Pictures of atomic arrangements should show a high degree of order for a solid, close packing but some disorder for a liquid, and large distances between atoms for a gas:

Ag(s) 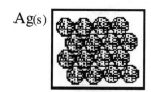 Ar(g) 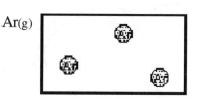 Hg(l)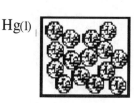

10.7 To determine deviation from ideal behavior, calculate the pressure using the ideal gas equation and compare the calculated result with the experimental value:

$$P_{ideal} = \frac{nRT}{V} = \frac{(1.00 \text{ mol})(8.206 \times 10^{-2} \frac{\text{L atm}}{\text{mol K}})(40.0 + 273.2 \text{ K})}{1.20 \text{ L}} = 21.4 \text{ atm}$$

$$\% \text{ deviation} = (100\%)\left(\frac{P_{actual} - P_{ideal}}{P_{ideal}}\right) = (100\%)\left(\frac{19.7 \text{ atm} - 21.4 \text{ atm}}{21.4 \text{ atm}}\right) = -7.9\%$$

163

10.9 To calculate pressure using the van der Waals equation, we must know n, V, T, a, and b:

$$P = \left(\frac{nRT}{V-nb}\right) - \left(\frac{n^2a}{V^2}\right)$$

Table 10-1 in your textbook lists values for Cl_2: $a = 6.493$ L^2 atm/mol^2, $b = 0.0562$ L/mol.

$$n_{Cl_2} = \frac{m}{MM} = 1.25\,kg\left(\frac{10^3\,g}{1\,kg}\right)\left(\frac{1\,mol}{70.906\,g}\right) = 17.63\,mol;$$

$$(V-nb) = 15.0\,L - 17.63\,mol\left(\frac{0.0562\,L}{1\,mol}\right) = 14.01\,L;$$

$$P = \frac{(17.63\,mol)(8.206 \times 10^{-2}\,\frac{L\,atm}{mol\,K})(295\,K)}{14.01\,L} - \frac{(17.63\,mol)^2(6.493\,\frac{L^2atm}{mol^2})}{(15.0\,L)^2}$$

$P = 30.46$ atm $- 8.97$ atm $= 21.5$ atm;

$$P_{ideal} = \frac{nRT}{V} = \frac{(17.63\,mol)(8.206 \times 10^{-2}\,\frac{L\,atm}{mol\,K})(295\,K)}{15.0\,L} = 28.5\,atm$$

$$\%\ deviation = (100\%)\left(\frac{P_{actual} - P_{ideal}}{P_{ideal}}\right) = (100\%)\left(\frac{21.5\,atm\ -\ 28.5\,atm}{28.5\,atm}\right) = -25\%$$

10.11 Ease of liquefaction depends on the magnitude of intermolecular attractions: The larger the attractions, the easier the substance is to liquefy. All these molecules are symmetric (tetrahedral), so the ranking depends entirely on dispersion forces. Ease of liquefaction will increase with molecular size: CH_4 (hardest to liquefy) $< CF_4 < CCl_4$ (easiest to liquefy).

10.13 Polarizability increases with the size of the highest occupied orbitals. The molecules with larger orbitals (CCl_4) should be shown with greater distortion than the molecules with smaller orbitals (CH_4). These molecules are close to spherical in the absence of polarization:

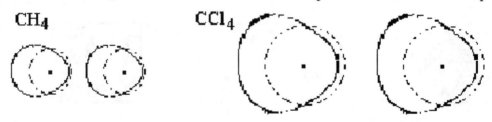

10.15 Boiling point depends on the magnitude of intermolecular attractions: The larger the attractions, the higher the boiling point. The size of intermolecular attractions depends on amount of dispersion forces (size of molecule and size of orbitals), presence of polar bonds, and hydrogen bonding. Among these substances, ethanol forms hydrogen bonds, so it has the highest boiling point. Propane, being smaller than n-pentane, has the lowest boiling point: Propane (lowest bp) $< n$-pentane $<$ ethanol (highest bp).

10.17 Hydrogen bonding capability requires the presence of an electronegative atom with lone pairs (O, F, or N) and an H-*X* bond to a highly electronegative atom (*X* = O, N, or F).
(b and d) hydrogen bonding; (a and c) no hydrogen bonding.

b)

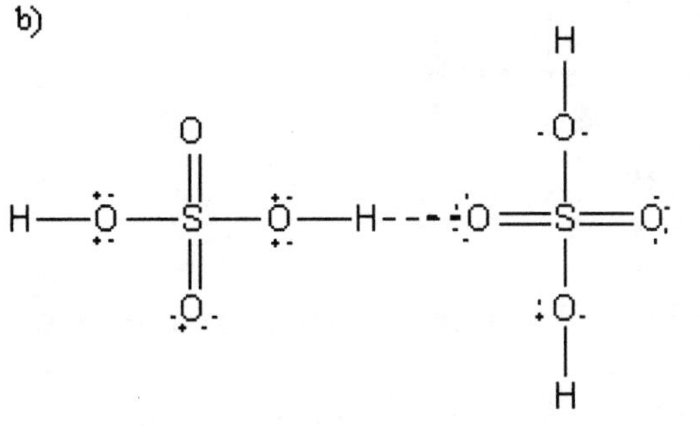

d)

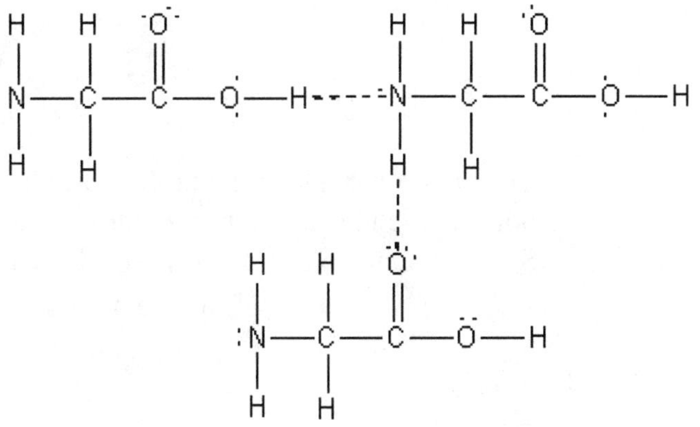

10.19 Hydrogen bonding capability requires the presence of an electronegative atom with lone pairs (O, F, or N) and an H–*X* bond to a highly electronegative atom (*X* = O, N, or F).
(a) Ammonia has one N atom with a lone pair that can form a hydrogen bond to any H atom of another ammonia molecule.

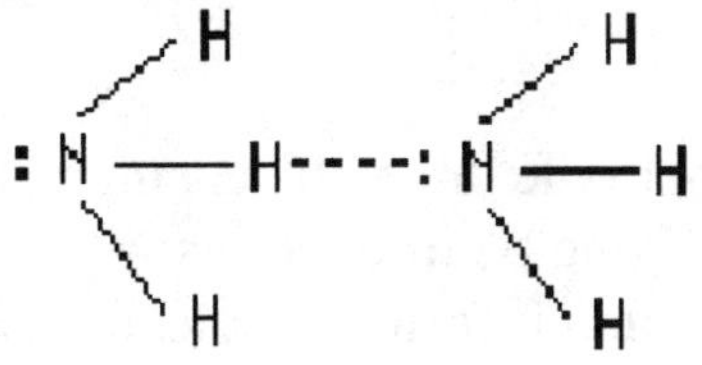

(b) Two types of hydrogen bonds are possible, between an N atom of NH_3 and an H atom of H_2O, and between an O atom of H_2O and an H atom of NH_3.

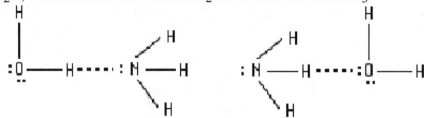

10.21 The position of elements in the periodic table, the chemical formula, and a knowledge of polyatomic ions all help in identifying types of solids:
Sn is a metal, so is a metallic solid; S_8 has a specific molecular formula, so is a molecular solid; Se is not a metal, so is a network solid; SiO_2 is not ionic, so is a network solid; and Na_2SO_4 contains Na^+ cations and SO_4^{2-} polyatomic anions, so is an ionic solid.

10.23 Consult your textbook for information about the differences between different types of solids.
(a) The bonding in metals comes from extended networks of delocalized electrons, while the bonding in network solids includes many individual covalent bonds.
(b) Metals conduct electricity, are malleable and ductile, and are shiny in appearance. Network solids are non-conductors or semiconductors, are brittle, and often have dull appearances.

10.25 The position of elements in the periodic table, the chemical formula, and knowledge of polyatomic ions all help in identifying types of solids:
(a) Br_2 is a discrete neutral molecule. It forms a molecular solid.
(b) KBr contains cations and anions. It forms an ionic solid.
(c) Ba is an alkaline earth metal. It forms a metallic solid.
(d) SiO_2 is a covalent molecule. It forms a network solid.
(e) CO_2 is a covalent molecule. It forms a molecular solid.

10.27 The density of a solid depends not only on the nature of the elements it contains but also on the tightness of bonding. A structure with more open bonding pattern has a lower density than one with a more compact bonding pattern. Graphite uses sp^2 hybrids and has an extended π bonding network, making it planar and leaving a relatively open structure between successive planes. Diamond uses sp^3 hybrids and has a compact network of bonds:

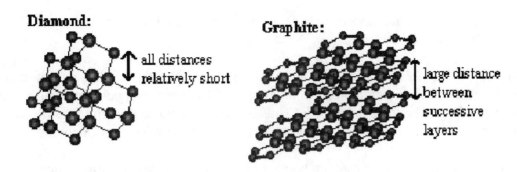

Diamond: all distances relatively short

Graphite: large distance between successive layers

10.29 To determine a chemical formula from a unit cell, count the number of atoms of each element contained within the unit cell. Atoms on a face count 1/2 (each is part of two unit cells), those on an edge count 1/4 (each is part of four unit cells), and those at a corner count 1/8 (each is part of eight unit cells).
There is one Ti atom within the cell.
Ca: 1/8 (8 corner atoms) = 1
O: 1/2 (6 face atoms) = 3.
The formula is $CaTiO_3$.

10.31 The problem states that carborundum is diamond-like. Diamond is composed entirely of tetrahedral carbon atoms (SN = 4), and the empirical formula of carborundum, SiC, indicates that alternate C atoms should be replaced with Si atoms. Each C atom bonds to four Si atoms and each Si atom bonds to 4 C atoms:

10.33 The viscosities of a set of similar substances increases with increasing chain length, because of increased dispersion forces (stronger intermolecular attractions make it harder for molecules to slide by one another) and increased "tangling" (longer chains are more entangled). Thus the order of increasing viscosity is:
pentane (C_5) < gasoline (C_8) < fuel oil (C_{12}).

10.35 The immiscibility of liquids can be explained using the concept that "like dissolves like." Molecules of salad oil are almost entirely hydrocarbons, which have dispersion-type intermolecular forces but low polarity and no hydrogen-bonding capability. A solution of acetic acid in water has high polarity and a large hydrogen-bonding capability. Thus these two liquids are not at all like each other and do not mix. In terms of balance of forces, hydrogen bonds would have to be broken for the two liquids to mix, making mixing energetically unfavorable.

10.37   The types of solids that dissolve in a solvent can be predicted using the concept that "like dissolves like." Liquid ammonia is similar to liquid water: It is a polar molecule with a strong hydrogen-bonding capability. Thus liquid ammonia should dissolve the same sorts of substances as water, ionic salts and polar organic materials such as alcohols.

10.39   To determine how many hydrogen bonds to water a molecule can form, count its polar H–X bonds and its electronegative atoms with lone pairs of electrons. Glycerol has three of each, so it can form six hydrogen bonds to water molecules:

10.41   In a chromatography column, components that are attracted to the support material move more slowly than those that are not attracted, so the least polar substance moves fastest through a polar column. The areas under chromatographic peaks are a measure of relative amounts.
(a) Heptane is least polar and comes off first, followed by methyl pentyl ether. The most polar compound, 1-hexanol, comes off last.
(b) The last peak is largest, so 1-hexanol is present in largest amount. The middle peak is smallest, so methyl pentyl ether is present in smallest amount.

10.43   Freezing points are calculated using Equation 10-2 from your textbook: $\Delta T_f = K_f \, c_m$. For water, K_f is 1.858 °C kg/mol. First, the molality of the solution must be calculated:

$$c_m = \frac{\text{mol solute}}{\text{kg solvent}}$$

Consider 100 g of a solution that is 12% by mass ethanol. It contains 12 g ethanol and 88 g of solvent (water). Thus

$$\text{mol solute} = 12 \, g\left(\frac{1 \, \text{mol}}{46 \, g}\right) = 0.26 \, \text{mol} \quad \text{and} \quad \text{kg solvent} = 88 \, g\left(\frac{10^{-3} \, \text{kg}}{1 \, g}\right) = 8.8 \times 10^{-2} \, \text{kg}.$$

$$c_m = \frac{0.26 \text{ mol}}{8.8 \times 10^{-2} \text{ kg}} = 2.95 \text{ mol/kg}$$

$$\Delta T_f = K_f \, c_m = \left(\frac{1.858 \text{ }^\circ\text{C kg}}{1 \text{ mol}}\right)\left(\frac{2.95 \text{ mol}}{1 \text{ kg}}\right) = 5.5 \text{ }^\circ\text{C}.$$

This is the depression in freezing point, so the new freezing point is $0.0 - 5.5 = -5.5 \text{ }^\circ\text{C}$.

10.45 Boiling points can be calculated for solutions of non-volatile solutes, using the boiling point elevation constant. However, ethanol is volatile, so the equation does not apply and the boiling point of the solution cannot be calculated from the information provided. To calculate boiling points, vapor pressure data would be required for both water and ethanol.

10.47 Molar masses can be calculated from osmotic pressures using Equation 10-5 in your textbook.

First convert torr into atm: $P = (64.8 \text{ torr})\left(\frac{1 \text{ atm}}{760 \text{ torr}}\right) = 8.53 \times 10^{-2} \text{ atm}$.

(a) $MM = \dfrac{mRT}{\Pi V} = \dfrac{(1.00 \text{ g})(8.206 \times 10^{-2} \frac{\text{L atm}}{\text{mol K}})(25 + 273 \text{ K})}{(8.53 \times 10^{-2} \text{ atm})(1.00 \text{ L})} = 287 \text{ g/mol}$

(b) The non-polar tail contains 11 C atoms and 2 H atoms for every C except the first, which has 3 H atoms, giving 23 H atoms altogether. It contributes:
$11(12.01 \text{ g/mol}) + 23(1.008 \text{ g/mol}) = 155 \text{ g/mol}$, leaving for the polar head group, $MM = 287 - 155 = 132 \text{ g/mol}$.

10.49 Surfactants must have a highly polar (or ionic) end and a non-polar end. Of these three substances, propanoic acid has too short a non-polar end and would be the worst. Lauryl alcohol and sodium lauryl sulfate have similar non-polar portions, but the ionic portion of sodium lauryl sulfate makes it the best surfactant of these three substances:

10.51 Surfactants must have a highly polar (or ionic) end and a non-polar end. The polar end of the surfactant molecule associates with a polar substance, while the non-polar end of a surfactant molecule associates with a non-polar substance. The figure shows the non-polar ends of sodium stearate immersed in the solvent, indicating that the solvent molecules are non-polar.

10.53 Gas samples are conveniently analyzed using gas chromatography. A gas sample mixed with a carrier gas passes through a column containing a stationary phase that attracts different gases to different extents. The more strongly attracted a particular component is to the stationary phase, the slower it passes through the column. Thus the column separates a gas mixture into a series of "peaks" that emerge from the column at different times. Each component can be identified by comparing its time of passage through the column with those of standard samples of known substances.

10.55 This problem deals with the freezing point depression of water using ethyl glycol. Use the following equation to solve this problem: $\Delta T_f = K_f c_m$

$$20°C = \left(\frac{1.858°C\,kg}{1\,mol}\right)(c_m)$$

$$c_m = 20\,°C\left(\frac{1\,mol}{1.858°C\,kg}\right) = 11\,mol/kg$$

10.57 The solubilities of organic substances in water depends on the presence of polar groups and groups that are capable of forming hydrogen bonds with water. Alcohols contain OH groups that enhance solubility. Among these compounds, benzene (C_6H_6) is non-polar and has the lowest solubility, while erythritol has four OH groups, making it highly soluble:

benzene (lowest solubility) < pentanol < erythritol (highest solubility).

10.59 Hydrophilicity is conveyed by the presence of ionic, polar, or hydrogen-bonding groups. DDT contains none of these, so it is highly hydrophobic. Thus it is stored in fatty tissues (which are non-polar) rather than being excreted from the body. Consequently, the effects of DDT are cumulative, especially for organisms high up on the "food chain," such as predator birds.

10.61 Differential solubilities can be explained using the concept that "like dissolves like." Here, iodine is a non-polar solid which will differentially dissolve in a non-polar liquid. Water is highly polar and CCl_4 is non-polar, so these two liquids are immiscible and I_2 preferentially dissolves in CCl_4.

10.63 All substances are subject to dispersion forces. In addition, look for polarity and hydrogen-bonding ability:
(a) NH_3 is polar and forms hydrogen bonds in addition to its dispersion interactions;
(b) $CHCl_3$ is a polar molecule in addition to having dispersion interactions;
(c) CCl_4 is symmetric (tetrahedral) so has only dispersion interactions; and
(d) CO_2 is symmetric (linear) so has only dispersion interactions.

10.65 All substances are subject to dispersion forces. In addition, look for polarity and hydrogen-bonding ability:
(a) CH_3OH can hydrogen bond, so it boils at a higher temperature than CH_3OCH_3;
(b) SiO_2 is a network solid with covalent bonds that must break for it to boil, so it has a higher boiling point than SO_2;
(c) HF can hydrogen bond, so it boils at a higher temperature than HCl; and

(d) I_2 has larger orbitals, giving it higher polarizability, larger dispersion forces, and a higher boiling point than Br_2.

10.67 When two substances with otherwise similar structures have different boiling points, look for differences in polarity and/or hydrogen-bonding behavior. The Lewis structures of the two isomers, shown with bond polarities drawn in, indicate why the isomers have different boiling points:

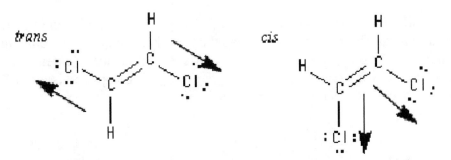

In the *trans* isomer, the two polar bonds oppose each other and cancel, making this a non-polar compound. In the *cis* isomer, the two polar bonds add, making this a polar compound. Thus, *cis* 1,2-dichloroethylene has the higher boiling point.

10.69 The notion of "like dissolves like" can be extended to the wetting behavior of one substance for another. Solder is a metal that will adhere well to another metal only if unlike substances have been removed from the surface. Non-polar substances such as oils and greases as well as network solids such as metal oxides must be removed by flux.

10.71 Deviations from ideal gas behavior can be predicted based on the strengths of intermolecular interactions. The larger the intermolecular interactions, the greater will be the deviations from ideality. Both pairs of substances are non-polar, so dispersion interactions dominate their intermolecular forces. The substance with the higher occupied n value will have larger dispersion forces and deviate more from ideality: Cl_2.

10.73 The boiling point for HCl is higher than that expected using the periodic trends (it should have a bp between PH_3 and SiH_4). This would suggest that H-bonding occurs. Hydrogen bonding would occur between the H atom on one HCl molecule and the electronegative Cl atom of another.

10.75 If water were linear rather than bent, the two polar O–H bonds would oppose each other and the molecule would be non-polar like CO_2. Non-polar water molecules could not associate with ions through ion-dipole interactions, so linear water would not be a good solvent for salts.

10.77 The magnitudes of dispersion forces depend on how extended the valence electrons are. In a molecule, valence electrons are spread over a larger volume because they are shared between atoms. This feature is illustrated by sketches of the electron clouds of H_2 and He:

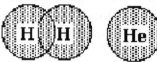

10.79 Pentane boils at a higher temperature than dimethylpropane, indicating that dispersion forces are greater for the linear chain than the branched chain. Both compounds have the same number of atoms and electrons, so the effect must be due to extension of the electron clouds. Line structures show that the branched compound is more compact than the linear compound, accounting for the difference:

10.81 Fish blood is isotonic with sea water, meaning that the overall concentrations of the two are equal. The molality of sea water can be calculated from its freezing point. To calculate the osmotic pressure, assume that molality and molarity differ negligibly for aqueous solutions:

$$\Delta T_f = 0\ °C - (-2.3\ °C) = 2.3\ °C$$

$$c_m = \frac{\Delta T_f}{K_f} = 2.3°C\left(\frac{1\ mol}{1.858°C\ kg}\right) = 1.24\ mol/kg.$$

$$M \cong c_m = 1.24\ mol/L$$

$$\Pi = MRT = \left(\frac{1.24\ mol}{1\ L}\right)\left(\frac{0.8206\ L\ atm}{1\ mol\ K}\right)(273+15K) = 29\ atm.$$

10.83 A surfactants has a polar (or ionic) end and a non-polar end. The polar end of the surfactant molecule associates with a polar substance, while the non-polar end of a surfactant molecule associates with a non-polar substance. A bilayer membrane between two aqueous layers will have its polar "heads" associated with the aqueous layers and its non-polar "tails" associated with one another, as the drawing shows:

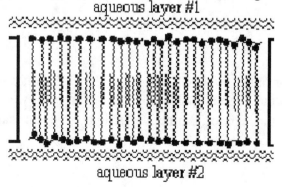

10.85 A gas molecule approaching the wall of a container is subject to intermolecular attractions from its neighboring molecules in the gas, which tend to pull it back into the gas volume. This slows the molecule so it exerts a lower force when it hits the wall. All other factors being the same, the more molecules that there are around a gas molecule, the greater this effect will be. Thus, the molecule in figure (a), which has fewer other molecules around it, will strike the wall with greater force than the molecule in figure (b).

10.87 Viscosity is determined by the strengths of intermolecular interactions and by the extent to which molecules get "tangled up" with one another. The ranking is as follows: 2,2-dimethylpropane (smallest viscosity) < *n*-pentane < butanol < propane-1,3-diol. Hydrogen bonding makes the two alcohols more viscous than the two hydrocarbons, with the diol most viscous because of multiple hydrogen-bonding interactions. The more compact branched hydrocarbon is least viscous because of near-spherical shape which prevents "tangling."

11.1 Convert line drawings to structural formulas following the standard rules, adding a C atom at each vertex and line end and adding –H bonds until each C atom has 4 bonds. Identify functional groups from memory:

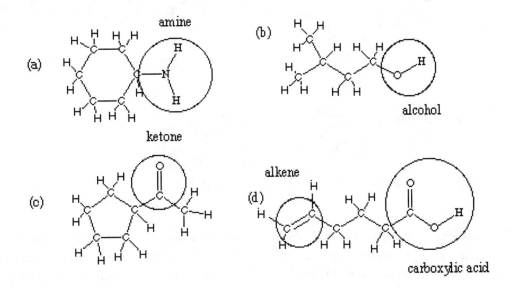

11.3 Identify functional groups from memory and convert structures to line drawings following the standard rules:

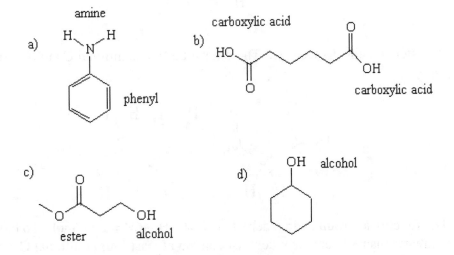

11.5 Convert line and ball-and-stick drawings to structural formulas following the standard rules, adding a C atom at each vertex and line end and adding –H bonds until each C atom has 4 bonds. Identify linkage groups from memory:

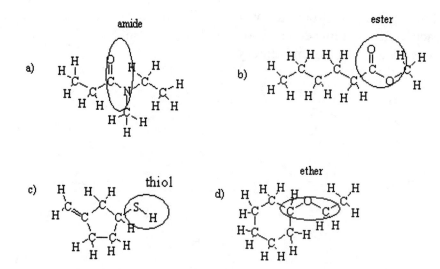

11.7   To draw examples of compound types, first identify the chemical formulas of its functional groups. Remember that carbon always forms four bonds. There are many possible correct answers to this problem. We provide just one example for each (a) an amine contains N bonded to C. The simplest amines contain $-NH_2$ units:

(b) An ester contains $C-CO_2-C$. There must be five additional C atoms, arranged in any fashion:

(c) The functional group in an aldehyde is $O=C-H$, $MM = 29$ g/mol. To have a molar mass greater than 80, an aldehyde must contain at least four additional C atoms:

(d) The ether linkage is $-O-$, and phenyl is the benzene ring:

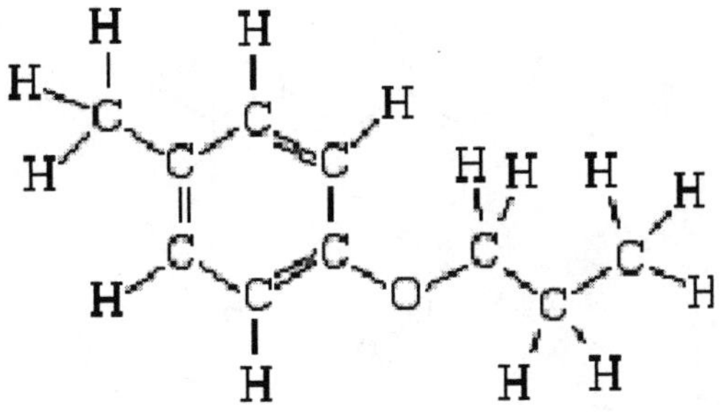

11.9 In a condensation reaction, two monomers react to form a larger unit, eliminating a small molecule such as H_2O in the process.

(a) The N–C linkage in the center of this molecule forms in a condensation reaction between a carboxylic acid and an amine:

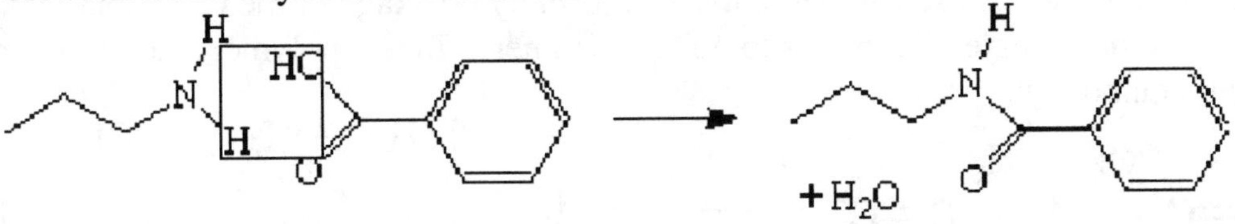

(b) The ester linkage in this molecule forms in a condensation reaction between a carboxylic acid and an alcohol:

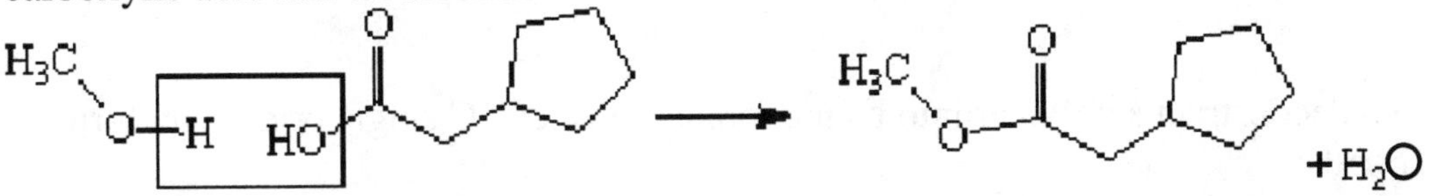

(c) The –O– linkage in this molecule forms in a condensation reaction between two alcohols:

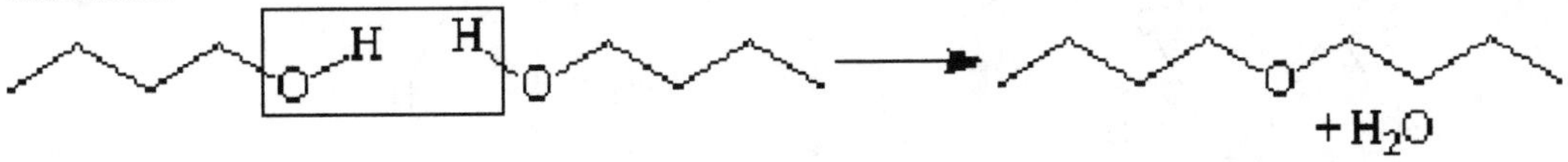

11.11 All amino acids can condense with each other in two ways. In addition, the extra carboxylic acid of glutamic acid could undergo condensation, so there are three possible products:

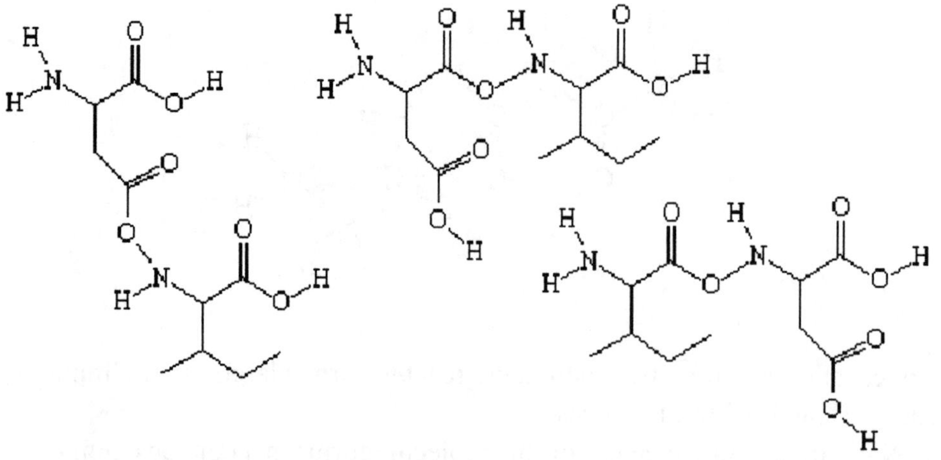

11.13 Polymers made from substituted ethylenes form by breakage of the C=C double bond to form two new single C–C bonds to other monomers. In a copolymer, each monomer occurs randomly:

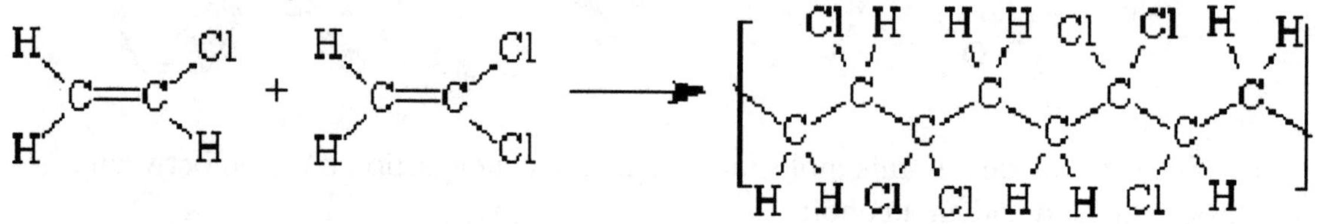

11.15 To deconstruct a polymer into its monomers, break C–C single bonds and form double bonds:

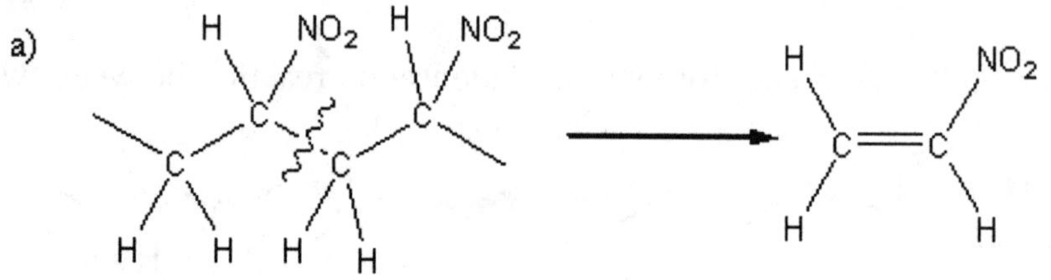

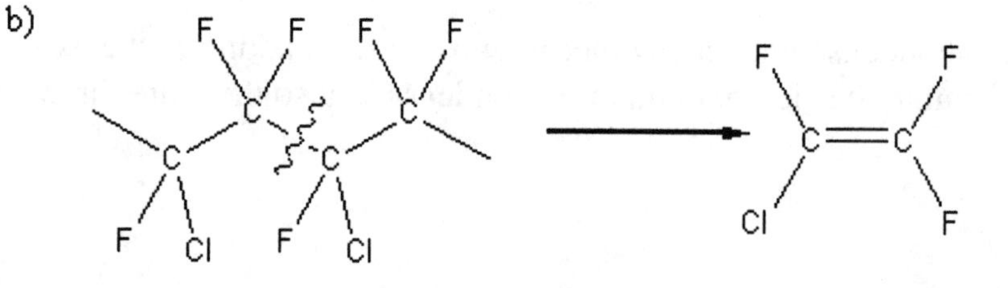

a)

b)

11.17 Polybutadiene forms from butadiene by a combination of double bond breakage and migration. The resulting polymer differs from polyethylene in having one C=C double bond in each repeat unit, whereas polyethylene is entirely CH_2 units connected by C–C single bonds (each arrow represents shifting one electron):

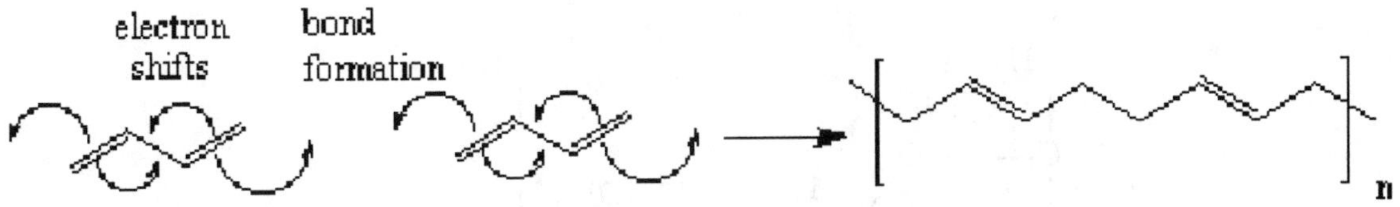

electron bond
shifts formation

11.19 To determine the monomers from which a condensation polymer has formed, decompose the polymer into its monomeric parts and add components of a small molecule, for example –H and –OH. Here, the structure shows a single component whose repeat unit is $(CH_2)_{10}$, and the linkage is an amide, which breaks down into an amine and a carboxylic acid:

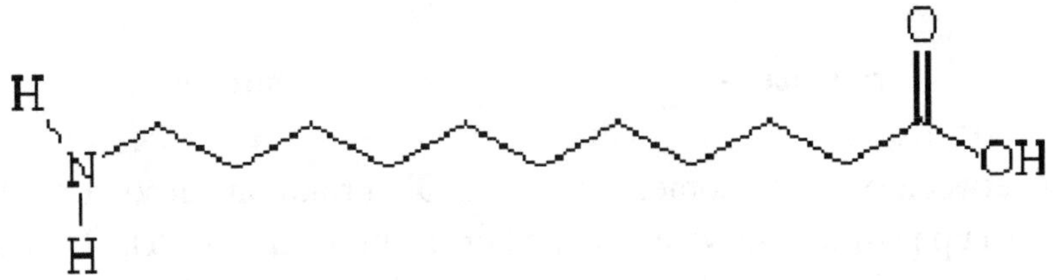

11.21 Construct polyethylene oxide by breaking a C–O bond and linking the fragments:

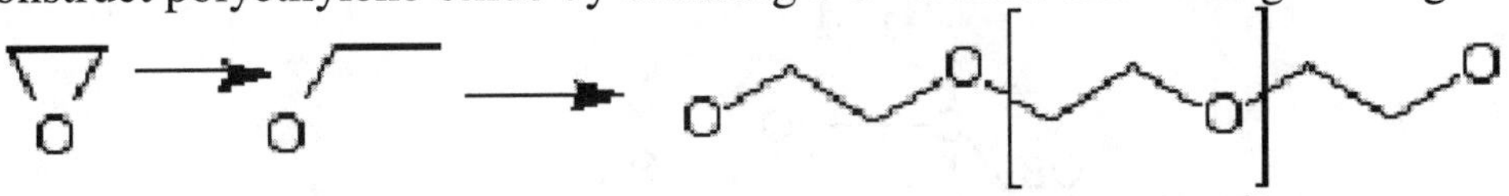

repeat unit

11.23 Cross-linking leads to increased rigidity, because chemical bonds between polymer chains restrict the ability of polymer chains to slide past one another. Thus, relatively rigid tires have more extensive cross-linking than flexible surgeon's gloves.

11.25 The categories of polymers and their characteristic properties are as follows: plastics, which exist as blocks or sheets; fibers, which can be drawn into long threads; and elastomers, which can be stretched without breaking.
(a) Balloons must stretch, so they are made of elastomers;
(b) rope is made of fibers;
(c) camera cases are rigid, so they are made of plastics.

11.27 Dioctylphthalate is an example of a liquid plasticizer, which reduces the amount of cross-linking in a polymer as well as adding a fluid component; both these changes result in improved flexibility of the polymer.

11.29 To draw the structures of sugars that are related to glucose, start with the glucose structure and make modifications as needed. For α-talose, move the OH group on carbons 2 and 4 from down positions to up positions.

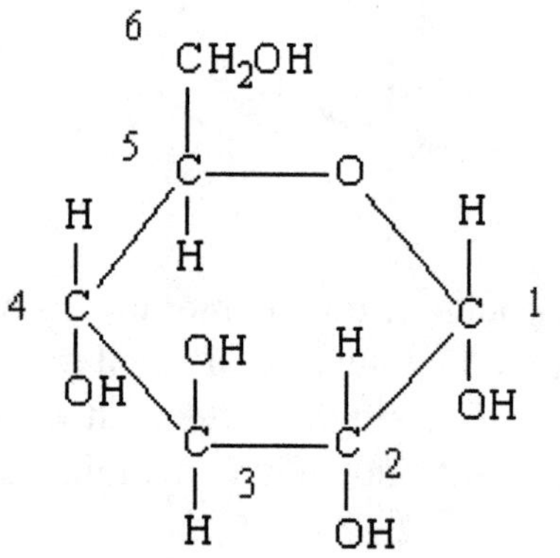

α-glucose 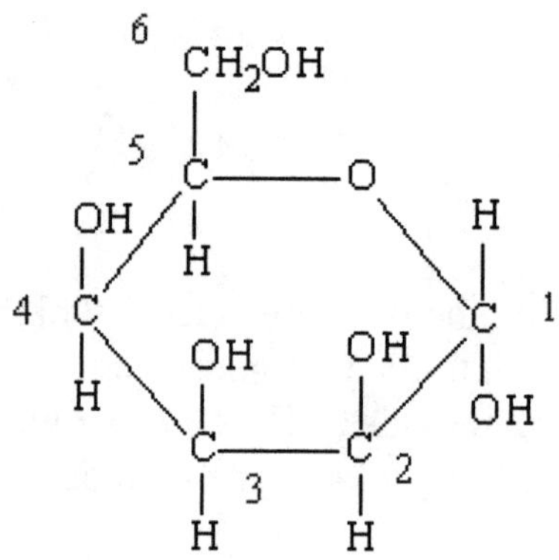   α-talose

11.31 To switch between α and β isomers move the OH group on carbon one from a down position to an up position (or vice versa when going from β to α). To draw β-talose, start with the structure of α-talose from problem 11.29 and move the OH on carbon one from a down position to an up position.

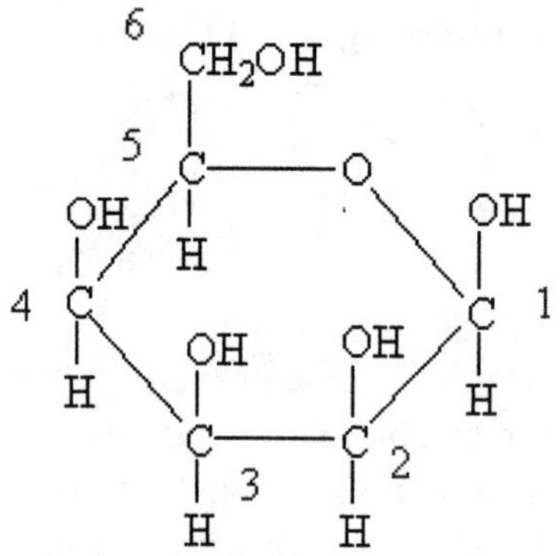

β-talose

11.33 As described in your textbook, polymers of β-glucose coil upon themselves. Glycogen is this type of polymer. Polymers of α-glucose, of which cellulose is an example, form planar sheets. It is easier to see the distinction between the two polymers by looking at them from a side-on view (as shown). In the drawings, the dark solid lines indicate the direction of the continuing chain:

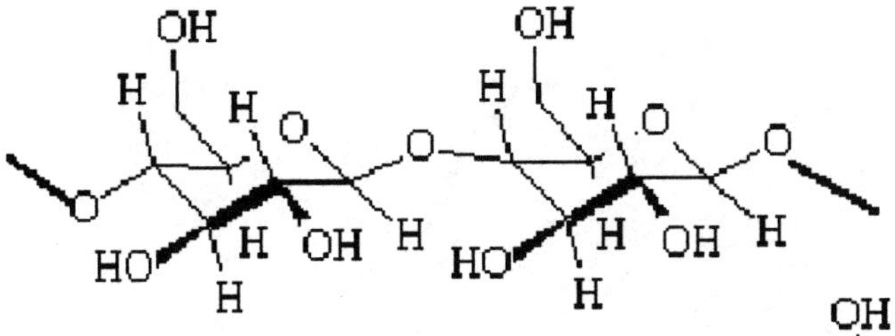

cellulose

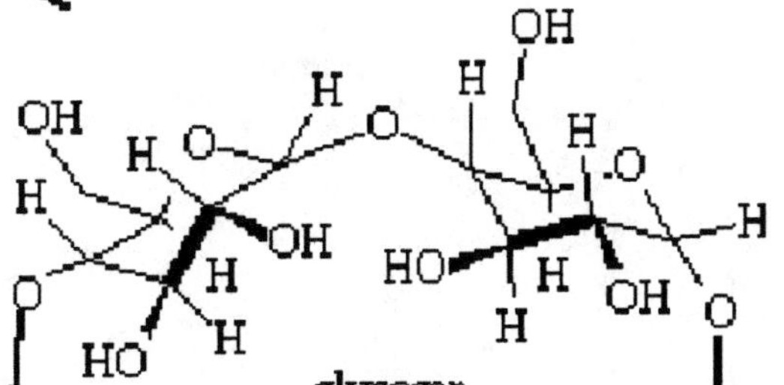

glycogen

11.35   The complementary strands of DNA form from hydrogen-bond linkages between specific nucleic acids. A pairs with T and G pairs with C. Thus, the complementary sequence of
A-A-T-G-C-A-C-T-G   is   T-T-A-C-G-T-G-A-C.

11.37   The structure of DNA consists of a phosphate-sugar-nucleic acid trio bonded to others through its phosphate groups. Apart from the structure of the nucleic acid, each unit is identical.

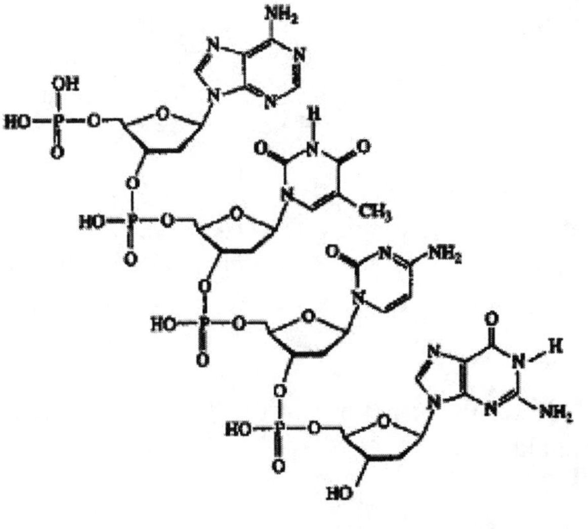

11.39   The strands of DNA fit together as a result of the hydrogen-bonding interactions illustrated in Figure 11-32. The complementary sequence for A-T-C is T-A-G. The backbones of complementary strands run in opposite directions. For clarity in showing the hydrogen-bonding interactions, we represent the backbone with a solid line rather than showing its details:

11.41 Consult Figure 11-36 for the structures of the different amino acids, all of which have the same amino acid backbone.

11.43 Hydrophilic side chains are characterized by the presence of N or O atoms that generate polar bonds and hydrogen-bonding capability, or an S–H bond that is polar. Among the

structures shown in Figure 11.41, Tyr (O–H bond) and Glu (CO$_2$H group) are hydrophilic, while Phe and Met are hydrophobic.

11.45 To identify an amino acid, examine the side chain attached to the carbon atom between the N atom and the C=O bond in the amino acid backbone. (a) the "side chain" is –H (hydrophobic side chain), making this glycine, the simplest amino acid; (b) the side chain is –CH$_2$OH (hydrophilic side chain), so this is serine; (c) the side chain is –CH$_2$SH (hydrophilic side chain), so this is cysteine.

11.47 When two amino acids condense, the amino end of either molecule can link to the carboxylic acid end of the other. A water molecule is eliminated, creating a C–N bond. Three amino acids can combine in six different ways, A–B, B–A, A–C, C–A, B–C, and C–B:

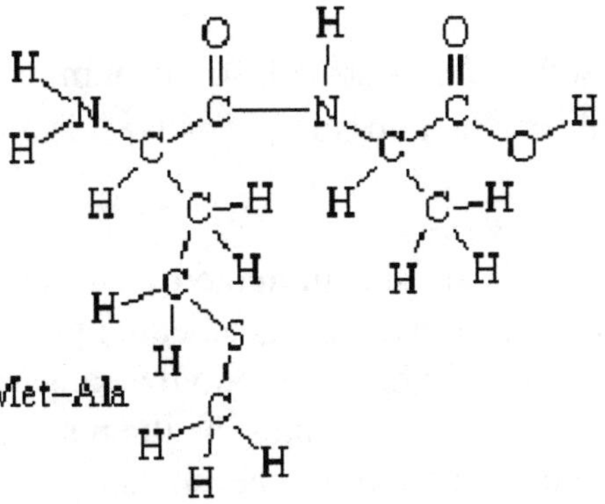

Met–Ala

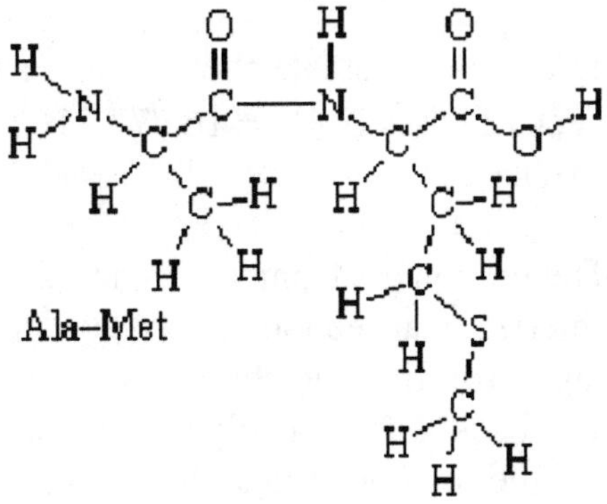

Ala–Met

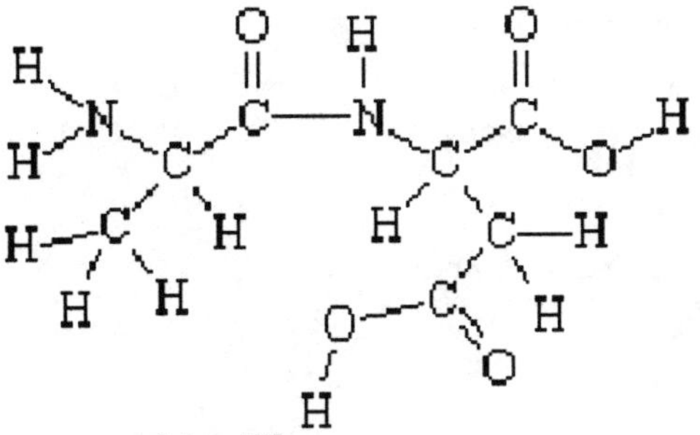

Ala–Glu

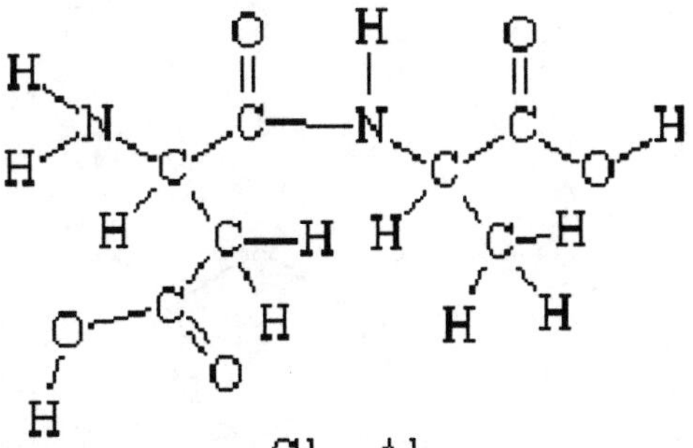

Glu –Ala

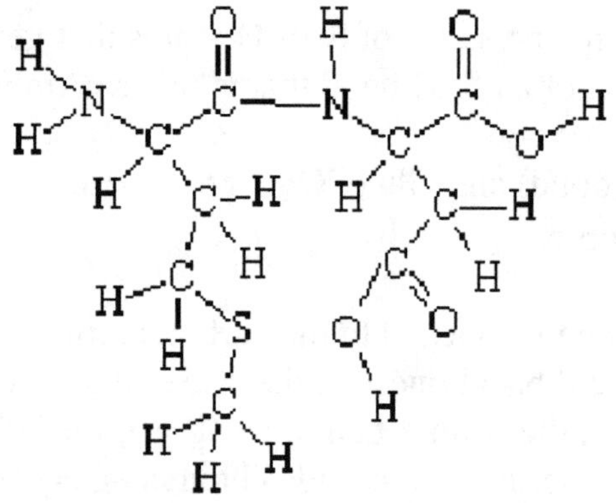

Met–Glu 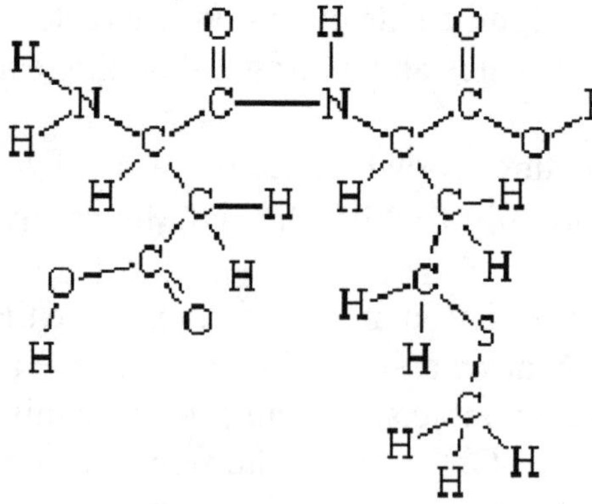   Glu–Met

11.49   The repeat unit in polyethylene is the ethylene molecule, C_2H_4, $MM = 28.1 \text{g/mol}$, so a polymer with 744 repeat units has

$$MM = 744\left(\frac{28.1\,\text{g}}{1\,\text{mol}}\right) = 2.09 \times 10^4 \text{ g/mol}.$$

11.51   There are four possible choices for each base in a DNA strand, so the number of ways to connect 12 bases is $4^{12} = 16{,}777{,}216$ (this is an exact number, because 4 is an exact number).

11.53   (a) The monomers from which hair spray is made are substituted ethylenes, which polymerize in a free radical process in which each ethylene has an equal probability of adding to the growing chain. Thus, this polymer will have a random arrangement.
(b) The backbone of a polyethylene is a carbon chain with the substituents connected to every other carbon atom. Here is a line structure of six monomer units:

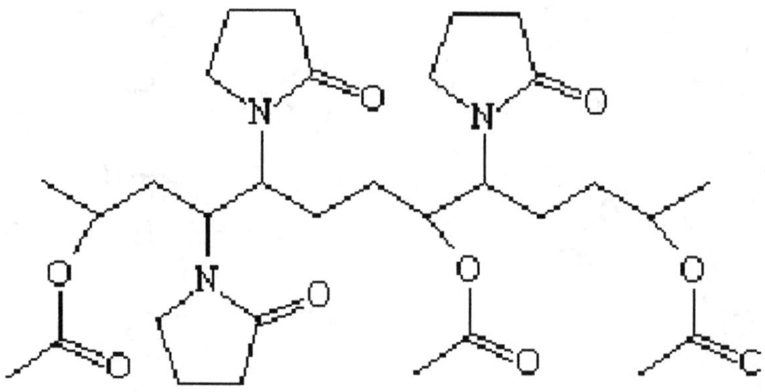

(c) The backbone of this polymer is hydrophobic, but its side chains contain highly polar C=O groups that interact with one another and with polar and hydrogen-bonding groups on hair.

11.55   A nucleotide is a combination of 1 base, 1 sugar, and 1 phosphate group. A duplex is a pair of nucleotides bound together by hydrogen-bonding interactions. The second nucleotide in a guanine duplex is cytosine.

184

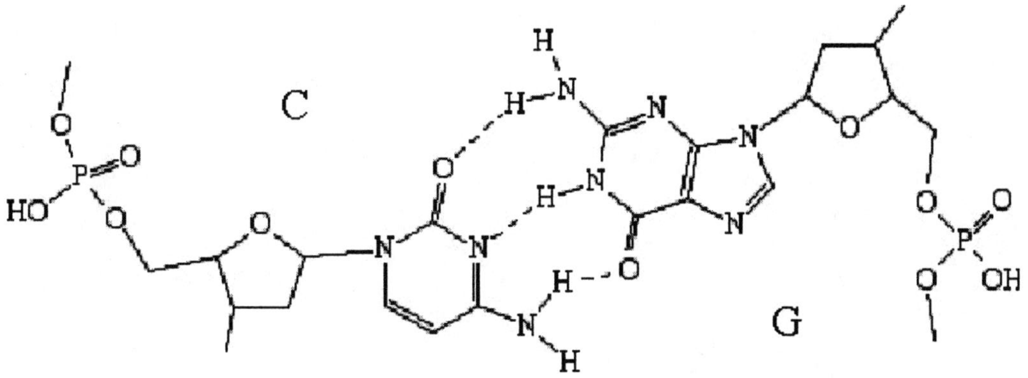

11.57 The three steps of free radical polymerization are initiation, which generates a free radical; propagation, in which monomer units add to the free radical end of the polymer chain; and termination, in which two free radical chains link together. The functional group on an ethylene monomer does not participate in any of these processes:

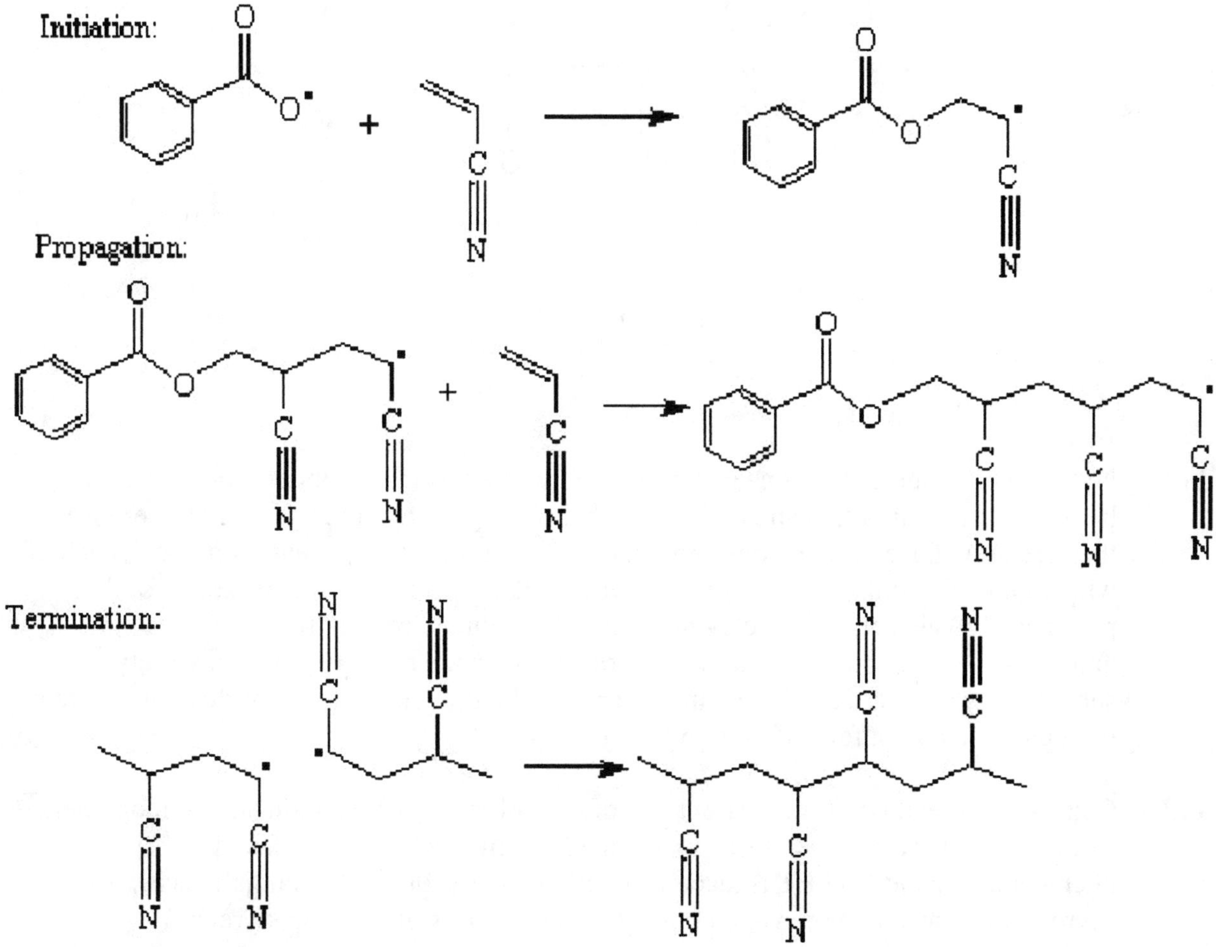

Initiation:

Propagation:

Termination:

11.59 All sugar molecules have multiple –OH functional groups, each of which can participate in hydrogen bond formation with water molecules. α-glucose has 5 –OH groups; in addition, its ring O atom can participate in formation of hydrogen bonds.

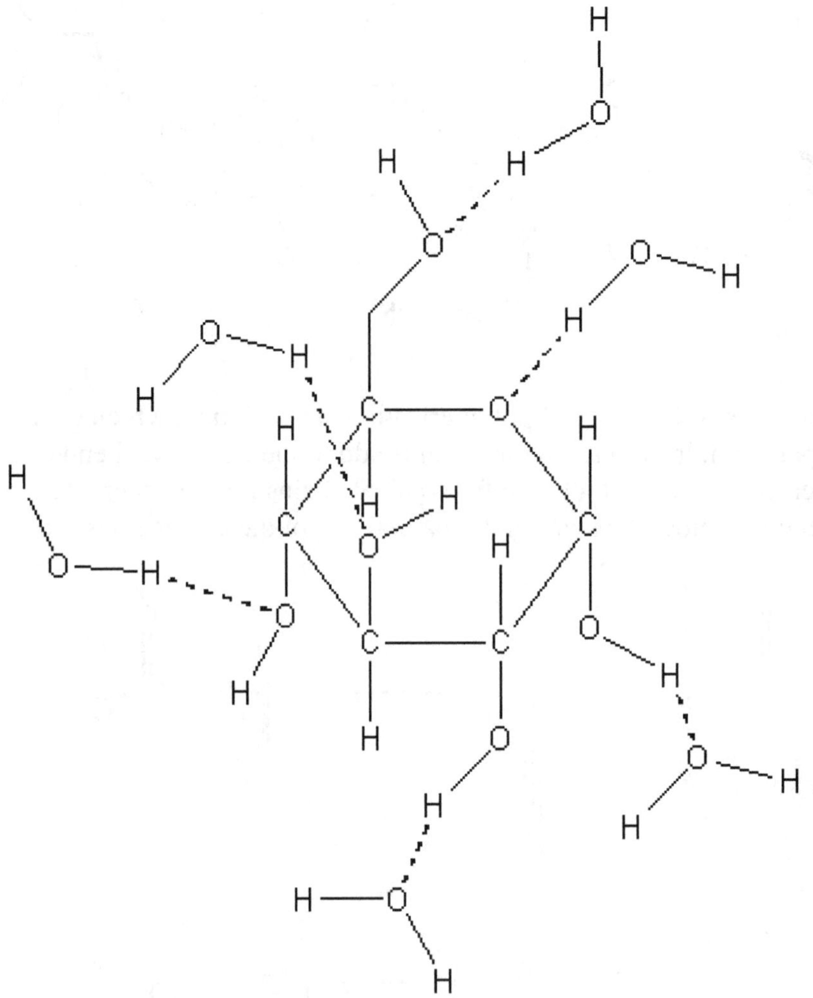

11.61 Nylons and proteins share one common feature: they form by condensation reactions between carboxylic acids and amines, so their linkages are amide groups. Otherwise, they are quite different. Nylons contain one or at most two different monomers, each of which typically contains several carbon atoms that form part of the backbone of the polymer. Proteins contain backbones that are absolutely regular repetitions of amide – C – amide bonding, but the carbon atoms in the backbone have a variety of substituent groups attached to them, generating the immense variety of different proteins (compared to only a few different nylons).

11.63 Consult your textbook for the structures of the polymers, which indicate the monomers from which they are made. (a)Kevlar is made from terephthalic acid and phenylenediamine; (b) PET is made from ethylene glycol and terephthalic acid; (c) Styrofoam is the common name for polystyrene, so it is made from styrene.

11.65 All polyethylene has the same empirical formula, $(CH_2)_n$, but whereas high-density polyethylene is all straight chains that nest together readily, low-density polyethylene has many side chains that cannot nest easily. Thus low-density polyethylene has more open space, accounting for the lower amount of CH_2 groups per unit volume. See Figure 11-10 for a visual representation.

11.67 The bases of the mRNA molecule must be complementary to the bases of the template DNA on which they are modeled, meaning A generates U, C generates G, G generates C, and T generates A (remember that in RNA, U appears rather than T). Thus the sequence generated by this strand is the same as that of Strand B except for U appearing in place of T:

1 = cytosine, 2 = uracil, 3 = adenine, and 4 = guanine.

11.69 The Watson-Crick model of DNA requires that there be a complementary base for each base in a given strand: one G for every C, one C for every G, one A for every T, and one T for every A. This requires that the molar ratios of A to T and G to C be 1.0. Chargoff's observations indicate that this relationship holds even though DNA from different sources has different sequences, some A, T-rich and others G, C-rich. These differences give rise to differences in the relative amounts of A, T vs. G, C, but pairing always results in 1:1 mole ratios of A:T and G:C.

11.71 All proteins contain both hydrophobic and hydrophilic amino acids. When a protein is in contact with a hydrophilic medium such as aqueous solution, its hydrophilic amino acids are most stable when facing outward, in contact with the solvent, while its hydrophobic amino acids are most stable when facing inward, in contact with the protein backbone. The opposite is the case when a protein is immersed in a hydrophobic medium such as a cell wall. The tertiary structure of a protein is the manner in which the individual amino acids orient themselves, so solvent interactions are the primary determinants of this tertiary structure.

11.73 To draw the structure of a condensation product, start with the structures of the two molecules undergoing condensation and then connect them together, eliminating a small molecule such as water:

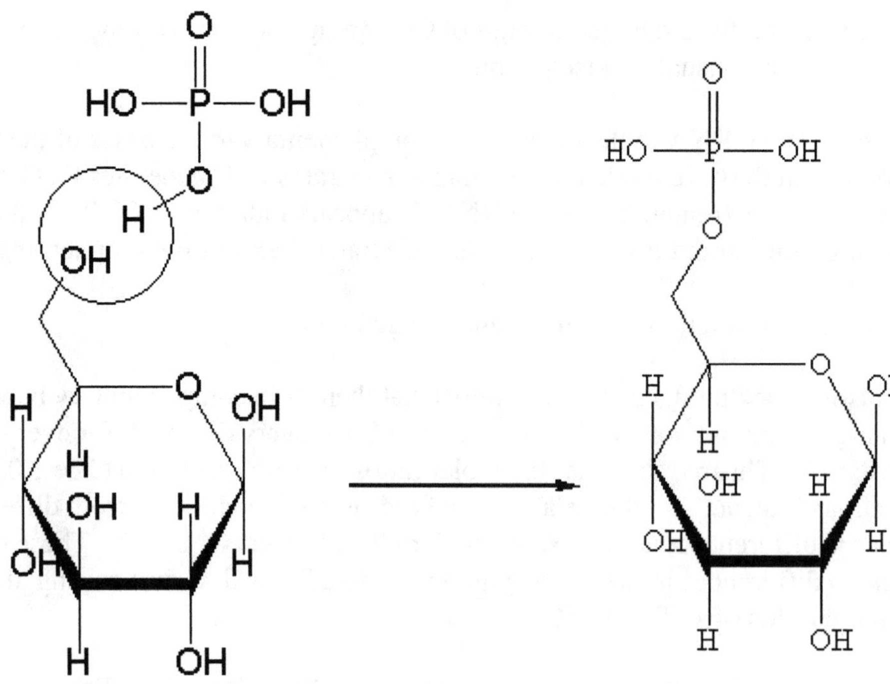

11.75 Polystyrene has the polyethylene backbone of carbon atoms, with a benzene ring attached to every second carbon atom. A divinylbenzene monomer reacts with the growing chain by means of one of its C=C double bonds, leaving the second C=C double bond available to cross-link by becoming incorporated into another polystyrene chain. Thus the benzene rings form "bridges" between polystyrene chains:

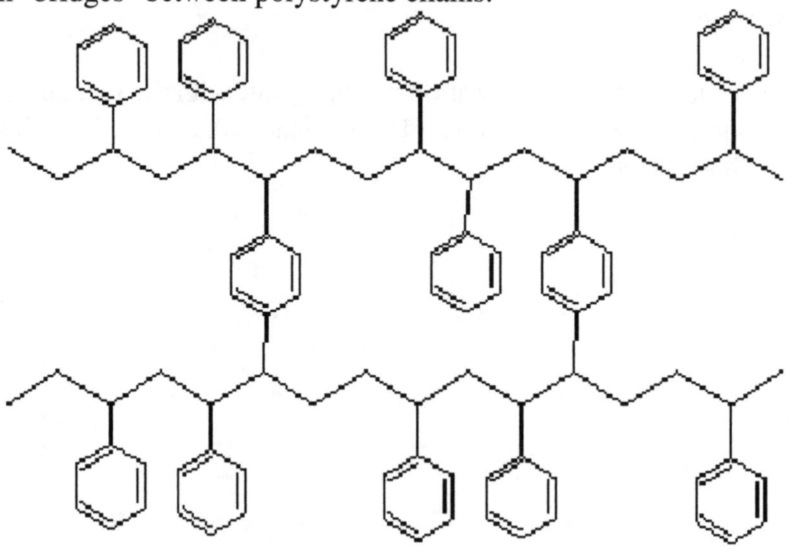

11.77 This problem is a "simple" stoichiometric problem that addresses the composition of a polymer. Use the percent by mass composition of copper to determine the mass of copper in 1 mol of the enzyme; convert to moles to obtain the molar ratio of copper to enzyme, which also gives the number of copper atoms per molecule:

$$64,000 \text{ g} \left(\frac{0.40\%}{100\%} \right) = 256 \text{ g of Cu in 1 mol of enzyme;}$$

$$256 \text{ g Cu} \left(\frac{1 \text{ mol}}{63.55 \text{ g}} \right) = 4.0 \text{ mol Cu/mol enzyme.}$$

There are 4 copper atoms in each molecule of this enzyme.

11.79 The description of the glucose-UDP molecule provides the information needed to construct it from the appropriate individual molecules. Uridine is uracil-ribose, and diphosphate indicates two phosphate units. Glucose is attached to the phosphate end via the specified –OH group:

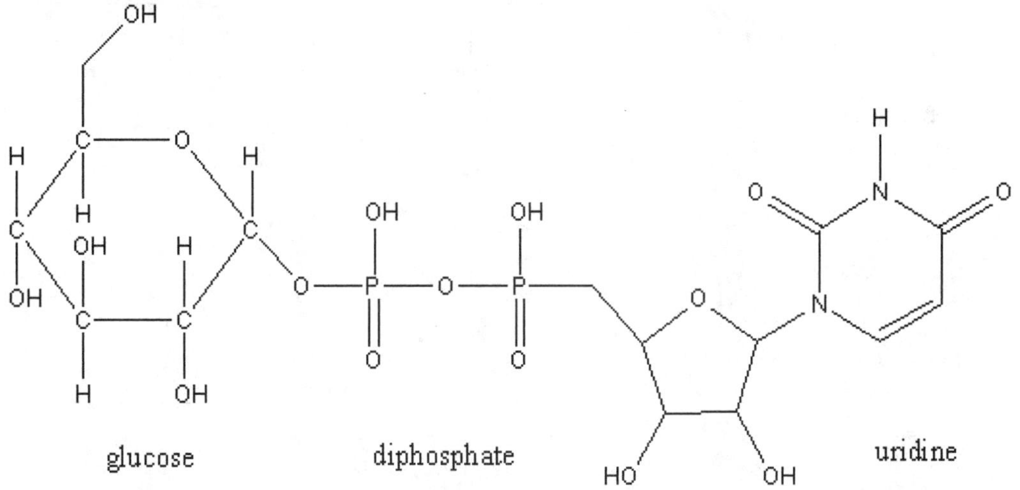

glucose   diphosphate   uridine

11.81 Trigonal planar geometry is best described by sp^2 hybrid orbitals. The simple Lewis structure of the peptide linkage indicates SN = 4, which normally requires sp^3 hybrid orbitals. The trigonal planar geometry indicates that the lone pair of electrons on the N atom is delocalized in a π bonding network, which can be illustrated using the Lewis structure of the gly-gly dimer. There is a resonance structure which places a π bond between C and N. Studies of bond lengths and strengths indicate that the peptide linkage is stronger than a single bond; this along with the trigonal planar geometry about the N atom confirms this bonding arrangement:

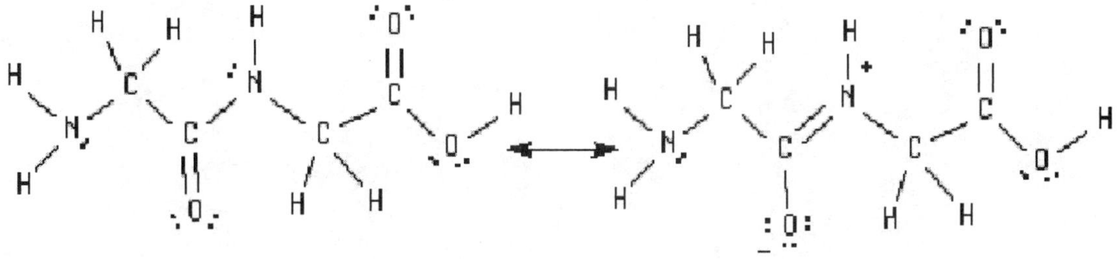

11.83 To draw the structure of a disaccharide, start with the appropriate monosaccharide and make the indicated connections. The structure of β-glucose appears in Figure 11-17. The linkage points are indicated **1** and **2**:

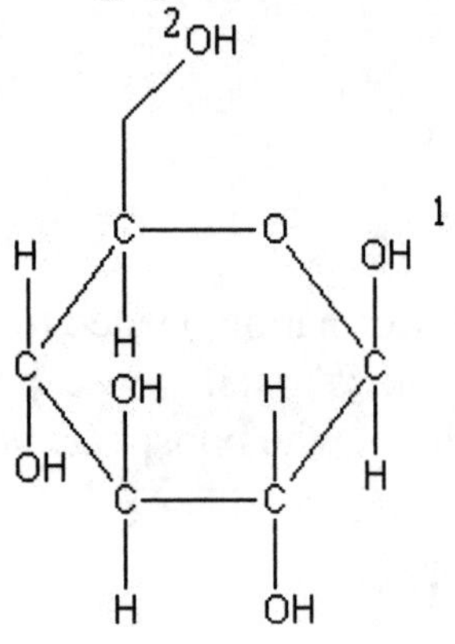

β-glucose

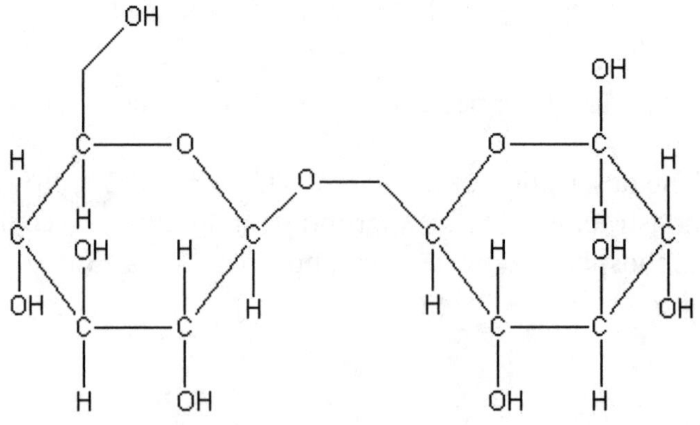

gentobiose

12.1   A system is isolated if it exchanges neither energy nor matter with its surroundings, closed if it exchanges energy but not matter, and open if it exchanges both:
(a) Open, energy and matter are constantly entering and leaving the system;
(b) (almost) isolated, because the thermos minimizes energy transfer to the coffee;
(c) open, energy and matter can flow in and out; and
(d) closed, the balloon isolates the helium from the outside.

12.3   State variables are properties that are independent of the past history of the system.
(a) not a state variable; (b) state variable; (c) state variable (mass is conserved); and
(d) not a state variable.

12.5   A system is any part of the universe that can be defined with specified boundaries. Among possible systems are the automobile and driver, the driver, the automobile, the cooling system, the engine, a cylinder, the battery, etc.

12.7   Calculate the temperature change resulting from an energy input using Equation 12-1 and C values from Table 12-1 of your textbook: $q = nC\Delta T$, which can be rearranged to

$$\Delta T = \frac{q}{nC}$$

(a) $n = \dfrac{m}{MM} = (10.0 \text{ g})\left(\dfrac{1 \text{ mol}}{26.98 \text{ g}}\right) = 0.3706 \text{ mol};$

$\Delta T = \dfrac{25.0 \text{ J}}{(0.3706 \text{ mol})(24.35 \text{ J mol}^{-1} \text{ K}^{-1})} = 2.77 \text{ K} = 2.77 \text{ °C}.$

$T_f = T_i + \Delta T = 15.0 + 2.77 = 17.8 \text{ °C}.$

(b) $n = \dfrac{m}{MM} = (25.0 \text{ g})\left(\dfrac{1 \text{ mol}}{26.98 \text{ g}}\right) = 0.9266 \text{ mol};$

$\Delta T = \dfrac{25.0 \text{ J}}{(0.9266 \text{ mol})(24.35 \text{ J mol}^{-1} \text{ K}^{-1})} = 1.108 \text{ K}.$

$T_f = T_i + \Delta T = 295 + 1.108 = 296 \text{ K}.$

(c) $n = \dfrac{m}{MM} = (25.0 \text{ g})\left(\dfrac{1 \text{ mol}}{107.9 \text{ g}}\right) = 0.2317 \text{ mol};$

$\Delta T = \dfrac{25.0 \text{ J}}{(0.2317 \text{ mol})(25.351 \text{ J mol}^{-1} \text{ K}^{-1})} = 4.256 \text{ K}.$

$T_f = T_i + \Delta T = 295 + 4.256 = 299 \text{ K}.$

(d) $n = \dfrac{m}{MM} = (25.0 \text{ g})\left(\dfrac{1 \text{ mol}}{18.02 \text{ g}}\right) = 1.387 \text{ mol};$

$\Delta T = \dfrac{25.0 \text{ J}}{(1.387 \text{ mol})(75.291 \text{ J mol}^{-1} \text{ K}^{-1})} = 0.2394 \text{ K} = 0.2394 \text{ °C}.$

$T_f = T_i + \Delta T = 22.0 + 0.2394 = 22.2\ °C.$

12.9 Calculate an energy change accompanying a temperature change using Equation 12-1 and C values from Table 12-1 of your textbook: $q = nC\Delta T$.
$\Delta T = 95.0 - 23.0 = 72.0\ °C = 72.0\ K;$

$$n_{kettle} = 1.35\ kg\left(\frac{1000\ g}{1\ kg}\right)\left(\frac{1\ mol}{55.85\ g}\right) = 24.2\ mol\ Fe$$

$q_{kettle} = (24.2\ mol)(25.10\ J\ mol^{-1}\ K^{-1})(72.0\ K) = 4.37\ x\ 10^4\ J;$

$$n_{water} = 2.75\ kg\left(\frac{1000\ g}{1\ kg}\right)\left(\frac{1\ mol}{18.02\ g}\right) = 153\ mol\ water$$

$q_{water} = (153\ mol)(75.291\ J\ mol^{-1}\ K^{-1})(72.0\ K) = 8.29\ x\ 10^5\ J.$

12.11 When two objects at different temperatures are placed in contact, energy flows from the warmer to the cooler object until the two objects are at the same temperature. Let that temperature be T. The energy lost by the warmer object equals the energy gained by the cooler object: $q_{cool} = -q_{warm}$, and $q = nC\Delta T$.

Water: $n = 37.5\ g\left(\frac{1\ mol}{18.02\ g}\right) = 2.081\ mol\ water;$

$q = (2.081\ mol)(75.351\ J\ mol^{-1}\ K^{-1})(T - 20.5\ °C) = (156.8\ J\ K^{-1})(T - 20.5\ °C).$

Coin: $n = (27.4\ g)\left(\frac{1\ mol}{107.9\ g}\right) = 0.2539\ mol\ Ag;$

$q = (0.2539\ mol)(25.351\ J\ mol^{-1}\ K^{-1})(T - 100.0\ °C) = (6.437\ J\ K^{-1})(T - 100.0\ °C).$
Substitute and solve for T:
$(156.8\ J\ K^{-1})(T - 20.5\ °C) = -(6.437\ J\ K^{-1})(T - 100.0\ °C).$
$156.8\ T - 3214.5 = -6.437\ T + 643.7;$
$163.2\ T = 3858.2,$ and
$T = 23.6\ °C.$

12.13 Calculate the energy change for a temperature change using Equation 12-1 and C values from Table 12-1 of your textbook: $q = nC\Delta T,$ and
$\Delta T = 87.6 - 21.5 = 66.1\ °C = 66.1\ K;$

$$n_{water} = \frac{m}{MM} = 475\ mL\left(\frac{1.00\ g}{1\ mL}\right)\left(\frac{1\ mol}{18.02\ g}\right) = 26.36\ mol\ ;$$

$$q_{water} = (26.36\ mol)\left(\frac{75.291\ J}{1\ mol\ K}\right)(66.1\ K) = 1.31\ x\ 10^5\ J = 131\ kJ.$$

12.15 Expansion work can be calculated using Equation 12-3, $w_{sys} = -P_{ext}\Delta V_{sys}$. The external pressure opposing inflation of a balloon is atmospheric pressure, 1.00 atm:
$V_f = 2.5\ L,\ V_i = 0.0\ L$
$\Delta V = 2.5\ L - 0.0\ L = 2.5\ L$

Chapter 12

$w_{sys} = -P_{ext}\Delta V_{sys} = -(1.00 \text{ atm})(2.5 \text{ L}) = -2.5 \text{ L atm}.$
Convert to joules:
$w_{sys} = -2.5 \text{ L atm}\left(\dfrac{101.325 \text{ J}}{1 \text{ L atm}}\right) = -2.5 \times 10^2 \text{ J}.$

12.17 To work this problem, use data in Chemistry and Life Box. First determine the energy difference(ΔE) between chicken and beef:

Chicken: $250 \text{ g}\left(\dfrac{6.0 \text{ kJ}}{1 \text{ g}}\right) = 1500 \text{ kJ};$

beef: $250 \text{ g}\left(\dfrac{16 \text{ kJ}}{1 \text{ g}}\right) = 4000 \text{ kJ};$

$\Delta E = 4000 \text{ kJ} - 1500 \text{ kJ} = 2500 \text{ kJ};$
According to data in Chemistry and Life Box, a 55-kg person walking 6.0 km/hr consumes 1090 kJ/hr:

$t = 2500 \text{ kJ}\left(\dfrac{1 \text{ hr}}{1090 \text{ kJ}}\right) = 2.3 \text{ hr}.$

Therefore, to consume the additional energy a person must walk:

$distance = 2.3 \text{ hr}\left(\dfrac{6.0 \text{ km}}{1 \text{ hr}}\right) = 1.4 \text{ km}.$

12.19 The heat capacity of a calorimeter can be found from energy and temperature data using Equation 12-5, $q_{calorimeter} = C_{cal}\,\Delta T.$
Here, the heat released by the combustion process is absorbed by the calorimeter so:

$q_{calorimeter} = -q_{glucose} = -1.7500 \text{ g}\left(\dfrac{-15.57 \text{ kJ}}{1 \text{ g}}\right) = 27.25 \text{ kJ}$

$\Delta T = 23.34 - 21.45 = 1.89 \text{ °C}$

$C_{cal} = \dfrac{q_{calorimeter}}{\Delta T} = \dfrac{27.25 \text{ kJ}}{1.89°\text{C}} = 14.4 \text{ kJ/°C}.$

12.21 (a) In a combustion reaction, the products are CO_2 and H_2O:
$$C_7H_6O_2 + O_2 \rightarrow CO_2 + H_2O \text{ (unbalanced)}$$

Follow standard procedures to balance the equation. Give CO_2 a coefficient of 7 to balance C, H_2O a coefficient of 3 to balance H:
$$C_7H_6O_2 + O_2 \rightarrow 7\,CO_2 + 3\,H_2O$$
$$7C + 6H + 4O \rightarrow 7C + 6H + 17O$$

Balance O by giving O_2 a coefficient of 15/2, then multiply by 2 to clear fractions:
$$2\,C_7H_6O_2 + 15\,O_2 \rightarrow 14\,CO_2 + 6\,H_2O$$

(b) Find energy per mole from energy per gram using the molar mass (122.1 g/mol):

$$\Delta E \text{ (kJ/mol)} = [\Delta E \text{ (kJ/g)}][MM \text{ (g/mol)}] = \left(\frac{-35.61\,\text{kJ}}{1.350\,\text{g}}\right)\left(\frac{122.1\,\text{g}}{1\,\text{mol}}\right) = -3.221 \times 10^3 \text{ kJ/mol}$$

(c) 15 moles O_2 is consumed for each 2 mol of benzoic acid, so the energy released per mol of O_2 is:

$$E = \left(\frac{-3.221 \times 10^3\,\text{kJ}}{1\,\text{mol acid}}\right)\left(\frac{2\,\text{mol acid}}{15\,\text{mol }O_2}\right) = -4.295 \times 10^2 \text{ kJ/mol.}$$

12.23 Calculate the energy released during combustion from the temperature increase and the total heat capacity of the calorimeter. Then convert to molar energy using molar mass.
$q = C\Delta T = (7.85 \text{ kJ/K})(302.04 \text{ K} - 297.65 \text{ K}) = 34.46 \text{ kJ}$
$\Delta E = -q = -34.46 \text{ kJ}$

$$\Delta E \text{ (kJ/mol)} = [\Delta E \text{ (kJ/g)}][MM \text{ (g/mol)}] = \left(\frac{-34.46\,\text{kJ}}{1.35\,\text{g}}\right)\left(\frac{194.2\,\text{g}}{1\,\text{mol}}\right) = -4.96 \times 10^3 \text{ kJ/mol}$$

12.25 Standard enthalpy changes are calculated from Equation 12-10 using standard enthalpies of formation, which can be found in Appendix D of your textbook:

$$\Delta H^{\circ}_{\text{reaction}} = \sum \text{coeff}_p\, \Delta H^{\circ}_f \text{ (products)} - \sum \text{coeff}_r\, \Delta H^{\circ}_f \text{(reactants)}$$

(a) $\Delta H^{\circ}_{\text{reaction}} = [2 \text{ mol}(-393.5 \text{ kJ/mol}) + 2 \text{ mol}(-285.83 \text{ kJ/mol})]$
$- [1 \text{ mol}(52.4 \text{ kJ/mol}) + 3 \text{ mol}(0 \text{ kJ/mol})] = -1411.1 \text{ kJ};$
(b) $\Delta H^{\circ}_{\text{reaction}} = 1 \text{ mol }(0 \text{ kJ/mol}) + 3 \text{ mol}(0 \text{ kJ/mol}) - 2 \text{ mol}(-45.9 \text{ kJ/mol}) = 91.8 \text{ kJ};$
(c) $\Delta H^{\circ}_{\text{reaction}} = [1 \text{ mol}(-2984.0 \text{ kJ/mol}) + 5 \text{ mol}(0\text{kJ/mol})]$
$- [5 \text{ mol}(-277.4 \text{ kJ/mol}) + 4 \text{ mol}(0 \text{ kJ/mol})] = -1597.0 \text{ kJ};$
(d) $\Delta H^{\circ}_{\text{reaction}} = [1 \text{ mol}(-910.7 \text{ kJ/mol}) + 4 \text{ mol}(-92.3 \text{ kJ/mol})]$
$- [1\text{mol}(-687.0 \text{ kJ/mol}) + 2 \text{ mol}(-285.83 \text{ kJ/mol})] = -21.2 \text{ kJ.}$

12.27 Reaction energies are related to reaction enthalpies (see Prob. 12.25) through Equation 12-9:
$$\Delta H_{\text{reaction}} \cong \Delta E_{\text{reaction}} + \Delta(nRT)_{\text{gases}}$$
Since temperature does not change, we can rearrange the equation to:
$$\Delta E_{\text{reaction}} \cong \Delta H_{\text{reaction}} - RT\Delta(n)_{(\text{gases})}$$

Begin by calculating the change in moles of *gases*, then use the rearranged equation to determine the $\Delta E_{\text{reaction}}$
(a) $\Delta n_{\text{gases}} = 2 - (3 + 1) = -2 \text{ mol};$

$$\Delta E_{\text{reaction}} \cong -1411.1 \text{ kJ} - (-2 \text{ mol})\left(\frac{8.314\,\text{J}}{1\,\text{mol K}}\right)(298\text{K})\left(\frac{10^{-3}\,\text{kJ}}{1\,\text{J}}\right)$$
$$= -1411.1 + 4.96 = -1406.1 \text{ kJ};$$

(b) $\Delta n_{\text{gases}} = 3 + 1 - 2 = 2 \text{ mol};$

$$\Delta E_{\text{reaction}} \cong 91.8 \text{ kJ} - (2 \text{ mol})\left(\frac{8.314 \text{ J}}{1 \text{ mol K}}\right)(298\text{K})\left(\frac{10^{-3}\text{kJ}}{1 \text{ J}}\right) = 91.8 - 4.96 = 86.8 \text{ kJ};$$

(c) all compounds are solid, $\Delta n_{\text{gases}} = 0$ mol;

$\Delta E_{\text{reaction}} \cong -1597.0$ kJ;

(d) $\Delta n_{\text{gases}} = 4$ mol;

$$\Delta E_{\text{reaction}} \cong -21.2 \text{ kJ} - (4 \text{ mol})\left(\frac{8.314 \text{ J}}{1 \text{ mol K}}\right)(298\text{K})\left(\frac{10^{-3}\text{kJ}}{1 \text{ J}}\right) = -21.2 + 9.91 = -11.3 \text{ kJ}$$

12.29   A formation reaction has elements in their standard states as reactants and 1 mol of a single product:

(a) $3 \text{ K(s)} + \text{P(s)} + 2 \text{ O}_2(g) \rightarrow \text{K}_3\text{PO}_4(s)$;

(b) $2 \text{ C(graphite)} + 2 \text{ H}_2(g) + \text{O}_2(g) \rightarrow \text{CH}_3\text{CO}_2\text{H(l)}$;

(c) $3 \text{ C(graphite)} + 9/2 \text{ H}_2(g) + 1/2 \text{ N}_2(g) \rightarrow (\text{CH}_3)_3\text{N(g)}$;

(d) $2 \text{ Al(s)} + 3/2 \text{ O}_2(g) \rightarrow \text{Al}_2\text{O}_3(s)$.

12.31   Standard enthalpy changes are calculated from Equation 12-10 using standard enthalpies of formation, which can be found in Appendix D of your textbook:

$$\Delta H^\circ_{\text{reaction}} = \Sigma \text{ coeff}_p \Delta H^\circ_f \text{ (products)} - \Sigma \text{ coeff}_r \Delta H^\circ_f \text{ (reactants)}$$

(a) $\Delta H^\circ_{\text{reaction}} = [4 \text{ mol}(91.3 \text{ kJ/mol}) + 6 \text{ mol}(-285.83 \text{ kJ/mol})]$
$\qquad\qquad\qquad - [4 \text{ mol}(-45.9 \text{ kJ/mol}) + 5 \text{ mol}(0 \text{ kJ/mol})] = -1166.2 \text{ kJ}$;

(b) $\Delta H^\circ_{\text{reaction}} = [2 \text{ mol}(0 \text{ kJ/mol}) + 6 \text{ mol}(-285.83 \text{ kJ/mol})]$
$\qquad\qquad\qquad - [4 \text{ mol}(-45.9 \text{ kJ/mol}) + 3 \text{ mol}(0 \text{ kJ/mol})] = -1531.4 \text{ kJ}$.

12.33   Calculate the energy released during the dissolving process from the temperature increase and the total heat capacity of the calorimeter. Then convert to molar energy using molar mass.

Assume that the heat capacity of the calorimeter is the heat capacity of its water contents:

$$C_{\text{cal}} = 110.0 \text{ g}\left(\frac{1 \text{ mol}}{18.02 \text{ g}}\right)\left(\frac{75.291 \text{ J}}{1 \text{ mol K}}\right) = 459.6 \text{ J/K} = 459.6 \text{ J/°C};$$

$$q = C_{\text{cal}}\Delta T = (459.6 \text{ J/°C})(29.7°\text{C} - 22.0°\text{C}) = 3.54 \times 10^3 \text{ J};$$

remember that the heat absorbed by the calorimeter is the heat lost by the solution process:

$$\Delta H = \frac{-q}{m} = \frac{-3.54 \times 10^3 \text{ J}}{4.75 \text{ g}} = -745 \text{ J/g}$$

$$\Delta H \text{ (J/mol)} = [\Delta H \text{ (J/g)}][MM \text{ (g/mol)}] = \left(\frac{-745 \text{ J}}{1 \text{ g}}\right)\left(\frac{111.0 \text{ g}}{1 \text{ mol}}\right) = -8.3 \times 10^4 \text{ J/mol}.$$

12.35 (a) The positive value for ΔH signals that reaction to form solid $MgSO_4$ is endothermic. Dissolving, the reverse reaction, is exothermic, releasing energy, which is absorbed by the water.

(b) Use the molar enthalpy and number of moles dissolving to calculate the energy released:

$$n = \frac{m}{MM} = 2.55 \text{ g}\left(\frac{1 \text{ mol}}{120.367 \text{ g}}\right) = 2.119 \times 10^{-2} \text{ mol } MgSO_4;$$

$$q_{water} = -q_{salt} = -n \, \Delta H = -(2.119 \times 10^{-2} \text{ mol})\left(\frac{-91.3 \text{ kJ}}{1 \text{ mol}}\right) = 1.93 \text{ kJ}.$$

(c) Use Equation 12-1, $q = nC\Delta T$, to determine the temperature change:

$$\Delta T = \frac{q}{nC};$$

$$n = \frac{m}{MM} = (5.00 \times 10^2 \text{ mL})\left(\frac{1.00 \text{ g}}{1 \text{ mL}}\right)\left(\frac{1 \text{ mol}}{18.02 \text{ g}}\right) = 27.747 \text{ mol};$$

$$\Delta T = \frac{1.93 \times 10^3 \text{ J}}{(27.747 \text{ mol})(75.291 \text{ J mol}^{-1} \text{ K}^{-1})} = 0.924 \text{ K}$$

12.37 The formation of a solid from ions in solution involves two opposing energy effects. Energy is released because of the coulombic attraction between cations and anions, but energy is absorbed to overcome the ion-dipole attractions between ions and water molecules. Solid formation is endothermic when the sum of all ion-dipole attractions in solution is greater than the ion-ion interactions in the solid.

12.39 The heats of phase changes indicate the strengths of intermolecular forces; the larger the intermolecular forces, the larger the heat of the phase change.
(a) Methane has a lower heat of vaporization than ethane because, being a smaller molecule, it has smaller dispersion forces.
(b) Ethanol has a significantly higher heat of vaporization than diethyl ether because of strong hydrogen bonding.
(c) Argon (18 electrons) has a higher heat of fusion than methane (10 electrons) because it has a higher polarizability due to its larger number of electrons.

12.41 Because E is a state function, the energy released (ΔE) when one gram of gasoline burns is the same regardless of the conditions. Work (w) is done when an automobile accelerates but no work is done when an automobile idles. Thus, $q_{idle} = \Delta E$ and $q_{accel} = \Delta E - w_{accel}$, so more heat must be removed under idling conditions (this is part of the reason why automobiles tend to overheat in traffic jams).

12.43 To work a problem involving heat transfers, it is useful to set up a block diagram illustrating the process. In this problem, a copper block transfers energy to ice:

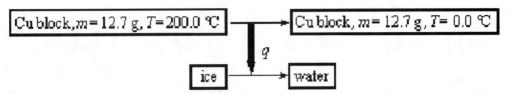

Thus, $q_{ice} = -q_{Cu}$,

$q_{Cu} = n_{Cu} C \Delta T$, and $q_{ice} = n_{ice} \Delta H_{fus}$.

Substituting gives:

$n_{ice} \Delta H_f = -n_{Cu} C \Delta T$

Here are the data needed for the calculation:

$$n_{Cu} = \frac{m}{MM} = 12.7 \text{ g} \left(\frac{1 \text{ mol}}{63.546 \text{ g}} \right) = 0.1999 \text{ mol};$$

$C = 24.435$ J/mol K,

$\Delta T = (0.0 \,^{\circ}\text{C} - 200.0 \,^{\circ}\text{C}) = -200 \,^{\circ}\text{C} = -200$ K;

$\Delta H_f = 6.01$ kJ/mol $= 6.01 \times 10^3$ J/mol;

Substitute and solve for the amount of ice that melts:

$$n_{ice} = \frac{-n_{Cu} C \Delta T}{\Delta H_f} = -\frac{(0.1999 \text{ mol})(24.435 \text{ J mol}^{-1}\text{ K}^{-1})(-200\text{K})}{6.01 \times 10^3 \text{ J mol}^{-1}} = 0.1625 \text{ mol}$$

Finally, convert to mass:

$$m_{ice} = n \, MM = 0.1625 \text{ mol} \left(\frac{18.02 \text{ g}}{1 \text{ mol}} \right) = 2.93 \text{ g of ice melts.}$$

12.45   (a) In a combustion reaction, a substance reacts with molecular oxygen to form CO_2 and H_2O:

$$C_6H_{12}O_6 + O_2 \rightarrow CO_2 + H_2O$$

Balance the equation using standard procedures. Give CO_2 a coefficient of 6 and H_2O a coefficient of 6 to balance C and H;

$$C_6H_{12}O_6 + O_2 \rightarrow 6\,CO_2 + 6\,H_2O$$
$$6C + 12H + 8\,O \rightarrow 6C + 12H + 18O$$

then give O_2 a coefficient of 6 to balance O:

$$C_6H_{12}O_6 + 6\,O_{2(g)} \rightarrow 6\,CO_{2(g)} + 6\,H_2O_{(l)}$$

(b) To determine the molar heat of combustion, multiply the heat released (negative, indicating an exothermic reaction) in burning one gram by the molar mass:

$$\Delta H_{molar} = \left(\frac{-15.7 \text{ kJ}}{1 \text{ g}} \right) \left(\frac{180.15 \text{g}}{1 \text{ mol}} \right) = -2.83 \times 10^3 \text{ kJ/mol};$$

(c) Use the molar heat of combustion along with Equation 12-10 and data from Appendix D to determine the heat of formation of glucose:

$$\Delta H^{\circ}_{reaction} = \sum coeff_p \, \Delta H^{\circ}_f \, (products) - \sum coeff_r \, \Delta H^{\circ}_f \, (reactants)$$

$$-2.83 \times 10^3 \text{ kJ/mol} = [6 \text{ mol}(-393.5 \text{ kJ/mol}) + 6 \text{ mol}(-285.83 \text{ kJ/mol})]$$

$$- [6 \text{ mol}(0 \text{ kJ/mol}) + \Delta H^{\circ}_f (glucose)]$$

$$\Delta H^{\circ}_f (glucose) = (2.83 \times 10^3 - 1.715 \times 10^3 - 2.361 \times 10^3) \text{ kJ/mol} = -1.25 \times 10^3 \text{ kJ/mol}$$

12.47 Living organisms require both matter and energy to carry out their life processes. An isolated system receives neither matter nor energy from its surroundings, so a living organism that is isolated will quickly die.

12.49 A molar heat capacity can be calculated from a temperature change using $q = n \, C\Delta T$:
$\Delta T = 299.6 \text{ K} - 280.0 \text{ K} = 19.6 \text{ K}$;

$$n = 52.5 \text{ g} \left(\frac{1 \text{ mol}}{207.2 \text{ g}} \right) = 0.2534 \text{ mol};$$

$$C = \frac{q}{n\Delta T} = \frac{(100.0 \text{ J})}{(0.2534 \text{ mol})(19.6 \text{ K})} = 20.1 \text{ J/mol K}.$$

12.51 In this process, neither P nor V is constant, so we cannot calculate q directly. Instead, we can calculate w and ΔE. An ideal gas has no intermolecular forces, so its energy does not change when it expands or contracts. Thus, for $\Delta T = 0$, $\Delta E = 0$.
Calculate w using $w = -P_{ext} \Delta V$:
$w = -(4.00 \text{ atm})(20.0 \text{ L} - 30.0 \text{ L}) = 40.0 \text{ L atm}$;
Convert to joules, recalling that 1 L atm = 101.325 J:

$$w = (40.0 \text{ L atm}) \left(\frac{101.325 \text{ J}}{1 \text{ L atm}} \right) = 4.05 \times 10^3 \text{ J}$$

$q = \Delta E - w = 0 - 4.05 \times 10^3 \text{ J} = -4.05 \times 10^3 \text{ J}$.

12.53 Standard enthalpy changes are calculated from Equation 12-10 using standard enthalpies of formation, which can be found in Appendix D of your textbook:

$$\Delta H^{\circ}_{reaction} = \sum coeff_p \, \Delta H^{\circ}_f \, (products) - \sum coeff_r \, \Delta H^{\circ}_f \, (reactants)$$

(a) $\Delta H^{\circ}_{reaction} = 2 \text{ mol}(-395.7 \text{ kJ/mol})$
$- [2 \text{ mol}(-296.8 \text{ kJ/mol}) + 1 \text{ mol}(0 \text{ kJ/mol})] = -197.8 \text{ kJ}$;
(b) $\Delta H^{\circ}_{reaction} = 1 \text{ mol}(11.1 \text{ kJ/mol}) - 2 \text{ mol}(33.2 \text{ kJ/mol}) = -55.3 \text{ kJ}$;
(c) $\Delta H^{\circ}_{reaction} = 1 \text{ mol}(-1675.7 \text{ kJ/mol}) + 2 \text{ mol}(0 \text{ kJ/mol})$
$- [1 \text{ mol}(-824.2 \text{ kJ/mol}) + 2 \text{ mol}(0 \text{ kJ/mol})] = -851.5 \text{ kJ}$.

12.55 An enthalpy of formation can be calculated from the heat of a reaction provided all other enthalpies of formation are known. For this reaction, formation enthalpies of three of the four reagents appear in Appendix D of your textbook:

$$\Delta H^{\circ}_{reaction} = \sum coeff_p \, \Delta H^{\circ}_f \, (products) - \sum coeff_r \, \Delta H^{\circ}_f \, (reactants)$$

$$-3677 \text{ kJ} = 1 \text{ mol}(-2984.0 \text{ kJ/mol}) + 3 \text{ mol}(-296.8 \text{ kJ/mol})$$

$$-[\Delta H_f^o(P_4S_3) + 8\ mol(0\ kJ/mol)]$$

$$\Delta H_f^o(P_4S_3) = 3677\ kJ - 2984.0\ kJ - 890.4\ kJ = -197\ kJ$$

12.57 To work a problem involving heat transfers, it is useful to set up a block diagram illustrating the process. In this problem, the unknown metal transfers energy to water:

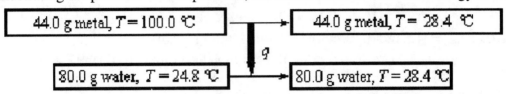

Thus, $q_{metal} = -q_{water} = -(nC\Delta T)_{water}$

$$n_{water} = \frac{m}{MM} = 80.0\ g\left(\frac{1\ mol}{18.02\ g}\right) = 4.44\ mol\ water;$$

$C = 75.291\ J/mol\ °C$,
and $\Delta T = (28.4 - 24.8) = 3.6\ °C$;

$$q_{metal} = -q_{water} = -(4.44\ mol)\left(\frac{75.291\ J}{1\ mol\ °C}\right)(3.6\ °C) = -1203\ J.$$

Without knowing the identity of the metal, we cannot determine n_{metal}, but heat capacity values are provided in units of J/g K, so we can use $q = mC\Delta T$:
$\Delta T = 28.4 - 100.0 = -71.6\ °C = -71.6\ K$;

$$C = \frac{q}{m\Delta T} = \frac{(-1203\ J)}{(44.0\ g)(-71.6\ K)} = 0.382\ J/g\ K$$

This matches the heat capacity of Cu (0.385 J/gK), so the metal is copper.

12.59 The process shown is condensation of a gas to form a solid, the reverse of sublimation. Sublimation always is endothermic, as energy must be provided to overcome intermolecular forces of attraction; thus the depicted process is exothermic.
(a) ΔH_{sys} is negative (exothermic process);
(b) ΔE_{surr} is positive (surroundings must absorb the energy released in the process);
(c) $\Delta E_{universe} = 0$ (total energy always is conserved).

12.61 The enthalpy of a reaction can be calculated from Equation 12-10 and standard heats of formation (see Appendix D):

$$\Delta H^o_{reaction} = \sum coeff_p\ \Delta H^o_f\ (products) - \sum coeff_r\ \Delta H^o_f\ (reactants)$$
For this "reaction," the calculation is simple:
$$\Delta H^o_{reaction} = 1\ mol(-241.83\ kJ/mol) - 1\ mol(-285.83\ kJ/mol) = 44.00\ kJ$$
Compare this value with $\Delta H_{vap} = 40.79\ kJ/mol$. The values differ because ΔH_{vap} refers to vaporization of liquid water at the normal boiling point, 373 K, while $\Delta H^o_{reaction}$ refers to vaporization of liquid water under standard thermodynamic conditions, 298 K.

12.63 The enthalpy of a reaction can be calculated from Equation 12-10 and standard heats of formation (see Appendix D):

$$\Delta H_{reaction} = \Sigma\ coeff_p\ \Delta H°_f\ (products) - \Sigma\ coeff_r\ \Delta H°_f\ (reactants)$$

(a) The question asks for ΔH when 1 mol of hydrazine burns, so the appropriate reaction is:

$$N_2H_{4(l)} + 1/2\ N_2O_{4(g)} \rightarrow 3/2\ N_{2(g)} + 2\ H_2O_{(g)}$$

$\Delta H_{reaction} = [1.5\ mol(0\ kJ/mol) + 2\ mol(-241.83\ kJ/mol)]$
$\qquad\qquad - [1\ mol(50.6\ kJ/mol) + 0.5\ mol(11.1\ kJ/mol)] = -539.8\ kJ/mol$

(b) When hydrazine burns in oxygen, the reaction is:

$$N_2H_{4(l)} + O_{2(g)} \rightarrow N_{2(g)} + 2\ H_2O_{(g)}$$

$\Delta H_{reaction} = [1\ mol(0\ kJ/mol) + 2\ mol(-241.83\ kJ/mol)]$
$\qquad\qquad - [1\ mol(50.6\ kJ/mol) + 1\ mol(0\ kJ/mol)] = -534.3\ kJ/mol$

The reaction with dinitrogen tetroxide is slightly more exothermic because that compound is slightly less stable than the elements from which it forms. Dinitrogen tetroxide is used in preference to molecular oxygen because the reaction occurs on contact, while oxygen would require an ignition device.

12.65 The difference between $\Delta H_{reaction}$ and $\Delta E_{reaction}$ can be estimated using Equation 12-9:

$\Delta H_{reaction} \cong \Delta E_{reaction} + \Delta(nRT)_{gases}$ so $\Delta H_{reaction} - \Delta E_{reaction} \cong \Delta(nRT)_{(gases)}$

(a) $\Delta n_{gases} = 0$, so $\Delta H_{reaction} - \Delta E_{reaction} \cong 0$;

(b) $\Delta n_{gases} = 1\ mol + 1\ mol - 0\ mol = 2\ mol$ gases;

$$\Delta H_{reaction} - \Delta E_{reaction} \cong (2\ mol)\left(\frac{8.314\ J}{1\ mol\ K}\right)(298\ K) = 4.96\ x\ 10^3\ J;$$

(c) The reaction is $C_4H_9OH_{(l)} + 6\ O_{2(g)} \rightarrow 4\ CO_{2(g)} + 5\ H_2O_{(l)}$;

$\Delta n_{gases} = 4\ mol - 6\ mol = -2\ mol$;

$$\Delta H_{reaction} - \Delta E_{reaction} \cong -(2\ mol)\frac{8.314\ J}{1\ mol\ K}(298\ K) = -4.96\ x\ 10^3\ J.$$

12.67 Use the definition of each type of process to identify the reaction associated with each process:

(a) $I_{2(s)} \rightarrow I_{2(g)}$;

(b) $1/2\ I_{2(s)} \rightarrow I_{(g)}$;

(c) $2\ C_{(graphite)} + 3/2\ H_{2(g)} + 1/2\ Cl_{2(g)} \rightarrow C_2H_3Cl_{(g)}$;

(d) $Na_2SO_{4(s)} \rightarrow 2\ Na^+_{(aq)} + SO_4^{2-}_{(aq)}$.

12.69 (a) This is a cooling process, for which $q = nC\Delta T$:

$$n = \frac{m}{MM} = (2.50 \text{ g})\left(\frac{1 \text{ mol}}{18.02 \text{ g}}\right) = 0.1387 \text{ mol};$$

$\Delta T = (37.5 - 100.0) \text{ °C} = -62.5 \text{ °C} = -62.5 \text{ K};$

$$q = (0.1387 \text{ mol})\left(\frac{75.291 \text{ J}}{1 \text{ mol °C}}\right)(-62.5 \text{ K}) = -653 \text{ J} .$$

(b) This is a phase change, for which $q = n\,\Delta H_{\text{phase}}$, followed by the same cooling process as in part (a):

$$q = (0.1387 \text{ mol})\left(\frac{-40.79 \text{ kJ}}{1 \text{ mol}}\right)\left(\frac{10^3 \text{ J}}{1 \text{ kJ}}\right) - 653 \text{ J} = -5658 \text{ J} - 653 \text{ J} = -6.31 \times 10^3 \text{ J}.$$

Steam at 100.0 °C releases nearly ten times as much heat as boiling water.

12.71 Compression work can be calculated using Equation 12-3, $w_{\text{sys}} = -P_{\text{ext}}\Delta V_{\text{sys}}$. The external pressure is the pressure exerted by the compressor, 15.0 atm. The air is initially at 1.00 atm and occupies a volume that can be calculated using $P_f V_f = P_i V_i$:

$$V_i = \frac{P_f V_f}{P_i} = \frac{(15.0 \text{ atm})(30.0 \text{ L})}{(1.00 \text{ atm})} = 450 \text{ L}$$

$\Delta V = V_f - V_i = 30.0 \text{ L} - 450 \text{ L} = -420 \text{ L};$

$w_{\text{sys}} = -P_{\text{ext}}\Delta V_{\text{sys}} = -(15.0 \text{ atm})(-420 \text{ L}) = 6.30 \times 10^3 \text{ L atm}.$
Convert to joules:

$$w_{\text{sys}} = (6.30 \times 10^3 \text{ L atm})\left(\frac{101.325 \text{ J}}{1 \text{ L atm}}\right) = 6.38 \times 10^5 \text{ kg m}^2/\text{s}^2 = 638 \text{ kJ}.$$

12.73 (a)The energy required in a heating process is calculated from Equation 12-1:

$$q = \frac{nC}{T} :$$

$$n_{\text{water}} = \frac{m}{MM} = (155 \text{ m}^3)\left(\frac{10^2 \text{ cm}}{1 \text{ m}}\right)^3\left(\frac{1.00 \text{ g}}{1 \text{ cm}^3}\right)\left(\frac{1 \text{ mol}}{18.02 \text{ g}}\right) = 8.60 \times 10^6 \text{ mol};$$

$C = 75.291 \text{ J/mol K},$
$\Delta T = (30 - 20) = 10\text{°C} = 10 \text{ K};$

$$q = (8.60 \times 10^6 \text{ mol})\left(\frac{75.291 \text{ J}}{1 \text{ mol K}}\right)(10 \text{ K}) = 6.48 \times 10^9 \text{ J}$$

(b) If heating is 80% efficient, the heat required is $(6.48 \times 10^9 \text{ J})\left(\frac{100\%}{80\%}\right) = 8.1 \times 10^9 \text{ J};$

Given that the heat of combustion of methane is -803 kJ/mol, the amount of methane

required is $(8.1 \times 10^9 \text{ J})\left(\frac{10^{-3} \text{ kJ}}{1 \text{ J}}\right)\left(\frac{1 \text{ mol}}{803 \text{ kJ}}\right) = 1.0 \times 10^4 \text{ mol};$

Multiply by MM to convert to grams: $(1.0 \times 10^4 \text{ mol})\left(\dfrac{16.04 \text{ g}}{1 \text{ mol}}\right) = 1.6 \times 10^5$ g methane

12.75 To work a problem involving heat transfers, it is useful to set up a block diagram illustrating the process. In this problem, a silver spoon absorbs energy from coffee (water):

Thus, $q_{\text{water}} = -q_{\text{Ag}}$,

$q_{\text{Ag}} = (nC_{\text{Ag}}\Delta T)_{\text{Ag}}$, and $q_{\text{water}} = (nC_{\text{H}_2\text{O}}\Delta T)_{\text{water}}$

For Ag,

$n = \dfrac{m}{MM} = 99 \text{ g}\left(\dfrac{1 \text{ mol}}{107.9 \text{ g}}\right) = 0.918$ mol Ag;

$C_{\text{Ag}} = 25.351$ J/mol K,

and $\Delta T = (x - 280)$ K;

For water,

$n = \dfrac{m}{MM} = 205 \text{ mL}\left(\dfrac{1.00 \text{ g}}{1 \text{ mL}}\right)\left(\dfrac{1 \text{ mol}}{18.02 \text{ g}}\right) = 11.38$ mol H_2O;

$C_{\text{H2O}} = 75.291$ J/mol K,

and $\Delta T = (x - 350)$ K;

Substitute and solve for x:

$(0.918 \text{ mol})\left(\dfrac{25.351 \text{ J}}{1 \text{ mol K}}\right)(x - 280) \text{ K} = -(11.38 \text{ mol})\left(\dfrac{75.291 \text{ J}}{1 \text{ mol K}}\right)(x - 350) \text{ K};$

$23.27 \, x - 6516 = -856.8 \, x + 299880,$

$880.1 \, x = 306390$

$x = 348$ K.

The final temperature of the coffee is 348 K.

To determine whether an Al spoon would be more or less effective, compare nC for the two spoons.

Al: $= 99 \text{ g}\left(\dfrac{1 \text{ mol}}{26.982 \text{ g}}\right)\left(\dfrac{24.35 \text{ J}}{1 \text{ mol K}}\right) = 89$ J/K;

Ag: $= 99 \text{ g}\left(\dfrac{1 \text{ mol}}{107.9 \text{ g}}\right)\left(\dfrac{25.351 \text{ J}}{1 \text{ mol K}}\right) = 23$ J/K;

An Al spoon requires more energy per degree temperature increase, so an Al spoon would be more effective in cooling the coffee.

12.77 The energy required in a heating process is calculated from Equation 12-1: $q = nC\Delta T$.

To determine q, we must first compute the amounts of N_2 and O_2 in the room, using the ideal gas equation: $n = \dfrac{PV}{RT}$ (assume no air escapes the room as it is heated):

$$V = (3.0 \text{ m})(5.0 \text{ m})(4.0 \text{ m})\left(\frac{10^2 \text{ cm}}{1 \text{ m}}\right)^3\left(\frac{1 \text{ L}}{10^3 \text{ cm}^3}\right) = 6.0 \times 10^4 \text{ L};$$

$p_{N_2} = (1 \text{ atm})(0.78) = 0.78 \text{ atm};$

$p_{O_2} = (1 \text{ atm})(0.22) = 0.22 \text{ atm};$

$T = 15 + 273 = 288 \text{ K};$

$$n_{N_2} = \frac{(0.78 \text{ atm})(6.0 \times 10^4 \text{ L})}{(0.08206 \frac{\text{L atm}}{\text{mol K}})(288 \text{ K})} = 1.98 \times 10^3 \text{ mol}$$

$$n_{O_2} = \frac{(0.22 \text{ atm})(6.0 \times 10^4 \text{ L})}{(0.08206 \frac{\text{L atm}}{\text{mol K}})(288 \text{ K})} = 0.559 \times 10^3 \text{ mol}$$

$\Delta T = 25 - 15 = 10 \text{ °C} = 10 \text{ K}.$

$$q_{N_2} = (1.98 \times 10^3 \text{ mol})\left(\frac{29.125 \text{ J}}{1 \text{ mol K}}\right)(10 \text{ K}) = 5.77 \times 10^5 \text{ J}$$

$$q_{O_2} = (0.559 \times 10^3 \text{ mol})\left(\frac{29.355 \text{ J}}{1 \text{ mol K}}\right)(10 \text{ K}) = 1.64 \times 10^5 \text{ J}$$

$q_{tot} = q_{N_2} + q_{O_2} = 5.77 \times 10^5 \text{ J} + 1.64 \times 10^5 \text{ J} = 7.4 \times 10^2 \text{ kJ}.$ (Value only has two sig. figs. because the gas pressures are known to only two sig. figs.)

To determine the amount of methane required, we need the heat of combustion of methane, which can be calculated from standard heats of formation using Equation 12-10 using data from Appendix D. However, the value is given in Problem 12.73: $\Delta H_{comb} = -803 \text{ kJ/mol}.$

Divide the energy required by the heat of combustion to determine amount of methane required: $(7.4 \times 10^2 \text{ kJ})\left(\dfrac{1 \text{ mol}}{803 \text{ kJ}}\right) = 0.92 \text{ mol}.$

Multiply by molar mass to convert to grams: $0.92 \text{ mol}\left(\dfrac{16.04 \text{ g}}{1 \text{ mol}}\right) = 15 \text{ g } CH_4$ required.

12.79 The process shows an expansion at constant temperature. As a gas expands, it does work on its surroundings, but if temperature remains constant, the energy of the gas does not change:
(a) w_{sys} is negative (system does work on surroundings);
(b) $\Delta E_{surr} = 0$ because $\Delta E_{sys} = 0$ and total energy is conserved;
(c) q_{sys} is positive because w_{sys} is negative and $\Delta E_{sys} = 0$.

12.81 First calculate the heat required to warm the bear. Then calculate the heat of combustion of arachadonic acid. Finally, determine the amount of arachidonic acid required to do the job:

$$q = mC\Delta T = 500 \text{ kg} \left(\frac{10^3 \text{ g}}{1 \text{ kg}}\right)\left(\frac{4.184 \text{ J}}{1 \text{ g °C}}\right)(25 \text{ °C} - 5 \text{ °C}) = 4.18 \times 10^7 \text{ J};$$

$$\Delta H_{comb} = [20 \text{ mol}(-393.5 \text{ kJ/mol}) + 16 \text{ mol}(-285.8 \text{ kJ/mol})]$$
$$- [1 \text{ mol}(-636 \text{ kJ/mol}) + 27 \text{ mol}(0 \text{ kJ/mol})]$$

$$\Delta H_{comb} = -1.18 \times 10^4 \text{ kJ}$$

$$\text{Mass required} = 4.18 \times 10^7 \text{ J}\left(\frac{1 \text{ kJ}}{10^3 \text{ J}}\right)\left(\frac{1 \text{ mol}}{1.18 \times 10^4 \text{ kJ}}\right)\left(\frac{304.45 \text{ g}}{1 \text{ mol}}\right) = 1.1 \times 10^3 \text{ g}.$$

12.83 To work a problem involving heat transfers, it is useful to set up a block diagram illustrating the process. In this problem, a copper block transfers energy to water:

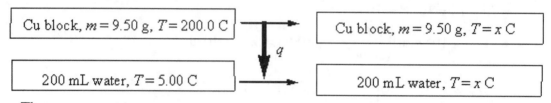

Thus, $q_{water} = -q_{Cu}$,

$q_{Cu} = (nC\Delta T)_{Cu}$, and $q_{water} = (nC\Delta T)_{water}$

For Cu, $n = \dfrac{m}{MM} = 9.50 \text{ g}\left(\dfrac{1 \text{ mol}}{63.546 \text{ g}}\right) = 0.1495 \text{ mol Cu};$
$C = 24.435 \text{ J/mol °C},$
and $\Delta T = (x - 200.0) \text{ °C};$

For water, $n = \dfrac{m}{MM} = 200 \text{ mL}\left(\dfrac{1.00 \text{ g}}{1 \text{ mL}}\right)\left(\dfrac{1 \text{ mol}}{18.02 \text{ g}}\right) = 11.10 \text{ mol water};$
$C = 75.291 \text{ J/mol °C},$
and $\Delta T = (x - 5.00) \text{ °C};$

Substitute and solve for x:

$$(0.1495 \text{ mol})\left(\frac{24.435 \text{ J}}{1 \text{ mol °C}}\right)(x - 200.0) \text{ °C} = -(11.10 \text{ mol})\left(\frac{75.291 \text{ J}}{1 \text{ mol °C}}\right)(x - 5.00) \text{ °C};$$

$$3.653 x - 730.6 = -835.7 x + 4179,$$
$$839.4 x = 4910,$$
$$x = 5.85 \text{ °C}$$

12.85 To determine the thermodynamic values first calculate the number of moles of the ammonia and the heat of vaporization:

moles $NH_3 = 275 \text{ mL} \left(\dfrac{0.81 \text{ g}}{1 \text{ mL}} \right) \left(\dfrac{1 \text{ mol}}{17.04 \text{ g}} \right) = 13.1 \text{ mol}$

$q = n \, \Delta H_{vap} = 13.1 \text{ mol} \left(\dfrac{23.2 \text{ kJ}}{1 \text{ mol}} \right) = 304 \text{ kJ}$

Since this system is at constant pressure, $\Delta H = q = 304 \text{ kJ}$

The only work on the system is expansion work. Thus, $w = -P\Delta V = -nRT_{vap}$

$w = -13.1 \text{ mol} \left(\dfrac{8.314 \text{ kJ}}{1 \text{ mol K}} \right) (240 \text{ K}) = -2.61 \times 10^4 \text{ J or } -26.1 \text{ kJ}$

Finally $\Delta E = \Delta H - P\Delta V = \Delta H - nRT_{vap} = 304\text{kJ} - 26.1 \text{ kJ} = 2.8 \times 10^2 \text{ kJ}$. (density of NH_3 is known to only two significant figures)

13.1 (a) A sand castle represents an ordered structure constructed by a person. Waves destroy that structure, returning the sand grains to a disordered arrangement.
(b) Two separate liquids represent an ordered arrangement (all molecules of one kind in one container). Upon mixing, the molecules in the liquids are distributed randomly throughout the container, an increase in disorder.
(c) Sticks in a bundle are ordered (all aligned in the same direction). When dropped, the sticks lose their alignment and become more disordered.
(d) Water in a puddle is relatively ordered, as it is confined to a small volume. When the water evaporates, the molecules spread over a much larger volume and become more disordered.

13.3 The molecules in a drop of ink are relatively ordered, because they occupy one particular part of the total volume. As they move about randomly, they spread throughout the volume of the container and become more disordered.

13.5 (a) The air molecules in a tire are relatively ordered, because they are confined to a specific, small volume. A puncture allows gas molecules to escape from the tire and fill a much larger volume, becoming less ordered in the process.
(b) Like air in a tire, the fragrant molecules in a perfume bottle are relatively ordered because they are confined to a specific, small volume. When the bottle is open, molecules escape from the confined volume into a much larger space of the room, becoming more disordered in the process.

13.7 Calculate W by determining how many ways each symbol can be placed. The first X can be placed in any of the nine compartments, the first O in any of the eight empty compartments, so $W = (9)(8) = 72$.

13.9 The entropy change accompanying a constant-temperature process is $\Delta S = \dfrac{q}{T}$.

(a) Melting involves absorption of heat:

$$q_{ice} = n\Delta H_{fus} = 13.8\ g\left(\frac{1\ mol}{18.02\ g}\right)\left(\frac{6.01\ kJ}{1\ mol}\right)\left(\frac{10^3\ J}{1\ kJ}\right) = 4.60 \times 10^3\ J;$$

$$\Delta S_{ice} = \frac{4.60 \times 10^3\ J}{273.15\ K} = 16.8\ J/K.$$

(b) The heat absorbed by the ice cube must be supplied by the pool water: $q_{pool} = -q_{ice}$;

$$\Delta S_{pool} = \frac{-4.60 \times 10^3\ J}{27.5 + 273.15\ K} = -15.3\ J/K.$$

(c) $\Delta S_{total} = \Delta S_{ice} + \Delta S_{pool} = (16.8\ J/K) + (-15.3\ J/K) = 1.5\ J/K.$

13.11 The sign of the entropy change will be the same as q.
(a) The system (water) absorbs heat to boil, so q_{sys} is positive and ΔS_{sys} is positive. The surroundings (stove) release heat to the system, so q_{surr} is negative and ΔS_{surr} is negative.

(b) The system (ice cubes) absorbs heat to melt, so q_{sys} is positive and ΔS_{sys} is positive. The surroundings (table top) release heat to the system, so q_{surr} is negative and ΔS_{surr} is negative.

(c) The system (coffee) absorbs heat, so q_{sys} is positive and ΔS_{sys} is positive. The surroundings (microwave oven) release heat to the system, so q_{surr} is negative and ΔS_{surr} is negative.

13.13 The entropy change accompanying a constant-temperature process is $\Delta S = \dfrac{q}{T}$.

Condensation involves release of heat:

$$q_{steam} = -n\Delta H_{condensation} = -15.5 \text{ g}\left(\frac{1 \text{ mol}}{18.02 \text{ g}}\right)\left(\frac{40.79 \text{ kJ}}{1 \text{ mol}}\right)\left(\frac{10^3 \text{ J}}{1 \text{ kJ}}\right) = -3.509 \times 10^4 \text{ J};$$

$$\Delta S_{steam} = \frac{-3.509 \times 10^4 \text{ J}}{373.15 \text{ K}} = -94.0 \text{ J/K}.$$

Knowing that ΔS_{total} must be positive, we can say without doing further calculations that $\Delta S_{surr} > 94.0$ J/K.

13.15 The molar entropy change accompanying fusion is $\Delta S_{molar} = \dfrac{\Delta H_{fus}}{T_{fus}}$.

(a) $\Delta S_{molar} = \left(\dfrac{1.3 \text{ kJ/mol}}{83 \text{ K}}\right)\left(\dfrac{10^3 \text{ J}}{1 \text{ kJ}}\right) = 16$ J/mol K;

(b) $\Delta S_{molar} = \left(\dfrac{0.84 \text{ kJ/mol}}{90 \text{ K}}\right)\left(\dfrac{10^3 \text{ J}}{1 \text{ kJ}}\right) = 9.3$ J/mol K;

(c) $\Delta S_{molar} = \left(\dfrac{7.61 \text{ kJ/mol}}{156 \text{ K}}\right)\left(\dfrac{10^3 \text{ J}}{1 \text{ kJ}}\right) = 48.8$ J/mol K;

(d) $\Delta S_{molar} = \left(\dfrac{23.4 \text{ kJ/mol}}{234 \text{ K}}\right)\left(\dfrac{10^3 \text{ J}}{1 \text{ kJ}}\right) = 1.00 \times 10^2$ J/mol K.

13.17 (a) Both are ionic solutions, but $MgCl_2$ produces three moles of ions per mole of substance, while NaCl produces two moles of ions per mole of substance, so $MgCl_2$ has the larger molar entropy;
(b) Both are ionic solids, but HgS has a higher molar mass than HgO, so HgS has the larger molar entropy;
(c) Both are diatomic molecules, but $Br_{2(l)}$ is liquid while $I_{2(s)}$ is solid, so $Br_{2(l)}$ has the larger molar entropy.

13.19 All these substances are small gaseous molecules. Methane is a highly symmetric molecule, so it has fewer distinguishable orientations in space. Thus it has lower entropy than the other two despite having five atoms per molecule. Ozone has more entropy than O_2 because it has three atoms per molecule while O_2 has only two.

13.21 Absolute entropies are tabulated in Appendix D of your textbook. To obtain the entropy per mole of atoms, divide each value by the number of atoms in one particle:

He: $S° = 126.153$ J/mol K$\left(\dfrac{1\,\text{mol He}}{1\,\text{mol atom}}\right) = 126.153$ J/mol K;

H$_2$: $S° = 130.680$ J/mol K$\left(\dfrac{1\,\text{mol H}_2}{2\,\text{mol atom}}\right) = 65.340$ J/mol K;

CH$_4$: $S° = 186.3$ J/mol K$\left(\dfrac{1\,\text{mol CH}_4}{5\,\text{mol atom}}\right) = 37.26$ J/mol K;

C$_3$H$_6$: $S° = 226.9$ J/mol K$\left(\dfrac{1\,\text{mol C}_3\text{H}_6}{9\,\text{mol atom}}\right) = 25.21$ J/mol K.

Entropy per mole of atoms decreases as the number of atoms in a species increases, because tying together atoms into a molecule increases the amount of order among those atoms.

13.23 Standard entropy changes can be calculated using tabulated values of absolute entropies (Appendix D of your textbook) and Equation 13-5:

$$\Delta S°_{\text{reaction}} = \Sigma\text{coeff}_p\, S°_{\text{(products)}} - \Sigma\text{coeff}_r\, S°_{\text{(reactants)}}$$

(a) $\Delta S°_{\text{reaction}} = 2$ mol(192.8 J/mol K)
$\quad\quad - [1\text{mol}(191.61\text{ J/mol K}) + 3\text{ mol}(130.680\text{ J/mol K})] = -198.1$ J/K;

(b) $\Delta S°_{\text{reaction}} = 2$ mol(238.9 J/mol K) $- 3$ mol(205.152 J/mol K) $= -137.7$ J/K;

(c) $\Delta S°_{\text{reaction}} = [1$ mol(64.8 J/mol K) $+ 2$ mol(37.99 J/mol K)]
$\quad\quad - [1$ mol(68.7 J/mol K) $+ 2$ mol(29.9 J/mol K)] $= 12.3$ J/K;

(d) $\Delta S°_{\text{reaction}} = [2$ mol(213.8 J/mol K) $+ 2$ mol(69.95 J/mol K)]
$\quad\quad - [1$ mol(219.3 J/mol K) $+ 3$ mol (205.152 J/mol K)] $= -267.3$ J/K.

13.25 Reactions have negative entropy changes when products are more ordered than reactants.
(a) There is a relatively large negative value because of the reduction in number of moles of gaseous substances: 4 moles of gaseous reactants are converted to 2 moles of gaseous products;
(b) There is a relatively large negative value because of the reduction in number of moles of gaseous substances: 3 moles of gaseous reactants are converted to 2 moles of gaseous products;
(c) This reaction has a near-zero entropy change because all reagents are relatively ordered solid substances;
(d) There is a relatively large negative value because of the reduction in number of moles of gaseous substances: 4 moles of gaseous reactants are converted to 2 moles of gaseous products.

13.27 Entropies of substances at pressures different from 1.00 atm or concentrations different from 1 M are calculated using Equation 13-4: $S_{(p \neq 1)} = S° - R \ln p$ or $S_{(c \neq 1)} = S° - R \ln c$. Take into account amounts different from 1 mol by multiplying by the number of moles.

(a) $S = (2.50 \text{ mol})[154.843 \text{ J/mol K} - (8.314 \text{ J/mol K})\ln(0.25 \text{ atm})] = 416 \text{ J/K}$;

(b) $S = (0.75 \text{ mol})[238.9 \text{ J/mol K} - (8.314 \text{ J/mol K})\ln(2.75 \text{ atm})] = 1.7 \times 10^2 \text{ J/K}$;

(c) Calculate the entropy of each component separately, using $n_i = X_i n_{tot}$ and $p_i = X_i p_{tot}$:

$S_{N_2} = (0.78)(0.45 \text{ mol})\{191.61 \text{ J/mol K} - (8.314 \text{ J/mol K})\ln[(0.78)(1.00 \text{ atm})]\} = 68 \text{ J/K}$;

$S_{O_2} = (0.22)(0.45 \text{ mol})\{205.152 \text{ J/mol K} - (8.314 \text{ J/mol K})\ln[(0.22)(1.00 \text{ atm})]\} = 22 \text{ J/K}$;

$S = S_{N_2} + S_{O_2} = 9.0 \times 10^1 \text{ J/K}$.

13.29 False statements can be made true in various ways; we choose the simplest change that corrects each statement:

(a) $\Delta G_{system} < 0$ for any spontaneous process at constant T and P.

(b) The free energy of a system decreases in any spontaneous process at constant T and P.

(c) $\Delta G = \Delta H - \Delta(TS)$.

13.31 Standard free energy changes are calculated from Equation 13-8 using standard free energies of formation, which can be found in Appendix D of your textbook:

$$\Delta G^{\circ}_{rxn} = \Sigma\text{coeff}_p \, \Delta G^{\circ}_f(\text{products}) - \Sigma \text{coeff}_r \, \Delta G^{\circ}_f(\text{reactants})$$

(a) $\Delta G^{\circ}_{reaction} = 2 \text{ mol}(-16.4 \text{ kJ/mol}) - [1 \text{ mol}(0 \text{ kJ/mol}) + 3 \text{ mol}(0 \text{ kJ/mol})] = -32.8 \text{ kJ}$;

(b) $\Delta G^{\circ}_{reaction} = 2 \text{ mol}(163.2 \text{ kJ/mol}) - 3 \text{ mol}(0 \text{ kJ/mol}) = 326.4 \text{ kJ}$.

(c) $\Delta G^{\circ}_{reaction} = [1 \text{ mol}(0) + 2 \text{ mol}(-211.7 \text{ kJ/mol})]$
$- [1 \text{ mol}(-217.3 \text{ kJ/mol}) + 2 \text{ mol}(0 \text{ kJ/mol})] = -206.1 \text{ kJ}$;

(d) $\Delta G^{\circ}_{reaction} = [2 \text{ mol}(-394.4 \text{ kJ/mol}) + 2 \text{ mol}(-237.1 \text{ kJ/mol})]$
$- [1 \text{ mol}(68.4 \text{ kJ/mol}) + 3 \text{ mol}(0 \text{ kJ/mol})] = -1331.4 \text{ kJ}$.

13.33 Use Equation 13-10 to estimate the standard free energy change at a temperature different from 298 K: $\Delta G^{\circ}_{reaction, T} \cong \Delta H^{\circ}_{reaction, 298} - T\Delta S^{\circ}_{reaction, 298}$. Standard entropy changes are calculated in Problem 13.23, but standard enthalpy changes need to be calculated from standard enthalpies of formation:

(a) $\Delta H^{\circ}_{reaction} = 2 \text{ mol}(-45.9 \text{ kJ/mol}) - [1 \text{ mol}(0 \text{ kJ/mol}) + 3 \text{ mol}(0 \text{ kJ/mol})] = -91.8 \text{ kJ}$;

$$\Delta G^{\circ}_{reaction, T} \cong -91.8 \text{ kJ} - 425 \text{ K}\left(-198.1 \text{ J/K}\right)\left(\frac{10^{-3} \text{ kJ}}{1 \text{ J}}\right) = -7.6 \text{ kJ};$$

(b) $\Delta H^{\circ}_{reaction} = 2 \text{ mol}(142.7 \text{ kJ/mol}) - 3 \text{ mol}(0 \text{ kJ/mol}) = 285.4 \text{ kJ}$;

$$\Delta G^{\circ}_{reaction, T} \cong 285.4 \text{ kJ} - 425 \text{ K}\left(-137.7 \text{ J/K}\right)\left(\frac{10^{-3} \text{ kJ}}{1 \text{ J}}\right) = 343.9 \text{ kJ};$$

(c) $\Delta H^{\circ}_{reaction} = [1 \text{ mol}(0 \text{ kJ/mol}) + 2 \text{ mol}(-239.7 \text{ kJ/mol})]$
$- [1 \text{ mol}(-277.4 \text{ kJ/mol}) + 2 \text{ mol}(0 \text{ kJ/mol})] = -202.0 \text{ kJ}$;

$$\Delta G^{\circ}_{reaction, T} \cong -202.0 \text{ kJ} - 425 \text{ K}\left(12.3 \text{ J/K}\right)\left(\frac{10^{-3} \text{ kJ}}{1 \text{ J}}\right) = -207.2 \text{ kJ};$$

(d) $\Delta H^{\circ}_{reaction} = [2 \text{ mol}(-393.5 \text{ kJ/mol}) + 2 \text{ mol}(-285.83 \text{ kJ/mol})]$
$- [1 \text{ mol}(52.4 \text{ kJ/mol}) + 3 \text{ mol}(0 \text{ kJ/mol})] = -1411.1 \text{ kJ}$;

$$\Delta G^{\circ}_{\text{reaction}, T} \cong -1411.1 \text{ kJ} - 425 \text{ K} \left(- 267.3 \text{ J/K}\right)\left(\frac{10^{-3} \text{ kJ}}{1 \text{ J}}\right) = -1297 \text{ kJ}.$$

13.35 Use Equations 12-10, 13-5, and 13-8 to calculate standard thermodynamic values:

$$\Delta H^{\circ}_{\text{reaction}} = \Sigma \text{ coeff}_p \, \Delta H^{\circ}_f(\text{products}) - \Sigma \text{ coeff}_r \, \Delta H^{\circ}_f(\text{reactants})$$

$$\Delta S^{\circ}_{\text{reaction}} = \Sigma \text{coeff}_p S^{\circ}_{(\text{products})} - \Sigma \text{coeff}_r S^{\circ}_{(\text{reactants})}$$

$$\Delta G^{\circ}_{\text{reaction}} = \Sigma \text{coeff}_p \, \Delta G^{\circ}_f(\text{products}) - \Sigma \text{ coeff}_r \, \Delta G^{\circ}_f(\text{reactants})$$

$$\Delta H^{\circ}_{\text{reaction}} = [1 \text{ mol}(-365.6 \text{ kJ/mol}) + 1 \text{ mol}(-285.83 \text{ kJ/mol})]$$
$$- [2 \text{ mol}(-45.9 \text{ kJ/mol}) + 2 \text{ mol}(0 \text{ kJ/mol})] = -559.6 \text{ kJ};$$

$$\Delta S^{\circ}_{\text{reaction}} = [1 \text{ mol}(151.1 \text{ J/mol K}) + 1 \text{ mol}(69.95 \text{ J/mol K})]$$
$$- [2 \text{ mol}(192.8 \text{ J/mol K}) + 2 \text{ mol}(205.152 \text{ J/mol K})] = -574.9 \text{ J/K};$$

$$\Delta G^{\circ}_{\text{reaction}} = [1 \text{ mol}(-183.9 \text{ kJ/mol}) + 1 \text{ mol}(-237.1 \text{ kJ/mol})]$$
$$- [2 \text{ mol}(-16.4 \text{ kJ/mol}) + 2 \text{ mol}(0 \text{ kJ/mol})] = -388.2 \text{ kJ}.$$

13.37 A phase diagram is a plot of P vs. T, and phase boundary lines meet at the triple point and cross the horizontal line at $P = 1$ atm at the normal freezing and boiling points:

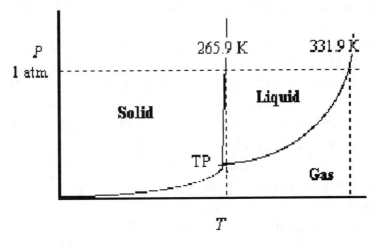

13.39 Phase diagrams provide "road maps" allowing us to determine what phase changes occur as temperature and pressure vary in particular ways.
(a) At $T = 400$ K, $P = 1.00$ atm, Br_2 is a gas. As it cools under a pressure of 1.00 atm, it liquefies at 331.9 K and solidifies at 265.9 K.
(b) At $T = 265.8$ K, $P = 1.00 \times 10^{-3}$ atm, Br_2 is a gas. Compressing it at this temperature causes it to liquefy at about $P = 6 \times 10^{-2}$ atm and solidify at about 0.5 atm.
(c) $P = 2.00 \times 10^{-2}$ atm is below the triple point of Br_2. Heating a solid at this pressure from 250 to 400 K causes sublimation to the vapor at about 265 K.

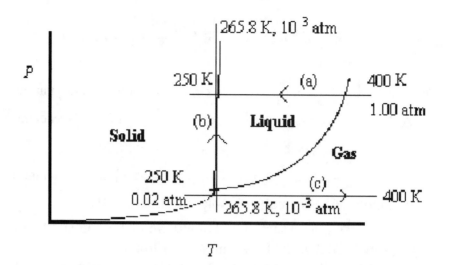

13.41 Coupled reactions share a common intermediate that transfers spontaneity (or energy) from one reaction to the other. In the case of the seesaw the first "reaction" would be child 1 starting on the ground and rising into the air. The second "reaction" would be child 2 starting in the air and falling to the ground. In this case child 2 would be the spontaneous reaction, child 1 would be non-spontaneous (things tend to fall down, not float up), so as child 2 is falling he will cause the first child to rise. When they are at equal heights (the common intermediate) they will switch spontaneity (or energy).

13.43 The reactions that are coupled in this example are the following:
$ATP + H_2O \rightarrow ADP + H_3PO_4$ $\Delta G° = -30.6$ kJ
$C_6H_{12}O_6$ (fructose) $+ C_6H_{12}O_6$ (glucose) $\rightarrow C_{12}H_{22}O_{11} + H_2O$ $\Delta G° = 23.0$ kJ;
Coupled reaction:
$ATP + C_6H_{12}O_6$ (fructose) $+ C_6H_{12}O_6$ (glucose) $\rightarrow ADP + C_{12}H_{22}O_{11}$
$\Delta G° = 23.0$ kJ $- 30.6$ kJ $= -7.6$ kJ.

13.45 (a) Energy is always conserved, so $\Delta E_{universe} = 0$;
(b) The teaspoon warms, so $\Delta E_{teaspoon} > 0$;
(c) This is a spontaneous process, so $\Delta S_{universe} > 0$;
(d) The water cools, so $\Delta S_{water} < 0$; and
(e) The teaspoon warms, so $q_{teaspoon} > 0$.

13.47 Entropy depends on amount, phase, temperature, and concentration. The order is:
(0.5 mol, l, 298 K) < (1 mol, l, 298 K) < (1 mol, l, 373 K)
< (1 mol, g, 1 atm, 373 K) < (1 mol, g, 0.1 atm, 373 K).

13.49 Spontaneity is determined by free energy, heat flow by enthalpy, and change in order by entropy: Use Equations 12-10, 13-5, and 13-8 to calculate standard thermodynamic values:

$$\Delta H°_{reaction} = \Sigma \, coeff_p \, \Delta H°_f \text{(products)} - \Sigma \, coeff_r \, \Delta H°_f \text{(reactants)}$$

$$\Delta S°_{reaction} = \Sigma coeff_p S°_{\text{(products)}} - \Sigma coeff_r S°_{\text{(reactants)}}$$

$$\Delta G^o_{reaction} = \Sigma coeff_p \; \Delta G^o_f (products) - \Sigma \; coeff_r \; \Delta G^o_f (reactants)$$

(a) $\Delta G^o_{reaction}$ = [2 mol(0 kJ/mol) + 3 mol(–237.1 kJ/mol)]

$\qquad$ – [1 mol(–1582.3 kJ/mol) + 3 mol(0 kJ/mol)] = 871.0 kJ.

The reaction is not spontaneous.

(b) $\Delta H^o_{reaction}$ = [2 mol(0 kJ/mol) + 3 mol(–285.83 kJ/mol)]

$\qquad$ – [1 mol(–1675.7 kJ/mol) + 3 mol(0 kJ/mol)] = 818.2 kJ.

The reaction absorbs heat.

(c) $\Delta S^o_{reaction}$ = [2 mol(28.3 J/mol K) + 3 mol(69.95 J/mol K)]

$\qquad$ – [1 mol(50.9 J/mol K) + 3 mol(130.680 J/mol K)] = –176.5 J/K.

Entropy decreases, so the products are more ordered than the reactants.

13.51 Use Equation 13-5 to calculate the entropy change of a reaction:

$$\Delta S^o_{reaction} = \Sigma coeff_p S^o_{(products)} - \Sigma coeff_r S^o_{(reactants)}$$

(a) $\Delta S^o_{reaction}$ = 2 mol(162 J/mol K)

$\qquad$ – [2 mol(101 J/mol K) + 1 mol(205.152 J/mol K)] = – 83 J/K;

(b) $\Delta S^o_{reaction}$ = [2 mol(87.4 J/mol K) + 6 mol(223.1 J/mol K)]

$\qquad$ – [4 mol(142.3 J/mol K) + 3 mol(205.152 J/mol K)] = 328.7 J/K;

(c) $\Delta S^o_{reaction}$ = [6 mol(210.8 J/mol K) + 6 mol(69.95 J/molK)]

$\qquad$ – [3 mol(121.2 J/mol K) + 4 mol(238.9 J/mol K)] = 365.3 J/K.

13.53 Use Equation 13-10 to calculate the standard free energy change of a reaction:

$$\Delta G^o_{reaction} = \Sigma coeff_p \; \Delta G^o_f (products) - \Sigma \; coeff_r \; \Delta G^o_f (reactants)$$

(a) $\Delta G^o_{reaction}$ = 2 mol(– 3 kJ/mol) – [2 mol(17 kJ/mol) + 1 mol(0 kJ/mol)] = – 40 kJ;

(b) $\Delta G^o_{reaction}$ = [2 mol(–742.2 kJ/mol) + 6 mol(0)]

$\qquad$ – [4 mol(–334.0 kJ/mol) + 3 mol(0 kJ/mol)] = – 148.4 kJ;

(c) $\Delta G^o_{reaction}$ = [6 mol(87.6 kJ/mol) + 6 mol(–237.1 kJ/mol)]

$\qquad$ – [3 mol(149.3 kJ/mol) + 4 mol(163.2 kJ/mol)] = – 1997.7 kJ.

13.55 To determine ΔG° at non-standard temperature, calculate ΔS° and ΔH° using Equations 13-5 and 12-10, then use Equation 13-10. The standard entropy changes are calculated in Problem 13.51.

$$\Delta S^o_{reaction} = \Sigma coeff_p S^o_{(products)} - \Sigma coeff_r S^o_{(reactants)}$$

$$\Delta H^o_{reaction} = \Sigma \; coeff_p \; \Delta H^o_f (products) - \Sigma \; coeff_r \; \Delta H^o_f (reactants)$$

$$\Delta G^o_{reaction, T} \cong \Delta H^o_{reaction, 298} - T \Delta S^o_{reaction, 298}$$

(a)

$\Delta H^o_{reaction}$ = 2 mol(– 104 kJ/mol) – [2 mol(–67 kJ/mol) + 1 mol(0 kJ/mol)] = – 74 kJ;

$$\Delta S^{\circ}{}_{reaction} = (623.2 \text{ K})\left(\frac{-83 \text{ J}}{1 \text{ K}}\right)\left(\frac{10^{-3} \text{ kJ}}{1 \text{ J}}\right) = -52 \text{ kJ};$$

$$\Delta G^{\circ}_{reaction, 350} = (-74 \text{ kJ}) - (-52 \text{ kJ}) = -22 \text{ kJ}.$$

(b)

$$\Delta H^{\circ}{}_{reaction} = [2 \text{ mol}(-824.2 \text{ kJ/mol}) + 6 \text{ mol}(0 \text{ kJ/mol})]$$
$$- [4 \text{ mol}(-399.5 \text{ kJ/mol}) + 3 \text{ mol}(0 \text{ kJ/mol})] = -50.4 \text{ kJ};$$

$$\Delta S^{\circ}{}_{reaction} = (623.2 \text{ K})\left(\frac{328.7 \text{ J}}{1 \text{ K}}\right)\left(\frac{10^{-3} \text{ kJ}}{1 \text{ J}}\right) = 204.8 \text{ kJ};$$

$$\Delta G^{\circ}_{reaction, 350} = (-50.4 \text{ kJ}) - (204.8 \text{ kJ}) = -255.2 \text{ kJ}.$$

(c)

$$\Delta H^{\circ}{}_{reaction} = [6 \text{ mol}(91.3 \text{ kJ/mol}) + 6 \text{ mol}(-285.83 \text{ kJ/mol})]$$
$$- [3 \text{ mol}(50.6 \text{ kJ/mol}) + 4 \text{ mol}(142.7 \text{ kJ/mol})] = -1889.8 \text{ kJ}.$$

$$\Delta S^{\circ}{}_{reaction} = (623.2 \text{ K})\left(\frac{-365.3 \text{ J}}{1 \text{ K}}\right)\left(\frac{10^{-3} \text{ kJ}}{1 \text{ J}}\right) = -227.7 \text{ kJ}$$

$$\Delta G^{\circ}_{reaction, 350} = (-1889.8 \text{ kJ}) - (-227.7 \text{ kJ}) = -1662.1 \text{ kJ}.$$

13.57 The figure shows a chemical reaction in which a set of diatomic molecules fragment into atoms with no change in volume or temperature.
(a) There is no volume change, so $w_{sys} = 0$;
(b) Energy must be supplied to break the chemical bonds, so $q_{sys} > 0$; and
(c) Because $q_{sys} > 0$, $q_{surr} < 0$, so $\Delta S_{surr} < 0$.

13.59 A reaction is thermodynamically feasible if $\Delta G < 0$.

(a) $\Delta G^{\circ} = [2 \text{ mol}(-73.5 \text{ kJ/mol}) + 1 \text{ mol}(87.6 \text{ kJ/mol})]$
$- [3 \text{ mol}(51.3 \text{ kJ/mol}) + 1 \text{ mol}(-237.1 \text{ kJ/mol})] = 23.8 \text{ kJ};$
This is not thermodynamically feasible under standard conditions.

(b) $\Delta G = \Delta G^{\circ} + RT \ln Q$, $\qquad Q = \left[\dfrac{(p_{HNO_3})^2(p_{NO})}{(p_{NO_2})^3}\right];$

$$\Delta G = 23.8 \text{ kJ} + (0.008314 \text{ kJ/mol K})(298 \text{ K}) \ln\left[\frac{(10^{-6})^2(10^{-6})}{(1)^3}\right] = 23.8 \text{ kJ} - 102.69 \text{ kJ}$$

$\Delta G = -78.9 \text{ kJ}$; thermodynamically feasible under these conditions.

13.61 Entropy changes for constant-temperature processes can be calculated using

Equation 13-1, $\Delta S = \dfrac{q_T}{T}$, and q for water freezing can be found using $q = -n\Delta H_{fus}$:

$$n_{water} = \frac{m}{MM} = 155 \text{ g} \left(\frac{1 \text{ mol}}{18.02 \text{ g}} \right) = 8.60 \text{ mol};$$

$$T_{water} = (0.0 + 273.15 \text{ K}) = 273.2 \text{ K};$$

$$q_{water} = -8.60 \text{ mol} \left(\frac{6.01 \text{ kJ}}{1 \text{ mol}} \right) \left(\frac{10^3 \text{ J}}{1 \text{ kJ}} \right) = -51.7 \times 10^3 \text{ J};$$

$$T_{surr} = (-20.0 + 273.15 \text{ K}) = 253.2 \text{ K};$$

(a) $\Delta S_{water} = \dfrac{-51.7 \times 10^3 \text{ J}}{273.2 \text{ K}} = -189 \text{ J/K}.$

(b) The energy released by the water will be absorbed by the surrounding so:

$$q_{surr} = -q_{water} = 51.7 \times 10^3 \text{ J/K}$$

$$\Delta S_{surr} = \frac{51.7 \times 10^3 \text{ J}}{253.2 \text{ K}} = 204 \text{ J/K};$$

$$\Delta S_{universe} = \Delta S_{water} + \Delta S_{surr} = -189 + 204 = 15 \text{ J/K}.$$

(c) Cooling the ice to –20 °C is an irreversible change (because the temperatures of ice and freezer are not the same except at the end of the process), for which the $\Delta S_{universe} > 0$. Thus, the cooling must generate an additional entropy increase for the universe.

13.63 Spontaneity is determined by free energy, heat flow by enthalpy, and change in order by entropy. Use Equations 12-10, 13-5, and 13-10 to calculate standard thermodynamic values:

$$\Delta H^\circ_{reaction} = \Sigma \text{ coeff}_p \Delta H_f^\circ (\text{products}) - \Sigma \text{ coeff}_r \Delta H_f^\circ (\text{reactants})$$

$$\Delta S^\circ_{reaction} = \Sigma \text{coeff}_p S^\circ_{(\text{products})} - \Sigma \text{coeff}_r S^\circ_{(\text{reactants})}$$

$$\Delta G^\circ_{reaction} = \Sigma \text{coeff}_p \Delta G_f^\circ (\text{products}) - \Sigma \text{ coeff}_r \Delta G_f^\circ (\text{reactants})$$

(a) $\Delta G^\circ_{reaction} = [1 \text{ mol}(-261.905 \text{ kJ/mol}) + 1 \text{ mol}(-131.0 \text{ kJ/mol})]$
    $- 1 \text{ mol}(-384.1 \text{ kJ/mol}) = -8.8 \text{ kJ}$. The reaction is spontaneous.
(b) $\Delta H^\circ_{reaction} = [1 \text{ mol}(-240.3 \text{ kJ/mol}) + 1 \text{ mol}(-167.1 \text{ kJ/mol})]$
    $- 1 \text{ mol}(-411.2 \text{ kJ/mol}) = 3.8 \text{ kJ}$. The reaction absorbs heat.
(c) $\Delta S^\circ_{reaction} = [1 \text{ mol}(58.5 \text{ J/mol K}) + 1 \text{ mol}(56.5 \text{ J/mol K})]$
    $- 1 \text{ mol}(72.1 \text{ J/mol K}) = 42.9 \text{ J/K}.$
Entropy increases, so the amount of order decreases during this reaction.

13.65 Energy "transactions" in the body are carried out using the ATP-ADP energy-storing reaction. When a cell needs energy, it "spends" ATP; when fat or carbohydrates ("capital") are consumed, the energy is stored by converting ADP to ATP ("buying" ATP). Like money, ATP is readily transported from place to place and is readily converted into "buying power."

13.67 A phase diagram provides a "road map" that allows us to determine the temperatures and pressures at which phase changes occur.
(a) A constant temperature process is a vertical line on the phase diagram. $T = 70$ K is in between the triple point (0.124 atm, 63 K) and normal boiling point (1 atm, 77 K) of nitrogen. At 70 K, 1 atm, nitrogen is liquid. When the pressure falls below about 0.5 atm, the liquid boils, and at 70 K, 0.1 atm, the substance is gaseous.
(b) A constant temperature process is a vertical line on the phase diagram. $T = 298$ K is within the gaseous region at all pressures, so compression from 1.00 atm to 50.0 atm leaves the substance gaseous.
(c) A constant pressure process is a horizontal line on the phase diagram, and 1 atm represents "normal" pressure, shown on the figure as a dotted line. When the temperature of the gas reaches 77 K, the gas liquefies, and when the temperature reaches 63 K, the liquid solidifies. At 50 K, N_2 is solid.

13.69 Ammonia is a toxic gas, while urea and ammonium nitrate both are relatively non-toxic solids. Thus the transport and application of ammonia entail significant risks to humans. Even in aqueous solution, ammonia is highly irritating, as anyone knows who has used strong ammonia as a cleanser.

13.71 According to your textbook, $\Delta G°_{glucose} = -2870$ kJ/mol and $\Delta G°_{palmitic\ acid} = -9790$ kJ/mol. Divide molar quantities by molar masses to obtain free energy per gram:

Glucose: $\Delta G_{per\ gram} = \left(\dfrac{-2870\,kJ}{1\,mol} \right)\left(\dfrac{1\,mol}{180\,g} \right) = -15.9$ kJ/g;

Palmitic acid: $\Delta G_{per\ gram} = \left(\dfrac{-9790\,kJ}{1\,mol} \right)\left(\dfrac{1\,mol}{256\,g} \right) = -38.2$ kJ/g;

Palmitic acid has the higher energy content per gram.

13.73 The process that takes place is vaporization of the liquid bromine to form bromine vapor.
(a) At the end of the process, all the bromine is in the gas phase, with the molecules well separated, and the piston has moved back:

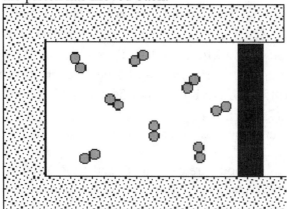

(b) The process is reversible, since the temperatures of system and surroundings are the same and the external and internal pressures both are 1.0 atm (bromine is at its boiling point). Thus, $\Delta S_{universe} = 0$.

13.75   The attractive forces in CaO are substantially larger than the attractive forces in KCl, so the individual Ca^{2+} and O^{2-} ions cannot as easily vibrate in the crystal. Thus the amount of vibrational disorder is substantially less in CaO than in KCl.

13.77   Make use of thermodynamic definitions to identify the state functions:
(a) $q_v = \Delta E$; (b) $q_p = \Delta H$; (c) $q_T = T\Delta S$.

13.79   (a) Whenever a substance cools, ΔS is negative.
(b) Whenever a gas is compressed (at constant T), ΔS is negative.
(c) Whenever two substances mix, disorder increases and ΔS is positive.

13.81   Use bond energy, intermolecular forces, and order-disorder to predict signs for ΔH and ΔS:
(a) Bonds form, so energy is released and ΔH is negative. The number of independent particles decreases, so order increases and ΔS is negative.
(b) When a solid melts, energy must be added; q is positive, ΔH is positive, and ΔS is positive.
(c) Combustion reactions release energy, so we expect ΔH to be negative. The number of molecules of gas remains the same during this reaction, so ΔS is expected to be small, but it is not possible to predict whether it is positive or negative without doing actual calculations.

13.83   Standard conditions refers to 1 atm, 298 K, all solutes at 1 M. In a cell, temperature is 37 °C (310 K) and no solutes are present at 1 M. In particular, hydronium ion concentration is around 10^{-7} M, far from standard conditions.

13.85   According to your textbook, the complete oxidation of 1 mol of palmitic acid produces 130 ATP molecules. The reactions are:

$$ADP + H_3PO_4 \rightarrow ATP + H_2O \quad \Delta G° = +30.6 \, kJ$$
$$\text{Palmitic acid: } C_{15}H_{31}CO_2H + 23\, O_2 \rightarrow 16\, CO_2 + 16\, H_2O \quad \Delta G° = -9790 \, kJ$$

Thus the amount of energy stored is (130)(30.6 kJ) = 3978 kJ and the efficiency is:

$$\text{Efficiency} = \frac{\text{energy stored}}{\text{energy released}} \times 100\% = \frac{3978 \, kJ}{9790 \, kJ} \times 100\% = 40.6 \, \%.$$

Metabolism of 1 mol of palmitic acid generates (9790 kJ) – (3978 kJ) = 5612 kJ of heat.

On a per gram basis, this is $q_{per\, gram} = \left(\dfrac{5612 \, kJ}{1 \, mol} \right) \left(\dfrac{1 \, mol}{256.42 \, g} \right) = 21.89 \, kJ/g$.

Evaporation of 75 g of water requires

$$q = n\Delta H_{vap} = 75\text{ g}\left(\frac{1\text{ mol}}{18.02\text{ g}}\right)\left(\frac{40.79\text{ kJ}}{1\text{ mol}}\right) = 1.7 \times 10^2\text{ kJ}.$$

$$\text{Mass required} = 1.7 \times 10^2\text{ kJ}\left(\frac{1\text{ g}}{21.89\text{ kJ}}\right) = 7.8\text{ g}.$$

13.87 Both of these processes are spontaneous. Thus, assuming that each of these processes occurs at constant T and P (perhaps more reasonable for the firefly than for lightning), both must have $\Delta G < 0$.

13.89 Your molecular picture should show all three phases simultaneously present, with transfers of atoms occurring among all phases:

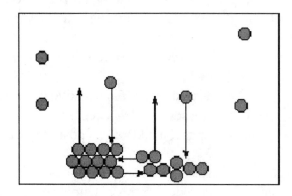

14.1 In any sequence of steps, the slowest one will be rate-determining. (a) Pouring the coffee from the urn into the cup; (b) Entering the items on the cash register (if the market has a good laser scanner, paying and receiving change may be rate-determining); (c) Preparing for the jump and passing through the door.

14.3 Every elementary reaction must depict actual molecular processes.
(a) $I_2 \rightarrow I + I$;
(b) $H_2 + I_2 \rightarrow H_2I_2$;
(c) $H_2 + I_2 \rightarrow H + HI_2$.

14.5 A molecular picture of an elementary reaction shows the reactants, the products, and (if necessary) the intermediate collision complex.

(a)

(b)

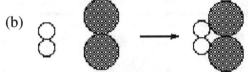

(c)

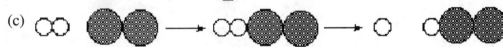

14.7 A satisfactory mechanism must consist entirely of reasonable elementary steps that sum to give the correct overall stoichiometry of the reaction.
(a) $I_2 \rightarrow I + I$
$I + H_2 \rightarrow HI + H$
$H + I \rightarrow HI$

(b) $H_2 + I_2 \rightarrow H_2I_2$
$H_2I_2 \rightarrow HI + HI$

(c) $H_2 + I_2 \rightarrow H + HI_2$
$H + HI_2 \rightarrow HI + HI$

14.9 The rate of a reaction can be expressed using the general expression, Rate $= -\dfrac{1}{a}\dfrac{\Delta[A]}{\Delta t}$.

(a) Rate $= -\dfrac{\Delta[Cl_2]}{\Delta t}$

(b) $-\dfrac{\Delta[Cl_2]}{\Delta t} = \dfrac{1}{2}\dfrac{\Delta[NOCl]}{\Delta t}$

(c) Use the result of (b) to calculate that NOCl appears at a rate of 94 M s^{-1}

14.11   In the reaction of NO and Cl_2, two NO molecules react for every Cl_2 that reacts, producing two NOCl molecules.

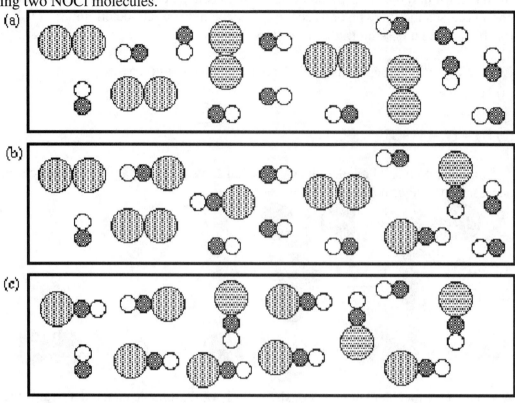

14.13 (a) To calculate the average rate of production, determine how many moles form during the time interval and divide by the time:

$$n = \frac{PV}{RT} = \frac{(0.15\,\text{atm})(5.0\,\text{L})}{(0.08206\,\frac{\text{L atm}}{\text{mol K}})(550 + 273.15\,\text{K})} = 1.11 \times 10^{-2}\,\text{mol};$$

$$\frac{\Delta[CO_2]}{\Delta t} = \frac{1.11 \times 10^{-2}\,\text{mol}}{5.0\,\text{min}} = 2.2 \times 10^{-3}\,\text{mol/min}$$

(b) The reaction is $CaCO_3 \rightarrow CaO + CO_2$, so the amount of $CaCO_3$ decomposing is the same as the amount of CO_2 produced, 1.1×10^{-2} mol.

14.15 (a) Before the reaction begins

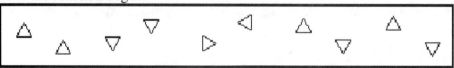

(b) After 20 minutes, the amount reacted is (20 min)(0.25 molecules/min) = 5 molecules

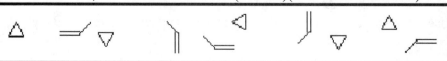

220

Chapter 14

14.17 (a) The number of decays is independent of the starting amount of the isotope in first-order kinetics, therefore there would be 60 decays in the 5.00 mol case.
(b) The fraction decaying is the same in both cases: 12 isotopes decay per 1 mole, or 12 decays/mol = 12 decays/6.022×10^{23} nuclei.
(c) In first-order kinetics the fraction reacting is independent of concentrations.

14.19 (a) The rate law for an elementary step contains the product of the reactant concentrations:
Rate = k[C][AB];
(b) The units of a rate constant have time in the denominator along with concentrations one power less than the number of reactants: units = (conc.)$^{-1}$ (time)$^{-1}$;
(c) The steps of the mechanism must sum to give the observed stoichiometry for the reaction. Intermediate A can react with AB, then B and C can react:
C + AB → BC + A
A + AB → B + A_2
B + C → BC
Net: 2 C + 2 AB → 2 BC + A_2

14.21 (a) The rate law for an elementary step contains the product of the reactant concentrations:
Rate = k[C][C][AB] = k[C]^2[AB];
(b) The units of a rate constant have time in the denominator along with concentrations one power less than the number of reactants: units = (conc.)$^{-2}$ (time)$^{-1}$;
(c) The steps of the mechanism must sum to give the observed stoichiometry for the reaction. A bimolecular reaction between the intermediate (AC) and AB is the simplest possibility:
2 C + AB → BC + AC
AC + AB → BC + A_2
Net: 2 C + 2 AB → 2 BC + A_2

14.23 According to the stated rate law, the rate of reaction is proportional to each concentration. A contains 6 molecules of each type, while B contains 10 NO and 2 O_3:
A: Rate = $k(6)(6) = 36\,k$;
B: Rate = $k(10)(2) = 20\,k$;
B will react slower by a factor of 20/36 = 0.56.

14.25 (a) One way to treat experimental data is by plotting: For first-order behavior, $\ln \frac{c_o}{c}$ vs. t is a straight line, and for second-order behavior, $\frac{1}{c} - \frac{1}{c_o}$ vs. t is a straight line. Here, the plot of $\ln c$ vs. t gives a straight line, so this reaction is first-order.

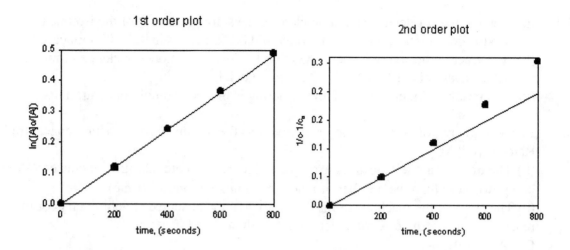

1st order plot

2nd order plot

(b) For a first-order reaction, k = slope of the graph

$$k = \text{slope} = \frac{\Delta y}{\Delta x} = \frac{0.491 - 0}{800\,\text{s} - 0\,\text{s}} = 6.14 \times 10^{-4}\ \text{s}^{-1}$$

(c) To find concentration at any particular time, use $\ln[A] = \ln[A]_o - kt$

$\ln[A] = \ln[2.50] - (6.14 \times 10^{-4}\ \text{s}^{-1})(1600\ \text{s}) = -0.0661$

$[A] = e^{-0.0661} = 0.936\ \text{atm}$

(d) To find the time at which concentration reaches a particular value, use

$$kt = \ln\left(\frac{[A]_o}{[A]}\right)$$

$$t = \frac{\ln\left(\frac{[A]_o}{[A]}\right)}{k} = \frac{\ln\left(\frac{2.50\,\text{atm}}{0.500\,\text{atm}}\right)}{6.14 \times 10^{-4}\,\text{s}^{-1}} = 2.62 \times 10^3\ \text{s}$$

14.27 This is stated to be a second-order reaction, so Rate = $k[\text{NOBr}]^2$ and Equation 14-5 applies:

$$\frac{1}{[A]} - \frac{1}{[A]_o} = kt\ \ ;$$

The problem states that $k = 25\ \text{M}^{-1}\ \text{min}^{-1}$.

(a) $kt = \dfrac{1}{(0.010\ \text{M})} - \dfrac{1}{(0.025\ \text{M})} = (100 - 40)\ \text{M}^{-1} = 60\ \text{M}^{-1};$

$$t = \frac{60\ \text{M}^{-1}}{25\ \text{M}^{-1}\ \text{min}^{-1}} = 2.4\ \text{min};$$

(b) $\dfrac{1}{[A]} = \dfrac{1}{(0.025\ \text{M})} + (25\ \text{M}^{-1}\ \text{min}^{-1})(125\ \text{min}) = (40 + 3125)\ \text{M}^{-1} = 3165\ \text{M}^{-1}$

$[A] = 3.2 \times 10^{-4}\ \text{M}.$

14.29 This is stated to be a first-order reaction, so Rate = $k[\text{C}_5\text{H}_{11}\text{Br}]$ and Equation 14.3 applies:

$$kt = \ln\left(\frac{[A]_o}{[A]}\right)$$

(a) $kt = \ln\left(\frac{0.125}{1.25 \times 10^{-3}}\right) = 4.61$

$t = \frac{4.61}{0.385\,\text{hr}^{-1}} = 12.0\,\text{hr}$

(b) $\ln[A] = \ln[0.125] - 3.5\,\text{hr}(0.385\,\text{hr}^{-1}) = -2.08 - 1.35 = -3.43$

$[A] = e^{-3.43} = 0.0324\,\text{M or } 3.24 \times 10^{-2}\,\text{M}$

14.31 To determine the order of a reaction from a set of experimental data, prepare plots of

$\ln\dfrac{c_o}{c}$ vs. t and $\dfrac{1}{c} - \dfrac{1}{c_o}$ vs. t:

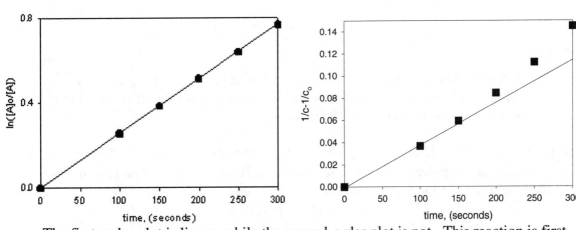

1st order plot

2nd order plot

The first-order plot is linear, while the second-order plot is not. This reaction is first-order. Determine the rate constant from

$k = \text{Slope} = \dfrac{\Delta y}{\Delta x} = \dfrac{0.766 - 0}{300\,\text{s} - 0\,\text{s}} = 2.55 \times 10^{-3}\,\text{s}^{-1}.$

14.33 The rate law should relate the rate of reaction to the concentration of the reactants. Reactive intermediates should not be shown in the rate law.

$2C + AB \rightleftarrows BC + AC$, followed by

$AC + AB \rightarrow BC + A_2$ (slow)

Net: $2\,C + 2\,AB \rightarrow 2\,BC + A_2$

The rate law is determined by the rate-determining step: Rate = $k_2[AC][AB]$. This is not satisfactory, however, because it contains the concentration of an intermediate (AC). Set the rates equal for the forward and reverse first step:

$$k_1[C]^2[AB] = k_{-1}[BC][AC];$$

Solve this equality for [AC]: $[AC] = \left(\dfrac{k_1}{k_{-1}}\right)\dfrac{[C]^2[AB]}{[BC]}$;

Substitute into the rate expression: $\quad \text{Rate} = k_2 \left(\dfrac{k_1}{k_{-1}} \right) \dfrac{[C]^2 [AB]^2}{[BC]}$

14.35 (a) The steps of a mechanism must sum to give the observed overall stoichiometry of the reaction. For ozone decomposition, this is $2\,O_3 \rightarrow 3\,O_2$. The two steps proposed by the student consume $1\,O_3$, produce $1\,O_2$, and generate an O atom which must be consumed:

$$O_3 + O \rightarrow 2\,O_2$$

(b) The rate law is determined by the rate-determining step: $\text{Rate} = k_2[O_5]$. This is not satisfactory, however, because it contains the concentration of an intermediate. Set the rates equal for the forward and reverse first step:

$$k_1[O_3][O_2] = k_{-1}[O_5];$$

Solve this equality for $[O_5]$: $\quad [O_5] = \dfrac{k_1}{k_{-1}}[O_3][O_2];$

Substitute into the rate expression:

$$\text{Rate} = k_2 \dfrac{k_1}{k_{-1}} [O_3][O_2]$$

(c) Atmospheric chemists would consider this mechanism to be molecularly unreasonable because fragmentation of O_5 in the second step (the breaking of two bonds simultaneously) is highly unlikely.

14.37 (a) The rate law is determined by the rate-determining step: $\text{Rate} = k_2[N_2O_2][O_2]$. This is not satisfactory, however, because it contains the concentration of an intermediate. Set the rates equal for the forward and reverse first step:

$$k_1[NO]^2 = k_{-1}[N_2O_2];$$

Solve this equality for $[N_2O_2]$: $[N_2O_2] = \left(\dfrac{k_1}{k_{-1}} \right)[NO]^2;$

Substitute into the rate expression:

$$\text{Rate} = k_2 \left(\dfrac{k_1}{k_{-1}} \right)[NO]^2[O_2]$$

(b) This rate expression has an overall order of $(2 + 1) = 3$, so the mechanism is consistent with third-order behavior.

(c) The intermediate species is N_2O_2. Two NO molecules could bind together in several ways:

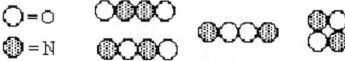

In the second step, O_2 collides with the intermediate and reacts to form two NO_2 molecules. The ONNO arrangement is the only one for which this rearrangement of bonds can easily occur:

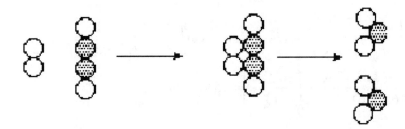

14.39 The rate constant for a reaction depends on temperature according to the Arrhenius equation (Equation 14-6): $k = Ae^{-E_a/RT}$. When $E_a = 0$, the exponent is $e^0 = 1$ and k is independent of temperature. A zero activation energy exists when a reaction can occur without first breaking any chemical bonds. The most common example is the combination of two free radicals, such as $H_3C\bullet + \bullet CH_3 \rightarrow H_3C–CH_3$.

14.41 An exothermic reaction has products lower in energy than reactants. In the activated complex A will be bonded to both B and C:

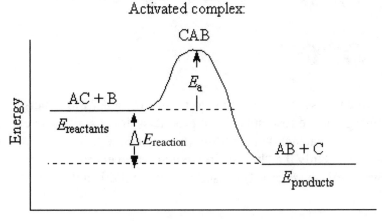

14.43 An exothermic reaction is "downhill" from reactants to products, and the activation energy plot should show this:

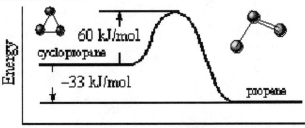

14.45 Assume that the ratio of the rate constants is proportional to the ratio of the number of flashes, then use the equation:

$$E_a = R\ln\left(\frac{k_2}{k_1}\right)\left(\frac{1}{T_1} - \frac{1}{T_2}\right)^{-1} = (8.314\,\text{J/molK})\ln\left(\frac{2.7}{3.3}\right)\left(\frac{1}{(29+273)K} - \frac{1}{(23+273)K}\right)^{-1}$$

$$E_a = \left(\frac{-1.67\,\text{J mol}^{-1}\,\text{K}^{-1}}{-6.7\times10^{-5}\,\text{K}^{-1}}\right)\left(\frac{10^{-3}\text{kJ}}{1\text{J}}\right) = 25\,\text{kJ/mol}$$

14.47 A catalyst, such as Pd metal from problem 14.49, would cause hydrogen gas to adsorb on the catalyst's surface as H atoms which will then react with the CO molecules much more easily (and quickly) than having the CO molecules break apart an H_2 molecule to get the hydrogens it needs. Bonds that must be broken for this reaction to occur (depicted by the squiggly lines) are two H-H single bonds and both the π bonds in the CO molecule. Bonds that are formed (dashed lines) are three C-H bonds and one O-H bond.

Bond Breakage   Bond Formation

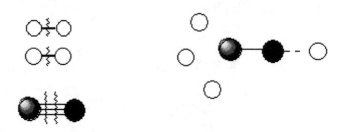

14.49 (a) The effect of a catalyst on a reaction is to reduce the activation energy, it has nothing to do with the energy of the reactants or the products. In an activation energy diagram the starting (energy of reactants) and ending (energy of the products) lines will be the same for both curves. Only the "bump" in the activation energy diagram will change, being smaller when Pd (the catalyst) is used than when Pd is not used.

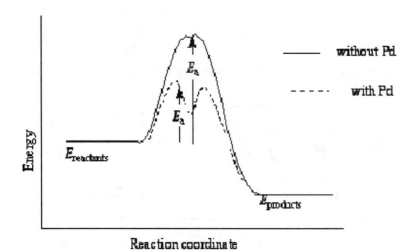

(b) A catalyst is anything that will speed up the rate of a reaction and is also never used up (it's a reactant as well as a product). The only thing that fits these criteria is the Pd metal. An intermediate is anything that is produced in one step of the mechanism and then used up in another step. The intermediates for this reaction are the H atoms that are formed when the H_2 gas is absorbed onto the Pd metal surface and then added to the ethylene (hence, the valley present between the 2 peaks).

(c)

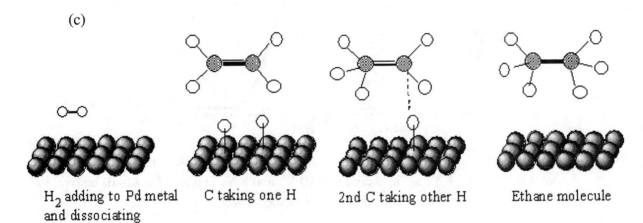

H$_2$ adding to Pd metal   C taking one H   2nd C taking other H   Ethane molecule
and dissociating

14.51   When the concentration of a reactant increases by a factor of three (triples), the rate of reaction changes by 3^n, where n is the order with respect to that concentration.
(a) nine-fold increase; (b) no change; (c) rate increases by 5.2 times.

14.53   (a)   The second step must use up the extra product, O, to form one of the reactants. Therefore one possible step would be, $O + O \rightarrow O_2$.
(b)

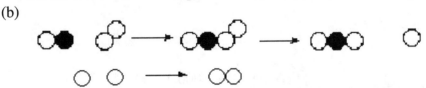

(The first step occurs twice for a net reaction of $2NO + O_2 \rightarrow 2NO_2$)

14.55   (a) A rate expression relates rate to Δc: $\text{Rate} = \dfrac{\Delta[C_6H_6]}{\Delta t} = -\dfrac{1}{3}\left(\dfrac{\Delta[C_2H_2]}{\Delta t}\right)$

(b) Rate laws must always be determined experimentally. Thus, there is insufficient information to write the rate law. Experiments would have to be carried out measuring the rate as a function of $[C_2H_2]$ and the data analyzed using techniques described in your textbook.

14.57   The essential units of information needed to construct an activation energy diagram are the energy change and activation energy. Here, the reaction is a formation reaction and the change in moles of gas is zero during the reaction, so $\Delta E \cong \Delta H_f^o$:

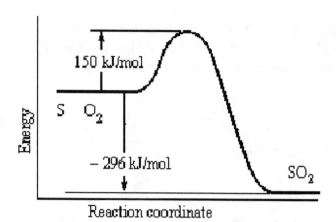

14.59 The mechanism has the rate law, Rate = $k[X_2]$, first-order in X_2 and independent of Y. A contains 5 X_2 and 8 Y, while B contains 10 X_2 and 8 Y. The rate for B will be twice that for A, because the concentration of X_2 is twice as great.

14.61 Reaction times for first-order reactions can be calculated using Equation 14-3, suitably rearranged: $t = \dfrac{\ln\left(\dfrac{[A]_o}{[A]}\right)}{k}$:

for 10.0%, $[A] = 0.900[A]_o$ and $\ln\left(\dfrac{[A]_o}{[A]}\right) = \ln\left(\dfrac{1.000}{0.900}\right) = 0.105$;

$t = \dfrac{0.105}{5.5 \times 10^{-4}\,s^{-1}} = 1.9 \times 10^2$ s;

for 50.0%, $[A] = 0.500[A]_o$ and $\ln\left(\dfrac{[A]_o}{[A]}\right) = \ln\left(\dfrac{1.000}{0.500}\right) = 0.693$;

$t = \dfrac{0.693}{5.5 \times 10^{-4}\,s^{-1}} = 1.3 \times 10^3$ s;

for 99.9%, $[A] = 0.001[A]_o$ and $\ln\left(\dfrac{[A]_o}{[A]}\right) = \ln\left(\dfrac{1.000}{0.001}\right) = 6.91$;

$t = \dfrac{6.91}{5.5 \times 10^{-4}\,s^{-1}} = 1.3 \times 10^4$ s.

14.63 (a) False (overall reaction would give fourth-order kinetics); (b) False (rate constants must be measured for at least two different temperatures to calculate E_a); (c) True (rates of reaction increase with increasing temperature); (d) True (a unimolecular step has first-order kinetics).

14.65 Reaction orders are given by the exponents on the concentrations appearing in the rate law. Overall order is the sum of those exponents.
(a) first-order in N_2O_5 and first-order overall;

(b) second-order in NO, first-order in H_2, and third-order overall;

(c) first-order in enzyme and first-order overall.

14.67 (a) Obtain the overall stoichiometry by adding the three steps, ignoring the reverse reaction of step one which leads to no net change: $2 NO + 2 H_2 \rightarrow N_2 + 2 H_2O$;

(b) The rate law is determined by the rate-determining step: Rate = $k_2[N_2O_2][H_2]$. This is not satisfactory, however, because it contains the concentration of an intermediate. Set the rates equal for the forward and reverse first step:

$$k_1[NO]^2 = k_{-1}[N_2O_2];$$

Solve this equality for $[N_2O_2]$: $\qquad [N_2O_2] = \dfrac{k_1}{k_{-1}}[NO]^2;$

Substitute into the rate expression: $\quad$ Rate $= k_2\dfrac{k_1}{k_{-1}}[NO]^2[H_2]$

14.69 Ultraviolet light causes chlorofluorocarbons to fragment, producing Cl atoms which catalyze the destruction of ozone:

$$CF_2Cl_2 \xrightarrow{h\upsilon} CF_2Cl + Cl$$
$$O_3 \xrightarrow{h\upsilon} O_2 + O$$
$$Cl + O_3 \rightarrow ClO + O_2$$
$$ClO + O \rightarrow Cl + O_2$$

Because the fourth reaction regenerates a Cl atom, the 2nd-4th reactions occur many hundreds of times for every CF_2Cl_2 molecule that fragments.

14.71 The speed of a chemical reaction refers to how fast it proceeds. The spontaneity of a chemical reaction refers to whether or not the reaction can go in the direction written without outside intervention. A spontaneous reaction may nevertheless have a very slow speed.

14.73 (a) At the molecular level, a catalyst binds to one or more of the reactants in a way that weakens chemical bonds and makes it easier for bonds to rearrange to form the products.
(b) When temperature increases, the average kinetic energies of the molecules increase, with the result that enough energy is present for reaction to occur in a larger fraction of molecular collisions.
(c) When concentration increases, the molecular density increases. There are more molecules to react, leading to a higher rate of molecular collisions. Both factors contribute to a greater rate of reaction.

14.75 Flask 1 contains 5 molecules, and Flask 2 contains 10 molecules, so the concentration is doubled. The problem states that Flask 2 reacts four times as fast. This is $(2)^2$, so the rate law is Rate = $k[A]^2$. At the molecular level, the rate-determining step could be a reaction between two A molecules. For this process, the rate increases because the higher concentration provides for a higher rate of collisions.

14.77 The hint for this problem suggests using the Arrhenius equation, $k = Ae^{-E_a/RT}$. Evaluate $\dfrac{E_a}{RT}$ for the catalyzed and uncatalyzed situations; $T = 21 + 273 = 294$ K:

$$\left(\frac{E_a}{RT}\right)_{uncat} = \frac{125\,kJ\,mol^{-1}}{(8.314\times10^{-3}\,kJ\,mol^{-1}\,K^{-1})(294\,K)} = 51.1;$$

$$\left(\frac{E_a}{RT}\right)_{cat} = \frac{46\,kJ\,mol^{-1}}{(8.314\times10^{-3}\,kJ\,mol^{-1}\,K^{-1})(294\,K)} = 18.8;$$

$k_{uncat} = Ae^{-51.1}$ and $k_{cat} = Ae^{-18.8}$;

Divide one of these by the other to eliminate A and find the ratio of rate constants:

$$\frac{k_{cat}}{k_{uncat}} = e^{(51.1-18.8)} = e^{32.3} = 1.1 \times 10^{14}.$$

14.79 Activation energies are calculated from the Arrhenius equation using Equation 14-8:

$$E_a = R\ln\left(\frac{k_2}{k_1}\right)\left(\frac{1}{T_1}-\frac{1}{T_2}\right)^{-1}.$$

Development time is inversely proportional to rate constant, so $\dfrac{k_2}{k_1} = \dfrac{t_1}{t_2}$.

(a) $\dfrac{t_1}{t_2} = 2$;

$T_1 = 20 + 273 = 293$ K;

$T_2 = 293 + 10 = 303$ K;

$$\left(\frac{1}{T_1}-\frac{1}{T_2}\right)^{-1} = (3.413\times10^{-3}\,K^{-1} - 3.300\times10^{-3}\,K^{-1})^{-1} = 8880.\,K;$$

$$E_a = \left(\frac{8.314\,J}{1\,mol\,K}\right)\left(\frac{10^{-3}\,kJ}{1\,J}\right)(\ln 2)(8880.\,K) = 51\,kJ/mol.$$

(b) To determine the time it takes at 25°C use $\ln\left(\dfrac{t_1}{t_2}\right) = \dfrac{E_a}{R}\left(\dfrac{1}{T_1}-\dfrac{1}{T_2}\right).$

$E_a = 51$ kJ/mol;

$T_1 = 20 + 273 = 293$ K;

$t_1 = 10$ min;

$T_2 = 25 + 273 = 298$ K;

$$\left(\frac{1}{T_1} - \frac{1}{T_2}\right) = (3.413 \times 10^{-3} - 3.356 \times 10^{-3}) = 5.726 \times 10^{-5} \ K^{-1};$$

$$\ln\left(\frac{10 \ min}{t_2}\right) = 5.726 \times 10^{-5} \ K^{-1}\left(\frac{1 \ mol \ K}{8.314 \times 10^{-3} \ kJ}\right)\left(\frac{51 \ kJ}{1 \ mol}\right) = 0.351;$$

$$\frac{10 \ min}{t_2} = 1.42;$$

$$t_2 = \frac{10 \ min}{1.42} = 7.0 \ min.$$

14.81 The only true statements are (d) and (f). Statement (a) is false because $\Delta E_{reaction} = C - A$. Here is an energy diagram showing the quantities involved:

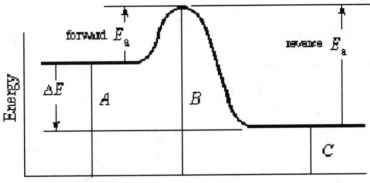

14.83 (a) The rate law is that for an elementary bimolecular reaction: Rate = $k[H_2][X_2]$;
(b) When a first step is rate-determining, it determines the rate law: Rate = $k[X_2]$;
(c) The rate law is determined by the rate-determining step: Rate = $k_2[X][H_2]$. This is not satisfactory, however, because it contains the concentration of an intermediate. Set the rates equal for the forward and reverse first step: $\qquad k_1[X_2] = k_{-1}[X]^2$;

Solve this equality for [X]: $[X] = \left(\dfrac{k_1}{k_{-1}}\right)^{1/2}[X_2]^{1/2}$;

Substitute into the rate expression: Rate = $k_2\left(\dfrac{k_1}{k_{-1}}\right)^{1/2}[H_2][X_2]^{1/2}$.

14.85 (a) The net reaction can be obtained by summing the three steps, because when the reverse step occurs there is no net change: $Cl_2 + CHCl_3 \rightarrow HCl + CCl_4$;
(b) Intermediates are produced in early steps and consumed in later steps: Cl and CCl_3;
(c) The rate law is determined by the rate-determining step: Rate = $k_2[CHCl_3][Cl]$. This is not satisfactory, however, because it contains the concentration of an intermediate. Set the rates equal for the forward and reverse first step: $\qquad k_1[Cl_2] = k_{-1}[Cl]^2$;

231

Solve this equality for [Cl]: $[Cl] = \left(\dfrac{k_1}{k_{-1}}\right)^{1/2} [Cl_2]^{1/2}$;

Substitute into the rate expression: $\text{Rate} = k_2 \left(\dfrac{k_1}{k_{-1}}\right)^{1/2} [CHCl_3][Cl_2]^{1/2}$.

14.87 Neither intermediates nor catalysts appear in the overall stoichiometry of the reaction, so any species that appears in the mechanism but not in the overall stoichiometry is either an intermediate or a catalyst. Catalysts are consumed in early steps and regenerated in later steps, while intermediates are produced in early steps and consumed in later steps. .

14.89 (a) Prepare first-order and second-order plots and look for linear behavior:

t	s	0	1000	2000	3000	4000
c	M	0.250	0.118	0.0770	0.0572	0.0455
$\ln(c_o/c)$		0.000	0.751	1.18	1.47	1.70
$1/c - 1/c_o$	M^{-1}	0.00	4.47	8.99	13.5	18.0

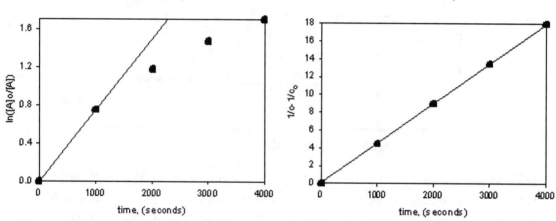

The second-order plot is linear, so $\text{Rate} = k[CH_3CHO]^2$;

(b) Determine the rate constant from the slope of the second-order plot:

$k = \text{Slope} = \dfrac{18.0\,\text{M}^{-1} - 0.00\,\text{M}^{-1}}{4000\,\text{s} - 0\,\text{s}} = 4.5 \times 10^{-3}\,\text{M}^{-1}\,\text{s}^{-1}$;

(c) Use Equation 14-5, suitably rearranged:

$$kt = \dfrac{1}{[A]} - \dfrac{1}{[A]_o}$$

$[A]_o = 0.250\,\text{M};\ [A] = 0.250\,\text{M}\left(\dfrac{100\% - 75\%}{100\%}\right) = 0.0625\,\text{M}$;

$t = \dfrac{16.0\,\text{M}^{-1} - 4.00\,\text{M}^{-1}}{4.5 \times 10^{-3}\,\text{M}^{-1}\,\text{s}^{-1}} = 2.7 \times 10^3\,\text{s}$.

15.1 The reaction proceeds, converting *cis*-butene into *trans*-butene, but as the concentration of *trans*-butene builds up, the reverse reaction becomes increasingly important until, when the concentration ratio is 3, the rates are equal. Then there is no net change. A plot of concentration vs. time appears similar to Figure 15-2, with 0.75 atm for the final pressure of *trans*-butene and 0.25 atm for the final pressure of *cis*-butene:

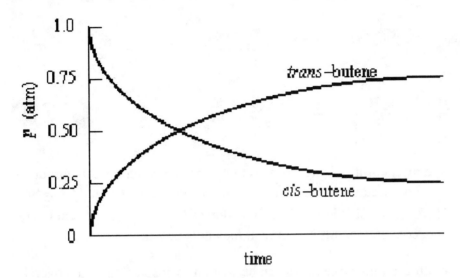

15.3 (a) According to the principle of reversibility, every elementary reaction can run in either direction, so as the system approaches equilibrium, the reverse reactions become important:

$$Cl^- \text{ (aq)} + ClO_2^- \text{ (aq)} \xrightarrow{\ k_1\ } ClO^- \text{ (aq)} + ClO^- \text{ (aq)}$$

$$Cl^- \text{ (aq)} + ClO_3^- \text{ (aq)} \xrightarrow{\ k_2\ } ClO^- \text{ (aq)} + ClO_2^- \text{ (aq)}$$

(b) The net reaction is $3 ClO^- \text{ (aq)} \rightleftharpoons 2 Cl^- \text{ (aq)} + ClO_3^- \text{ (aq)}$. By inspection, the

equilibrium constant expression is $K_{eq} = \dfrac{[ClO_3^-]_{eq}[Cl^-]_{eq}^2}{[ClO^-]_{eq}^3}$

(c) The net reaction is the sum of the two elementary steps, so the equilibrium constant is the product of the rate ratios of the two steps: $K_{eq} = \dfrac{k_1 k_2}{k_{-1} k_{-2}}$.

15.5 A molecular picture of an elementary reaction shows the reactants, the products, and (if necessary) the intermediate collision complex.

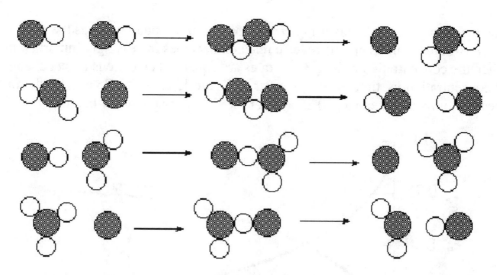

15.7 To test the reversibility of a reaction, set up a system containing the products and observe whether or not reactants form. Here, a solution containing Cl⁻ and ClO₃⁻ ions should react to form some ClO⁻ ions. (If the reaction of ClO⁻ to form Cl⁻ and ClO₃⁻ goes virtually to completion, this experiment may not succeed.)

15.9 Equilibrium constant expressions can be written by inspection of the overall stoichiometry, remembering that pure liquids and solids do not appear in the expression. Equilibrium constant expressions contain product concentrations over reactant concentrations raised to their stoichiometric coefficient:

(a) $K_{eq} = \dfrac{(p_{IF_5})^2_{eq}}{(p_{F_2})^5_{eq}}$; (b) $K_{eq} = \dfrac{1}{(p_{O_2})^5_{eq}}$; (c) $K_{eq} = (p_{CO})^2_{eq}$;

(d) $K_{eq} = \dfrac{1}{(p_{CO})_{eq}(p_{H_2})^2_{eq}}$; (e) $K_{eq} = \dfrac{[H_3O^+]^3_{eq}[PO_4^{3-}]_{eq}}{[H_3PO_4]_{eq}}$.

15.11 Equilibrium constant expressions can be written by inspection of the overall stoichiometry, remembering that pure liquids and solids do not appear in the expression:

(a) 2IF₅ (g) ⇌ I₂ (s) + 5 F₂ (g)　　　　$K_{eq} = \dfrac{(p_{F_2})^5_{eq}}{(p_{IF_5})^2_{eq}}$;

(b) P₄O₁₀(s) ⇌ P₄ (s) + 5 O₂ (g)　　　　　　　　　　　　　 ;
　　　　　　　　　　　　　　　　　　$K_{eq} = (p_{O_2})^5_{eq}$

(c) BaO (s) + 2 CO (g) ⇌ BaCO₃ (s) + C (s)　$K_{eq} = \dfrac{1}{(p_{CO})^2_{eq}}$;

(d) CH₃OH(l) ⇌ CO (g) + 2 H₂ (g)　　　　　　　　　　　　　 ;
　　　　　　　　　　　　　　　$K_{eq} = (p_{CO})_{eq}(p_{H_2})^2_{eq}$

234

(e) $PO_4^{3-}{}_{(aq)} + 3\ H_3O^+{}_{(aq)} \longrightarrow H_3PO_4{}_{(aq)} + 3\ H_2O_{(l)}$ $\quad K_{eq} = \dfrac{[H_3PO_4]_{eq}}{[H_3O^+]^3_{eq}[PO_4^{3-}]_{eq}}$.

15.13   The standard states of gases are gases at 1 atm, the standard states of solutes are solutions at 1 M, and the standard states of solvents and pure liquids and solids are unit mole fractions, $X = 1$.
(a) $p = 1$ atm for F_2 and IF_5, $X = 1$ for I_2;
(b) $p = 1$ atm for O_2, $X = 1$ for P_4 and P_4O_{10};
(c) $p = 1$ atm for CO, $X = 1$ for others;
(d) $p = 1$ atm for CO and H_2, $X = 1$ for CH_3OH;
(e) $c = 1$ M for H_3PO_4, H_3O^+, and PO_4^{3-}, $X = 1$ for H_2O.

15.15   A molecular picture of an equilibrium between phases shows the concentrations of the species and arrows indicating transfers between phases:

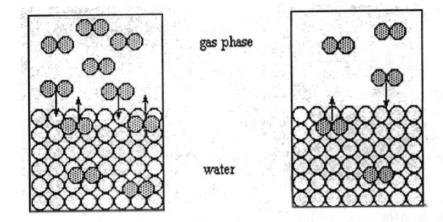

15.17   Table 15-1 indicates that the solubility of O_2 in water increases as the temperature falls from 25 °C to 0 °C. Use Henry's law to calculate the concentration at each temperature:

$$\frac{[O_{2(aq)}]_{eq}}{(p_{O_2})_{eq}} = K_H, \text{ so } [O_{2(aq)}]_{eq} = (p_{O_2})_{eq}K_H$$

The atmosphere is 22 % O_2, so the partial pressure of O_2 is 0.22 atm.

At 25 °C, $[O_{2(aq)}]_{eq} = (0.22 \text{ atm})(1.3 \times 10^{-3}) = 2.9 \times 10^{-4}$ M;

At 0 °C, $[O_{2(aq)}]_{eq} = (0.22 \text{ atm})(2.5 \times 10^{-3}) = 5.5 \times 10^{-4}$ M;

The percentage change is: $(100 \text{ \%})\dfrac{5.5 \times 10^{-4}\,\text{M} - 2.9 \times 10^{-4}\,\text{M}}{5.5 \times 10^{-4}\,\text{M}} = 47\%$.

15.19   To determine an equilibrium constant at standard temperature from thermodynamic tables, calculate $\Delta G^o{}_{reaction}$ from tabulated values for ΔG_f^o and then use Equation 15-3:

$$\Delta G^o = -RT \ln K_{eq}$$

(a) $\Delta G^o{}_{reaction} = [1 \text{ mol}(0 \text{ kJ/mol}) + 1 \text{ mol}(-394.4 \text{ kJ/mol})]$
$\qquad\qquad - [1 \text{ mol}(-137.2 \text{ kJ/mol}) + 1 \text{ mol}(-237.1 \text{ kJ/mol})] = -20.1$ kJ/mol;

$$\ln K_{eq} = -\frac{-2.01 \times 10^4 \text{ J mol}^{-1}}{(8.314 \text{ J mol}^{-1} \text{ K}^{-1})(298 \text{ K})} = 8.11;$$

$$K_{eq} = e^{8.11} = 3.3 \times 10^3;$$

(b) $\Delta G^o_{reaction} = 2 \text{ mol}(-394.4 \text{ kJ/mol})$
$$- [1 \text{ mol}(0 \text{ kJ/mol}) + 2 \text{ mol}(-137.2 \text{ kJ/mol})] = -514.4 \text{ kJ/mol};$$

$$\ln K_{eq} = -\frac{-5.144 \times 10^5 \text{ J mol}^{-1}}{(8.314 \text{ J mol}^{-1} \text{ K}^{-1})(298 \text{ K})} = 207.6;$$

$$K_{eq} = e^{207.6} = 1 \times 10^{90};$$

(c) $\Delta G^o_{reaction} = [1 \text{ mol}(-520.3 \text{ kJ/mol}) + 2 \text{ mol}(-137.2 \text{ kJ/mol})]$
$$- [1 \text{ mol}(0 \text{ kJ/mol}) + 1 \text{ mol}(-1134.4 \text{ kJ/mol})] = 339.7 \text{ kJ/mol};$$

$$\ln K_{eq} = -\frac{3.397 \times 10^5 \text{ J mol}^{-1}}{(8.314 \text{ J mol}^{-1} \text{ K}^{-1})(298 \text{ K})} = -137.1;$$

$$K_{eq} = e^{-137.1} = 3 \times 10^{-60};$$

(d) $\Delta G^o_{reaction} = [3 \text{ mol}(-237.1 \text{ kJ/mol}) + 1 \text{ mol}(74.62 \text{ kJ/mol})]$
$$- [6 \text{ mol}(0 \text{ kJ/mol}) + 3 \text{ mol}(-137.2 \text{ kJ/mol})] = -225.1 \text{ kJ/mol};$$

$$\ln K_{eq} = -\frac{-2.251 \times 10^5 \text{ J mol}^{-1}}{(8.314 \text{ J mol}^{-1} \text{ K}^{-1})(298 \text{ K})} = 90.86;$$

$$K_{eq} = e^{90.86} = 2.9 \times 10^{39}.$$

15.21 To estimate the equilibrium constant at a temperature different from 298 K, calculate $\Delta H^o_{reaction}$ and $\Delta S^o_{reaction}$ at 298 K and then use Equations 13-9 and 15-3:

$$\Delta G^o = \Delta H^o - T\Delta S^o \qquad \Delta G^o = -RT \ln K_{eq}$$

15.19 (b) $\Delta H^o_{reaction} = 2 \text{ mol}(-393.5 \text{ kJ/mol})$
$$- [1 \text{ mol}(0 \text{ kJ/mol}) + 2 \text{ mol}(-110.5 \text{ kJ/mol})] = -566.0 \text{ kJ/mol};$$
$\Delta S^o_{reaction} = 2 \text{ mol}(213.8 \text{ J/mol K})$
$$- [1 \text{ mol}(205.15 \text{ J/mol K}) + 2 \text{ mol}(197.7 \text{ J/mol K})] = -173.0 \text{ J/mol K};$$
$\Delta G^o_{reaction, 250K} = (-566.0 \text{ kJ/mol}) - (250 \text{ K})(-0.1730 \text{ kJ/mol K}) = -522.8 \text{ kJ/mol}$

$$\ln K_{eq} = -\frac{-5.228 \times 10^5 \text{ J mol}^{-1}}{(8.314 \text{ J mol}^{-1} \text{ K}^{-1})(250 \text{ K})} = 251.5;$$

$$K_{eq} = e^{251.5} = 2 \times 10^{109};$$

15.19 (c) $\Delta H^o_{reaction} = 1 \text{ mol}(-548.0 \text{ kJ/mol}) + 2 \text{ mol}(-110.5 \text{ kJ/mol})$
$$- [1 \text{ mol}(0 \text{ kJ/mol}) + 1 \text{mol}(-1213.0 \text{ kJ/mol})] = 444.0 \text{ kJ/mol};$$
$\Delta S^o_{reaction} = [1 \text{ mol}(72.1 \text{ J/mol K}) + 2 \text{ mol}(197.7 \text{ J/mol K})]$
$$- [1 \text{ mol}(5.7 \text{ J/mol K}) + 1 \text{ mol}(112.1 \text{ J/mol K})] = 349.7 \text{ J/mol K};$$
$\Delta G^o_{reaction, 250K} = (444.0 \text{ kJ/mol}) - (250 \text{ K})(0.3497 \text{ J/mol K}) = 356.6 \text{ kJ/mol}$

$$\ln K_{eq} = -\frac{356.6 \times 10^3 \text{ J mol}^{-1}}{(8.314 \text{ J mol}^{-1} \text{ K}^{-1})(250 \text{ K})} = -171.6;$$

$K_{eq} = e^{-171.6} = 3 \times 10^{-75}$

15.23 To estimate the equilibrium constant at a temperature different from 298 K, calculate $\Delta H^\circ_{reaction}$ and ΔS°_{rxn} at 298 K and then use Equations 13-9 and 15-3:

$$\Delta G^\circ = \Delta H^\circ - T\Delta S^\circ \qquad\qquad \Delta G^\circ = -RT \ln K_{eq}$$

15.19 (a)

$\Delta H^\circ_{reaction} = [1 \text{ mol}(-393.5 \text{ kJ/mol}) + 1 \text{ mol}(0 \text{ kJ/mol})]$
$\qquad - [1 \text{ mol}(-110.5 \text{ kJ/mol}) + 1 \text{ mol}(-241.83 \text{ kJ/mol})] = -41.2 \text{ kJ/mol};$

$\Delta S^\circ_{reaction} = [1 \text{ mol}(213.8 \text{ J/mol K}) + 1 \text{ mol}(130.680 \text{ J/mol K})]$
$\qquad - [1 \text{ mol}(197.7 \text{ J/mol K}) + 1 \text{ mol}(188.835 \text{ J/mol K})] = -42.1 \text{ J/mol K};$

$\Delta G^\circ_{reaction,\ 395K} = (-41.2 \text{ kJ/mol}) - (395 \text{ K})(-0.0421 \text{ kJ/mol K}) = -24.6 \text{ kJ/mol}$

$\ln K_{eq} = -\dfrac{-2.46 \times 10^4 \text{ J mol}^{-1}}{(8.314 \text{ J mol}^{-1} \text{ K}^{-1})(395 \text{ K})} = 7.49;$

$K_{eq} = e^{7.49} = 1.8 \times 10^3;$

15.19 (d)

$\Delta H^\circ_{reaction} = 1 \text{ mol}(20.0 \text{ kJ/mol}) + 3 \text{ mol}(-241.83 \text{ kJ/mol})$
$\qquad - [3 \text{ mol}(-110.5 \text{ kJ/mol}) + 6 \text{mol}(0 \text{ kJ/mol})] = -374.0 \text{ kJ/mol};$

$\Delta S^\circ_{reaction} = [1 \text{ mol}(226.9 \text{ J/mol K}) + 3 \text{ mol}(188.835 \text{ J/mol K})]$
$\qquad - [3 \text{ mol}(197.7 \text{ J/mol K}) + 6 \text{ mol}(130.680 \text{ J/mol K})] = -583.8 \text{ J/mol K};$

$\Delta G^\circ_{reaction,\ 395K} = (-374.0 \text{ kJ/mol}) - (395 \text{ K})(-0.5838 \text{ J/mol K}) = -143.4 \text{ kJ/mol}$

$\ln K_{eq} = -\dfrac{-1.434 \times 10^5 \text{ J mol}^{-1}}{(8.314 \text{ J mol}^{-1} \text{ K}^{-1})(395 \text{ K})} = 43.67;$

$K_{eq} = e^{43.67} = 9.2 \times 10^{18}$

15.25 Use Le Châtelier's principle to determine the effect on an equilibrium position caused by adding one reagent:
In reactions (a), (b), and (d), CO is a reactant, so adding CO causes the reaction to go to the right, forming products; in reaction (c), CO is a product, so adding CO causes the reaction to proceed to the left, forming reactants.

15.27 Use Le Châtelier's principle to determine the effect on an equilibrium position caused by a change in conditions:
(a) Solid reactants do not appear in the equilibrium constant expression, so adding $PbCl_2$ has no effect;
(b) Addition of water dilutes the solution, reducing the concentrations of the product ions, so more of the solid will dissolve;
(c) Addition of NaCl increases the concentration of one of the product ions, Cl^-, so some solid will precipitate;
(d) Addition of KNO_3 does not change the concentrations in the equilibrium constant expression, so adding this solid has no effect.

15.29 Use Le Châtelier's principle to determine what changes in conditions will drive an equilibrium in any given direction. This reaction can be driven to the left by removing SO_2, removing Cl_2, adding SO_2Cl_2, or increasing the temperature.

15.31 To calculate an equilibrium constant when amounts are available, convert amounts data into concentrations or pressures using stoichiometric reasoning, then complete an amounts table and substitute into the equilibrium constant expression. Generally, pressures in atmospheres are required, but in this problem, the equilibrium constant expression contains p^2 in the numerator and denominator, so units cancel and amounts can be used directly. Set up an amounts table, using stoichiometric reasoning and the fact that the change is 0.49 mol for CO:

Reaction:	H_2 +	CO_2 ⇌	H_2O +	CO
Initial (mol)	1.00	1.00	0	0
Change (mol)	−0.49	−0.49	+ 0.49	+ 0.49
Equilibrium (mol)	0.51	0.51	0.49	0.49

Now substitute into the equilibrium constant expression and evaluate K:

$$K_{eq} = \frac{(p_{H_2O})_{eq}(p_{CO})_{eq}}{(p_{H_2})_{eq}(p_{CO_2})_{eq}} = \frac{(n_{H_2O})_{eq}(n_{CO})_{eq}}{(n_{H_2})_{eq}(n_{CO_2})_{eq}} = \frac{(0.49)^2}{(0.51)^2} = 0.92$$

15.33 To calculate an equilibrium constant when experimental data concerning amounts are available, identify the reaction, convert amounts data into concentration using stoichiometric reasoning, then complete a concentration table and substitute into the equilibrium constant expression. For solutes, concentrations must be in mol/L.

$$[C_2H_5CO_2H]_{initial} = \frac{0.0500 \text{ mol}}{0.500 \text{ L}} = 0.100 \text{ M}$$

To complete the concentration table, use stoichiometric reasoning and the fact that the change is 1.15×10^{-3} M for H_3O^+:

Reaction:	H_2O +	$C_2H_5CO_2H$ ⇌	$C_2H_5CO_2^-$ +	H_3O^+
Initial (M)	----	0.100	0	0
Change (M)	----	-1.15×10^{-3}	$+ 1.15 \times 10^{-3}$	$+1.15 \times 10^{-3}$
Equilibrium (M)	----	0.099	1.15×10^{-3}	1.15×10^{-3}

Now substitute into the equilibrium constant expression and evaluate K:

$$K_a = \frac{[C_2H_5CO_2^-]_{eq}[H_3O^+]_{eq}}{[C_2H_5CO_2H]_{eq}} = \frac{(1.15 \times 10^{-3})^2}{(0.099)} = 1.3 \times 10^{-5}$$

15.35   To calculate concentrations at equilibrium from initial conditions, set up a concentration table. For a gas phase reaction, concentrations must be expressed in atm. Let x = change in $[H_2]$:

Reaction:	H_2 +	$Br_2 \rightleftharpoons$	2 HBr
Initial (atm)	0	0	10.0
Change (atm)	$+x$	$+x$	$-2x$
Equilibrium (atm)	x	x	$10.0 - 2x$

Now substitute into the equilibrium constant expression and solve for x:

$$K_{eq} = \frac{(p_{HBr})^2_{eq}}{(p_{H_2})_{eq}(p_{Br_2})_{eq}} = \frac{(10.0 - 2x)^2}{(x)(x)} = 1.6 \times 10^5$$

To simplify, assume that $2x \ll 10.0$:

$$1.6 \times 10^5 = \frac{(10.0)^2}{(x)^2} \qquad \text{so}$$

$$x^2 = \frac{100}{1.6 \times 10^5} = 6.25 \times 10^{-4}$$

$x = 2.5 \times 10^{-2} = (p_{H_2})_{eq} = (p_{Br_2})_{eq}$;

$(p_{HBr})_{eq} = 10.0 - 2(2.5 \times 10^{-2}) = 10.0$ atm

$2(2.5 \times 10^{-2}) = 0.050 \ll 10.0$, so the approximation is valid.

15.37   To calculate concentrations at equilibrium from initial conditions, set up a concentration table. For a gas phase reaction, concentrations must be expressed in atm. Let x = change in $[CO_2]$:

Reaction:	$FeO(s)$ +	$CO(g) \rightleftharpoons$	$CO_2(g)$ +	$Fe(s)$
Initial (M)	excess	5.0	0	----
Change (M)	----	$-x$	$+x$	----
Equilibrium (M)	----	$5.0 - x$	x	----

Now substitute into the equilibrium constant expression and solve for x:

$$K_{eq} = \frac{(p_{CO_2})_{eq}}{(p_{CO})_{eq}} = \frac{x}{(5.0 - x)} = 0.403$$

$x = (0.403)(5.0 - x) = 2.015 - 0.403\,x$ so
$1.403\,x = 2.015$ and $x = 1.436$

$(p_{CO_2})_{eq} = 1.4$ atm

$(p_{CO})_{eq} = 5.0 - 1.4 = 3.6$ atm.

15.39   To identify species in solution, first identify the nature of the solute. Strong acids, strong bases, and salts generate ions, while all other substances remain molecular:
(a) Weak acid, major species are H_2O and CH_3CO_2H;
(b) salt, major species are H_2O, NH_4^+ and Cl^-;
(c) salt, major species are H_2O, K^+ and Cl^-;
(d) salt, major species are H_2O, Na^+ and $CH_3CO_2^-$; and
(e) strong base, major species are H_2O, Na^+ and OH^-.

15.41   The equilibria among major species depend on the nature of the species. The proton transfer reaction of water always plays a role:
H_2O (l) + H_2O (l) $\rightleftharpoons$ H_3O^+ (aq) + OH^- (aq)
(a) weak acid equilibrium:
CH_3CO_2H (aq) + H_2O (l) $\rightleftharpoons$ CH_3CO^-(aq) + H_3O^+(aq)
(b) ammonium ion is the weak conjugate acid of NH_3:
NH_4^+ (aq) + H_2O (l) $\rightleftharpoons$ NH_3 (aq) + H_3O^+(aq)
(c) There are no equilibria other than the water equilibrium;
(d) acetate ion is the weak conjugate base of acetic acid:
$CH_3CO_2^-$(aq) + H_2O(l) $\rightleftharpoons$ CH_3CO_2H (aq) + OH^-(aq)
(e) There are no equilibria other than the water equilibrium.

15.43   To identify species in solution, first identify the nature of the solute. Strong acids, strong bases, and salts generate ions, while all other substances remain molecular:
(a) major species are $(CH_3)_2CO$ (acetone) and H_2O;
(b) salt, major species are H_2O, K^+, and Br^-;
(c) strong base, major species are H_2O, Li^+, and OH^-;
(d) strong acid, major species are H_2O, H_3O^+, and HSO_4^-.

15.45   Equilibrium constant expressions have the concentrations of the products in the numerator and the concentrations of the reactants in the denominator with each concentration raised to the power of its stoichiometric coefficient. Remember to omit liquids and solids from the expressions. If the reaction deals with a weak acid (or base) reacting with water it will be related to K_a (or K_b if a base). If the reaction deals with a solid it will be related to K_{sp}.

(a) $K = \dfrac{[ClO_2^-][H_3O^+]}{[HClO_2]} = K_a$;

Chapter 15

(b) $K = \dfrac{1}{[Fe^{3+}][OH^-]^3} = \dfrac{1}{K_{sp}}$;

(c) $K = \dfrac{[HCN]}{[CN^-][H_3O^+]} = \dfrac{1}{K_a}$.

15.47 To identify the spectator ions present first determine the reaction (if any) that occurs when the solutions are mixed. Whatever does not react will then be a spectator ion.

(a) $CH_3CO_2H + OH^- \rightleftharpoons CH_3CO_2^- + H_2O$; spectator ions : Na^+;

(b) $3\,Ca^{2+} + 2\,PO_4^{3-} \rightleftharpoons Ca_3(PO_4)_2$; spectator ions : Cl^- and K^+;

(c) $H_3O^+ + OH^- \rightleftharpoons 2\,H_2O$; spectator ions: K^+ and NO_3^-.

15.49 Equilibrium constant expressions are determined by the stoichiometry of the overall reaction, with pure liquids, solids, and solvent omitted from the expression. Equilibrium constant expressions have the concentrations of the products in the numerator and the concentration of the reactants in the denominator with each concentration raised to the power of its stoichiometric coefficient:

(a) $K_{eq} = (p_{CO_2})_{eq}(p_{H_2O})_{eq}$; (b) $K_{eq} = \dfrac{(p_{NH_3})^4_{eq}(p_{O_2})^3_{eq}}{(p_{N_2})^2_{eq}}$; (c) $K_{eq} = \dfrac{(p_{CH_3CHO})^2_{eq}}{(p_{C_2H_4})^2_{eq}(p_{O_2})_{eq}}$;

(d) $K_{eq} = [Ag^+]^2_{eq}[SO_4^{2-}]_{eq}$; (e) $K_{eq} = (p_{H_2S})_{eq}(p_{NH_3})_{eq}$.

15.51 This problem describes an equilibrium reaction. We are asked to determine the equilibrium pressures of all the gases. To calculate pressures at equilibrium from initial conditions, set up a concentration table, write the K expression, and solve for the pressures. For a gas phase reaction, concentrations must be expressed in atm.

Use $PV = nRT$ to calculate the initial pressure of phosgene:

$P = \dfrac{(0.31\,mol)(0.08206\,\frac{L\,atm}{mol\,K})(360 + 273\,K)}{2.55\,L} = 6.3\,atm$

Reaction:	COCl$_2$ $\rightleftharpoons$	CO +	Cl$_2$
Initial (atm)	6.3	0	0
Change (atm)	$-x$	$+x$	$+x$
Equilib(atm)	6.3 - x	x	x

Now substitute into the equilibrium constant expression and solve for x:

241

$K_{eq} = 8.3 \times 10^{-4} = \dfrac{(p_{CO})_{eq}(p_{Cl_2})_{eq}}{(p_{COCl_2})_{eq}} = \dfrac{(x)^2}{6.3 - x}$; Assume that $x \ll 6.3$:

$8.3 \times 10^{-4} = \dfrac{x^2}{6.3}$, so

$x^2 = (6.3)(8.3 \times 10^{-4}) = 5.23 \times 10^{-3}$;

$x = 7.2 \times 10^{-2}$ atm $= (p_{CO})_{eq} = (p_{Cl_2})_{eq}$;

$(p_{COCl_2})_{eq} = 6.3 - 7.2 \times 10^{-2} = 6.2$ atm;

7.23×10^{-2} is about 1.1% of 6.3, so x is $\ll 6.3$ and the approximation is valid.

15.53 Equilibrium exists when the rates of evaporation and condensation are equal. A test tube has a smaller surface area than a petri dish, so fewer molecules escape per unit time from the test tube. However, the rate of escape per unit surface area is the same for both samples, so equilibrium is established at the same pressure (molecules colliding per unit surface area) for both samples.

15.55 At equilibrium, a molecular picture should show the presence of both reactants and products, in relative amounts that are determined by the value of the equilibrium constant. Set up a "concentration" table to determine how many of each species are present at equilibrium:

Species:

	Initial	12	12	0	0
Change	$-x$	$-x$	$+x$	$+x$
Equilibrium	$12 - x$	$12 - x$	x	x

Substitute in the equilibrium constant expression and solve for x:

$K_{eq} = 25 = \dfrac{(x)^2}{(12 - x)^2}$; Taking the square root of each side gives $5 = \dfrac{x}{(12 - x)}$;

$(60 - 5x) = x$,

$6x = 60$ and $x = 10$

The molecular picture should show $(12 - 10) = 2$ of each reactant and 10 of each product:

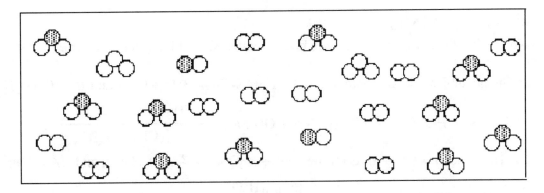

15.57 To calculate equilibrium pressures, set up a concentration table. In this problem, the table can be short, because two of the three equilibrium pressures are provided:

Reaction:	Br_2 (g) +	I_2 (g) $\rightleftharpoons$	2 IBr (g)
Equilib(atm)	0.512	0.327	x

Substitute into the equilibrium constant expression and solve for x:

$$322 = \frac{[IBr]^2_{eq}}{[Br_2]_{eq}[I_2]_{eq}} = \frac{x^2}{(0.512)(0.327)}, \text{ from which}$$

$x^2 = (322)(0.512)(0.327) = 53.91;$
$x = [IBr]_{eq} = 7.34$ atm.

15.59 To determine an equilibrium constant at standard temperature from thermodynamic tables, calculate ΔG^o_{rxn} from tabulated values for ΔG^o_f; calculate ΔH^o_{rxn} and ΔS^o_{rxn} at 298K and then use Equations 13-9 and 15-3 to estimate the equilibrium constant at a temperature different from 298 K:

$$\Delta G^o = \Delta H^o - T\Delta S^o \qquad \Delta G^o = -RT \ln K_{eq}$$

at 298 K:

$\Delta G^o_{reaction} = 1$ mol(99.8 kJ/mol) $- 2$ mol(51.3 kJ/mol) $= -2.8$ kJ/mol;

$$\ln K_{eq} = -\frac{-2.8 \times 10^3 \text{ J mol}^{-1}}{(8.314 \text{ J mol}^{-1} \text{ K}^{-1})(298 \text{ K})} = 1.1;$$

$K_{eq} = e^{1.1} = 3.0;$

At 525 K:

$\Delta H^o_{reaction} = 1$ mol(11.1 kJ/mol) $- 2$ mol(33.2 kJ/mol) $= -55.3$ kJ/mol;

$\Delta S^o_{reaction} = 1$ mol(304.4 J/mol K) $- 2$ mol(240.1 J/mol K) $= -175.8$ J/mol K;

$$\Delta G^o_{reaction} = (-55.3 \text{ kJ/mol}) - (525 \text{ K})\left(\frac{-175.8 \text{ J}}{1 \text{ mol K}}\right)\left(\frac{10^{-3} \text{ kJ}}{1 \text{ J}}\right) = 37.0 \text{ kJ/mol}$$

$$\ln K_{eq} = -\frac{3.70 \times 10^4 \text{ J mol}^{-1}}{(8.314 \text{ J mol}^{-1} \text{ K}^{-1})(525 \text{ K})} = -8.48;$$

$K_{eq} = e^{-8.48} = 2.1 \times 10^{-4}$

Notice that the exothermic reaction has a smaller K_{eq} at higher temperature.

15.61 (a) The general weak base reaction is $B + H_2O \rightleftharpoons BH^+ + OH^-$, and $B = (CH_3)_3N$:

$$(CH_3)_3N + H_2O \rightleftharpoons (CH_3)_3NH^+ + OH^-, \quad K_b = \frac{[(CH_3)_3NH^+]_{eq}[OH^-]_{eq}}{[(CH_3)_3N]_{eq}};$$

(b) The general weak acid reaction is $HA + H_2O \rightleftharpoons A^- + H_3O^+$, and $HA = HF$:

$$HF + H_2O \rightleftharpoons F^- + H_3O^+, \quad K_a = \frac{[F^-]_{eq}[H_3O^+]_{eq}}{[HF]_{eq}};$$

(c) A solubility reaction involves solid dissolving to produce ions:

$$CaSO_{4\,(s)} \rightleftharpoons Ca^{2+}{}_{(aq)} + SO_4{}^{2-}{}_{(aq)}, \quad K_{sp} = [Ca^{2+}]_{eq}[SO_4{}^{2-}]_{eq}.$$

15.63 To predict effects of changes on equilibrium position, apply Le Châtelier's principle: The system will respond in the direction that reduces the effect of the change. The reaction in example 15-15 is: $2\,NO_{(g)} + O_{2\,(g)} \rightleftharpoons 2\,NO_{2\,(g)}$,

(a) NO_2 is a product, so reducing its pressure causes the reaction to proceed to the right;

(b) There are more gases on the reactant side, thus by doubling the volume the reaction proceeds to the left;

(c) Adding Ar does not change any of the pressures in the equilibrium expression, so this change has no effect.

15.65 To determine the volume that will contain a single molecule of SnH_4 we first need to find the equilibrium pressure of the gas. To calculate equilibrium pressures, set up a concentration table:

Reaction:	$Sn_{(s)}$ +	$2\,H_{2\,(g)} \rightleftharpoons$	$SnH_{4\,(g)}$
Initial (atm)	----	200	0
Change (atm)	----	$-2x$	$+x$
Equilib(atm)	----	$200 - 2x$	x

Because K_{eq} is very small, we assume that $2x \ll 200$:

$$1.07 \times 10^{-33} = \frac{[SnH_4]_{eq}}{[H_2]_{eq}^2} = \frac{x}{(200)^2},$$

from which $x = (1.07 \times 10^{-33})(200)^2 = 4.28 \times 10^{-29}$

Thus $[SnH_4]_{eq} = 4.28 \times 10^{-29}$ atm (a very small pressure).

To calculate the volume that would be expected to contain a single molecule, use the ideal gas equation, rearranged to solve for V:

$$V = \frac{RT}{N_A p} = \frac{(0.0821 \frac{L\,atm}{mol\,K})(298\,K)}{(6.022 \times 10^{23}\,mol^{-1})(4.28 \times 10^{-29}\,atm)} = 9.49 \times 10^5\,L.$$

15.67 To calculate an equilibrium constant from initial and equilibrium conditions, set up a concentration table:

Reaction:	CCl_4 (g) $\rightleftharpoons$	2 Cl_2 (g) +	C (s)
Initial (atm)	1.00	0	----
Change (atm)	– x	+ 2x	----
Equilib(atm)	1.00 – x	2x	----

The problem gives the total equilibrium pressure, which is the sum of partial pressures: 1.35 atm = (1.00 – x) + 2 x = 1.00 + x, from which x = (1.35 – 1.00) = 0.35; Substitute into the equilibrium constant expression and calculate K:

$$K_{eq} = \frac{(p_{Cl_2})^2_{eq}}{(p_{CCl_4})_{eq}} = \frac{[(2)(0.35)]^2}{(1.00-0.35)} = \frac{0.70^2}{0.65} = 0.75.$$

15.69 (a) The equilibrium constant expression for a reaction can be written by inspection of the stoichiometry of the reaction, omitting pure liquids, solids, and solvents:

$$K_{eq} = \frac{[CO_3^{2-}]_{eq}}{(p_{CO_2})_{eq}[OH^-]^2_{eq}}$$

(b) Use Le Châtelier's Principle to predict the effect of changes on a system at equilibrium. Dissolving Na_2CO_3 leads to an increase in the concentration of CO_3^{2-}, so the equilibrium shifts to the left and the pressure of CO_2 increases.

(c) At first glance, it may appear that HCl will not affect this equilibrium, but recall that HCl is a strong acid which generates H_3O^+ in solution. This, in turn, will react with OH^-, reducing the concentration of a reactant. Again the equilibrium shifts to the left and the pressure of CO_2 increases.

15.71 To determine an equilibrium constant at standard temperature from thermodynamic tables, calculate $\Delta G^o_{reaction}$ from tabulated values for ΔG^o_f; to estimate the equilibrium constant at a temperature different from 298 K, calculate $\Delta H^o_{reaction}$ and $\Delta S^o_{reaction}$ at 298 K and then use Equations 13-9 and 15-3:

$$\Delta G^o = \Delta H^o - T\Delta S^o \qquad \Delta G^o = -RT \ln K_{eq}$$

(a) $\Delta G^o_{reaction} = $ 1 mol(–210.7 kJ/mol)

$\qquad$ – [1 mol(31.8 kJ/mol) + 1 mol(–178.6 kJ/mol)] = – 63.9 kJ/mol;

$$\ln K_{eq} = -\frac{-6.39 \times 10^4\,J\,mol^{-1}}{(8.314\,J\,mol^{-1}\,K^{-1})(298\,K)} = 25.8;$$

$K_{eq} = e^{25.8} = 1.6 \times 10^{11}$;

(b) When the equilibrium pressure is 1.00 atm, $K_{eq} = 1$, $\ln(K_{eq}) = 0$ and $\Delta G^\circ_{rxn} = 0$:

$$0 = \Delta H^\circ - T\Delta S^\circ, \qquad \text{so} \qquad T\Delta S^\circ = \Delta H^\circ \qquad \text{and} \qquad T = \frac{\Delta H^\circ}{\Delta S^\circ}$$

$\Delta H^\circ_{reaction} = 1\ \text{mol}(-265.4\ \text{kJ/mol})$
$\qquad\qquad - [1\ \text{mol}(61.4\ \text{kJ/mol}) + 1\ \text{mol}(-224.3\ \text{kJ/mol})] = -102.5\ \text{kJ/mol}$;
$\Delta S^\circ_{reaction} = 1\ \text{mol}(191.6\ \text{J/mol K})$
$\qquad\qquad - [1\ \text{mol}(175.0\ \text{J/mol K}) + 1\ \text{mol}(146.0\ \text{J/mol K})] = -129.4\ \text{J/mol K}$;

$$T = \left(\frac{-102.5\ \text{kJ}}{1\ \text{mol}}\right)\left(\frac{10^3\ \text{J}}{1\ \text{kJ}}\right)\left(\frac{1\ \text{mol K}}{-129.4\ \text{J}}\right) = 792\ \text{K}$$

(c) $\Delta G^\circ_{reaction,\ 1050K} = (-102.5\ \text{kJ/mol}) - (1050\ \text{K})\left(\frac{-129.4\ \text{J}}{1\ \text{mol K}}\right)\left(\frac{10^{-3}\ \text{kJ}}{1\ \text{J}}\right) = 33.4\ \text{kJ/mol}$

$$\ln K_{eq} = -\frac{3.34 \times 10^4\ \text{J mol}^{-1}}{(8.314\ \text{J mol}^{-1}\ \text{K}^{-1})(1050\ \text{K})} = -3.83;$$

$K_{eq} = e^{-3.83} = 2.2 \times 10^{-2}$

Notice that K_{eq} decreases as T increases for this exothermic reaction.

15.73 To find an equilibrium total pressure, it is necessary to calculate equilibrium partial pressures of all gaseous participants. First determine the initial pressures of the gases, using the ideal gas equation:

$$p = \frac{nRT}{V} = \frac{(0.500\ \text{mol})(0.08206\ \tfrac{\text{L atm}}{\text{mol K}})(1020 + 273\ \text{K})}{1.00\ \text{L}} = 53.1\ \text{atm}$$

Set up a concentration table, write the equilibrium expression and solve for the pressure:

Reaction:	$C_{(s)}$ +	$CO_2\ _{(g)} \rightleftharpoons$	$2\ CO_{(g)}$
Initial (atm)	----	53.1	53.1
Change (atm)	----	$-x$	$+2x$
Equilib(atm)	----	$53.1 - x$	$53.1 + 2x$

Substitute into the equilibrium constant expression and solve for x:

$$167.5 = \frac{(53.1 + 2x)^2}{(53.1 - x)}$$

$(167.5)(53.1 - x) = (53.1 + 2x)^2$

Chapter 15

$8894 - 167.5\,x = 2820 + 212.4\,x + 4\,x^2;$

$4\,x^2 + 379.9\,x - 6074 = 0$

$$x = \frac{-b \pm \sqrt{b^2 - 4ac}}{2a} = \frac{-379.9 \pm \sqrt{(379.9)^2 - 4(4)(-6074)}}{2(4)} = \frac{-379.9 \pm 491.4}{8} = 13.9;$$

Rule out the negative value, which would give a negative pressure;

$(p_{CO_2})_{eq} = 53.1 - 13.9 = 39.2$ atm;

$(p_{CO})_{eq} = 53.1 + 2(13.9) = 80.9$ atm;

$P_{total} = 39.2 + 80.9 = 1.20 \times 10^2$ atm.

15.75 This molecular picture illustrates starting conditions and equilibrium conditions. The symbols indicate BG_3 as the starting material and G_2 and GB as products. Count numbers of symbols to determine initial and equilibrium concentrations. Set up a concentration table using numbers of symbols:

Initial	15	0	0
Change	-12	+12	+12
Equilibrium	3	12	12

(a) From the amounts of change, the stoichiometry is 1:1, so the net reaction is

;

(b) Use numbers of symbols at equilibrium to calculate the equilibrium constant:

$$K_{eq} = \frac{[GB]_{eq}[G_2]_{eq}}{[BG_3]_{eq}} = \frac{(12)(12)}{(3)} = 48 \, .$$

16.1   HBr is a strong acid and will therefore give its hydrogen atom to a water molecule, making a hydronium cation and a bromine anion.

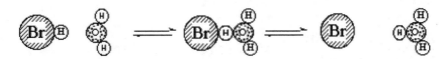

16.3 $HClO_4$ is a strong acid and dissolves in water to generate H_3O^+ cations and ClO_4^- anions. Since there are no bases present for the hydronium ion to react with, the only equilibrium occurring is proton transfer reaction between water molecules:

$$H_2O + H_2O \rightleftharpoons OH^- + H_3O^+ \qquad K_w = 1.00 \times 10^{-14}$$

To determine the concentrations of hydroxide and hydronium ions, set up a concentration table, write the equilibrium expression, and solve for the final concentrations.

The initial concentration of hydronium ions is the same as the concentration of perchloric acid, $[H_3O^+] = 1.25 \times 10^{-3}$M. The concentration table is:

Reaction:	H_2O +	$H_2O \rightleftharpoons$		H_3O^+ +	OH^-
Start (M)	----	----		1.25×10^{-3}	0
Change (M)	----	----		$+ x$	$+ x$
Final (M)	solvent	solvent		$1.25 \times 10^{-3} + x$	x

The equilibrium expression is:
$K_w = [H_3O^+][OH^-]$
$1.00 \times 10^{-14} = [0.00125 + x][x]$,

We can make the assumption that $x \ll 0.00125$M
$1.00 \times 10^{-14} = [0.00125][x]$,

$x = 8.00 \times 10^{-12}$ M $= [OH^-]$;
$[H_3O^+] = 1.25 \times 10^{-3}$ M $+ 8.00 \times 10^{-12}$ M $= 1.25 \times 10^{-3}$ M
Thus, the assumption that $x \ll 0.00125$ is valid.

16.5   We are asked to determine the final concentrations of all the ions in a final solution. Begin by analyzing the chemistry. HCl is strong acid and dissolves in water to generate H_3O^+ cations and Cl^- anions. Any water solution always has OH^- and H_3O^+ ions with the equilibrium:

$$H_2O + H_2O \rightleftharpoons OH^- + H_3O^+ \qquad K_w = 1.00 \times 10^{-14}$$

Therefore, the ions present in this solution are: H_3O^+, OH^-, and Cl^-

This is a dilution type problem. The first step is to determine the concentration of HCl in the flask after the dilution.

$$M_iV_i = M_fV_f$$

$$M_f = \frac{M_iV_i}{V_f} = \frac{(12.1\,M)(1.00\,mL)}{100.\,mL} = 0.121\,M\,HCl\text{ in final solution.}$$

Since Cl⁻ is a spectator ion, its concentration is the same as that of HCl, [Cl⁻] = 0.121 M. The rest of the ion concentrations are determined by the equilibrium. Set up a concentration table, write the equilibrium expression, and solve for the final concentrations.

Reaction:	H_2O +	$H_2O \rightleftharpoons$	H_3O^+ +	OH^-
Start (M)	----	----	0.121	0
Change (M)	----	----	+ x	+ x
Final (M)	solvent	solvent	0.121 + x	x

The equilibrium expression is:
$K_w = [H_3O^+][OH^-]$
$1.00 \times 10^{-14} = x(0.121 + x)$;
assume $x \ll 0.121$

$1.00 \times 10^{-14} = x(0.121)$,
$x = [OH^-] = 8.26 \times 10^{-14}\,M$
$[H_3O^+] = 0.121\,M + 8.26 \times 10^{-14}\,M = 0.121\,M$

Here are the final concentrations:
$[Cl^-] = [H_3O^+] = 0.121\,M$
$[OH^-] = 8.26 \times 10^{-14}\,M$

16.7 We are asked to determine the concentrations of hydroxide and hydronium ions in the solution. Begin by analyzing the chemistry. HCl is a strong acid and dissolves in water to generate H_3O^+ cations and Cl⁻ anions. Any water solution always has OH⁻ and H_3O^+ ions with the equilibrium:

$$H_2O + H_2O \rightleftharpoons OH^- + H_3O^+ \qquad\qquad K_w = 1.00 \times 10^{-14}$$

The first step is to determine the initial number of moles of HCl gas and convert that to concentration of HCl dissolved in the solution:
$MM_{HCl} = 1.008\,g/mol + 35.453\,g/mol = 36.461\,g/mol$

$$0.488\,g\left(\frac{1\,mol}{36.461\,g}\right) = 0.01338\,mol\,HCl\text{ dissolved.}$$

$$[HCl] = \frac{0.01338\,mol}{0.325\,L} = 0.0412\,M$$

To determine the concentrations of the ions, construct a concentration table, write the equilibrium expression, and solve for the final concentrations. The initial concentration of hydronium ions will be the same as the concentration of HCl, $[H_3O^+] = 0.0412$ M.

Reaction:	H_2O +	H_2O ⇌	H_3O^+ +	OH^-
Start (M)	----	----	0.0412	0
Change (M)	----	----	$+x$	$+x$
Final (M)	solvent	solvent	$0.0412 + x$	x

The equilibrium expression is:
$K_w = [H_3O^+][OH^-]$
$1.00 \times 10^{-14} = (0.0412 + x)x$;
Assume $x \ll 0.0412$

$1.00 \times 10^{-14} = (0.0412)x$;
$x = 2.43 \times 10^{-13}$ M $= [OH^-]$;   the assumption is vaild

$[H_3O^+] = 0.0412$ M $+ 2.43 \times 10^{-13}$ M $= 0.0412$ M

16.9   Conversion from hydronium ion molarity to pH is accomplished by taking logarithm to base ten and changing sign, pH $= - (\log[H_3O^+])$:
(a) -0.60; (b) 5.426; (c) 2.32; and (d) 3.593.

16.11 Because pH + pOH = 14.00, pH = 14.00 − pOH. Convert from hydroxide ion molarity to pOH by taking logarithm to base ten and changing sign, pOH $= - (\log[OH^-])$.
Thus, pH $= 14 + \log[OH^-]$:
(a) 14.60 ; (b) 8.574; (c) 11.68; and (d) 10.407.

16.13 Take 10^{-pH} to convert pH into hydronium ion concentration:
(a) 0.22 M; (b) 1.4×10^{-8} M; (c) 2.1×10^{-4} M; (d) 4.7×10^{-15} M.

16.15 Use pH + pOH = 14.00 to convert pH to pOH. Then take 10^{-pOH} to convert pOH into hydroxide ion concentration:
(a) pOH = 13.34, $[OH^-] = 4.6 \times 10^{-14}$ M;
(b) pOH = 6.15, $[OH^-] = 7.1 \times 10^{-7}$ M;
(c) pOH = 10.32, $[OH^-] = 4.8 \times 10^{-11}$ M;

(d) pOH = – 0.33, [OH⁻] = 2.1 M.

16.17 To calculate the pH of a solution, it is necessary to determine either the hydronium ion concentration or the hydroxide ion concentration.

For each species determine the initial concentrations, construct a concentration table, write the equilibrium expression, and solve for the concentrations.

(a) Strong base, carry out the water equilibrium to determine H_3O^+

Reaction:	H_2O +	H_2O ⇌	OH^- +	H_3O^+
Start (M)	----	----	1.5	0
Change (M)	----	----	+ x	+ x
Final (M)	solvent	solvent	1.5 + x	x

The equilbrium expression is:
$K_w = 1.0 \times 10^{-14} = [H_3O^+][OH^-] = x(1.5 + x)$; assume $x \ll 1.5$

$1.0 \times 10^{-14} = x(1.5)$; from which $x = 6.7 \times 10^{-15}$ M $= [H_3O^+]$; assumption valid

pH = -log(6.7 x 10^{-15}) = 14.18.

(b) weak base, carry out equilibrium calculation to determine [OH⁻]:

Reaction:	H_2O +	C_5H_5N ⇌	$C_5H_5NH^+$ +	OH^-
Start (M)	----	1.5	0	0
Change (M)	----	-x	+x	+x
Final (M)	Solvent	1.5 – x	x	x

The equilibrium expression is:

$K_b = 1.7 \times 10^{-9} = \dfrac{[C_5H_5NH^+]_{eq}[OH^-]_{eq}}{[C_5H_5N]_{eq}} = \dfrac{x^2}{1.5 - x}$; Assume $x \ll 1.5$:

$1.7 \times 10^{-9} = \dfrac{x^2}{1.5}$, from which $x^2 = 2.55 \times 10^{-9}$ and $x = 5.0 \times 10^{-5}$; assumption is valid

[OH⁻] = 5.0 x 10^{-5} M,
pOH = -log (5.0 x10^{-5}) = 4.30, and
pH = 14.00 – 4.30 = 9.70;

(c) weak base, carry out equilibrium calculation to determine [OH⁻]:

Reaction:	H_2O +	$NH_2OH \rightleftharpoons$	NH_3OH^+ +	OH^-
Start (M)	----	1.5	0	0
Change (M)	----	$-x$	$+x$	$+x$
Final (M)	solvent	$1.5 - x$	x	x

The equilibrium expression is:

$$K_b = 8.7 \times 10^{-9} = \frac{[NH_3OH^+]_{eq}[OH^-]_{eq}}{[NH_2OH]_{eq}} = \frac{x^2}{1.5 - x}; \quad \text{Assume } x \ll 1.5:$$

$8.7 \times 10^{-9} = \dfrac{x^2}{1.5}$, from which $x^2 = 1.31 \times 10^{-8}$ and $x = 1.1 \times 10^{-4}$; assumption valid

[OH⁻] = 1.1×10^{-4} M,
pOH = $-\log(1.1 \times 10^{-4}) = 3.96$, and
pH = $14.00 - 3.96 = 10.04$;

(d) Weak acid, carry out equilibrium calculation to determine [H₃O⁺]:

Reaction:	H_2O +	$HCO_2H \rightleftharpoons$	HCO_2^- +	H_3O^+
Start (M)	----	1.5	0	0
Change (M)	----	$-x$	$+x$	$+x$
Final (M)	solvent	$1.5 - x$	x	x

The equilibrium expression is:

$$K_a = 1.8 \times 10^{-4} = \frac{[HCO_2^-]_{eq}[H_3O^+]_{eq}}{[HCO_2H]_{eq}} = \frac{x^2}{1.5 - x}; \quad \text{Assume } x \ll 1.5:$$

$1.8 \times 10^{-4} = \dfrac{x^2}{1.5}$, from which $x^2 = 2.7 \times 10^{-4}$ and $x = 1.6 \times 10^{-2}$; assumption is valid

[H₃O⁺] = 1.6×10^{-2} M, and
pH = $-\log(1.6 \times 10^{-2}) = 1.80$

16.19 HONH$_2$ is a weak base and will therefore take a proton from water to form hydroxide ions.

HCO$_2$H is a weak acid and will give a proton to water to form hydronium ions.

16.21 Follow standard procedures for dealing with equilibrium problems:
(a) HN$_3$ is a weak acid. Major species: HN$_3$ and H$_2$O, Minor species: N$_3^-$ and H$_3$O$^+$, and OH$^-$
(b) Construct the concentration table, write the equilibrium expression, and solve for the concentrations.

Reaction:	H$_2$O +	HN$_3$ ⇌	N$_3^-$ +	H$_3$O$^+$
Start (M)	----	1.50	0	0
Change (M)	----	- x	+ x	+ x
Final (M)	solvent	1.50 - x	x	x

$$K_a = 2.5 \times 10^{-5} = \frac{[N_3^-]_{eq}[H_3O^+]_{eq}}{[HN_3]_{eq}} = \frac{x^2}{1.50 - x} ; \text{ assume that } x << 1.50$$

$$2.5 \times 10^{-5} = \frac{x^2}{1.50};$$

$$x^2 = 3.75 \times 10^{-5} \text{ and } x = 6.1 \times 10^{-3}$$

[H$_3$O$^+$] = [N$_3^-$] = 6.1 $\times$ 10^{-3} M, and [HN$_3$] = 1.50 M – 6.1 $\times$ 10^{-3} M = 1.50 M

$$[OH^-] = \frac{1.0 \times 10^{-14}}{6.1 \times 10^{-3}} = 1.6 \times 10^{-12} \text{ M}$$

(c) pH = – log (6.1 $\times$ 10^{-3}) = 2.21;

(d) The dominant equilibrium is proton transfer from water to HN$_3$:

16.23 Follow standard procedures for dealing with equilibrium problems:
(a) $N(CH_3)_3$ is a weak base. Major species: $N(CH_3)_3$, H_2O, Minor species:
$HN(CH_3)_3^+$, OH^-, and H_3O^+;
(b) Construct the concentration table, write the equilibrium expression, and solve for the concentrations.

Reaction:	H_2O +	$N(CH_3)_3 \rightleftharpoons$	OH^- +	$HN(CH_3)_3^+$
Start (M)	----	0.350	0	0
Change (M)	----	$-x$	$+x$	$+x$
Final (M)	solvent	$0.350 - x$	x	x

The equilibrium expression is:

$$K_b = 6.5 \times 10^{-5} = \frac{[HN(CH_3)_3^+]_{eq}[OH^-]_{eq}}{[N(CH_3)_3]_{eq}} = \frac{x^2}{0.350 - x}; \quad \text{assume } x \ll 0.350$$

$6.5 \times 10^{-5} = \dfrac{x^2}{0.350};$

$x^2 = 2.28 \times 10^{-5}$ and $x = 4.8 \times 10^{-3}$; assumption is valid

$[OH^-] = [HN(CH_3)_3^+] = 4.8 \times 10^{-3}$ M,
$[N(CH_3)_3] = 0.350 - 4.8 \times 10^{-3} = 0.345$ M;

$[H_3O^+] = \dfrac{1.0 \times 10^{-14}}{4.8 \times 10^{-3}} = 2.1 \times 10^{-12}$ M

(c) pH $= -\log(2.1 \times 10^{-12}) = 11.68$;
(d) The dominant equilibrium is proton transfer from water to trimethylamine:

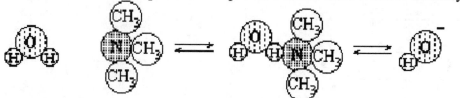

16.25 Examine the chemical formula to determine the nature of a compound: (a) weak base;
(b) weak acid, (c) weak acid; (d) strong base.

16.27 The conjugate base will have one less H and one less charge, a conjugate acid will have one more H and one higher charge:

Problem 16.17

C_5H_5N conjugate acid is $C_5H_5NH^+$;

$HONH_2$ conjugate acid is $HONH_3^+$;

HCO_2H conjugate base is HCO_2^-;

16.29 Conjugate pairs are connected. Any aqueous solution always has OH^- and H_3O^+ ions with the equilibrium:

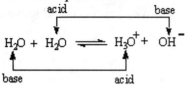

(a) NH_3 is a weak base:

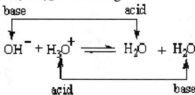

(b) HCNO is a weak acid:

base   acid
$H_2O + HCNO \rightleftharpoons H_3O^+ + CNO^-$
acid   base

(c) HClO is a weak acid:

base   acid
$H_2O + HClO \rightleftharpoons H_3O^+ + ClO^-$
acid   base

(d) $Ba(OH)_2$ is a strong base:

base   acid
$OH^- + H_3O^+ \rightleftharpoons H_2O + H_2O$
acid   base

16.31 Follow standard procedures for dealing with equilibrium problems:

(a) The compound is a salt, so major species are Na^+, SO_3^{2-}, and H_2O;

(b) The species with acid-base properties are SO_3^{2-} (a weak base) and H_2O, so the dominant equilibrium is: $H_2O_{(l)} + SO_3^{2-}{}_{(aq)} \rightleftharpoons HSO_3^-{}_{(aq)} + OH^-{}_{(aq)}$

(c) Construct the concentration table, write the equilibrium expression, and carry out an equilibrium calculation to determine $[OH^-]$:

Reaction:	H_2O +	$SO_3^{2-} \rightleftharpoons$	HSO_3^- +	OH^-
Start (M)	---	0.45	0	0
Change (M)	---	$-x$	$+x$	$+x$
Final (M)	solvent	$0.45 - x$	x	x

The equilibrium reaction is a weak base proton transfer. HSO_3^- is the species resulting from the gain of a proton from SO_3^{2-}, so use K_{a2} to determine $pK_b = 14.00 - pK_a$:
$pK_a = 7.20$, $pK_b = 6.80$;

The equilibrium expression is:

$$K_b = 1.6 \times 10^{-7} = \frac{[HSO_3^-]_{eq}[OH^-]_{eq}}{[SO_3^{2-}]_{eq}} = \frac{x^2}{0.45 - x}; \quad \text{Assume } x \ll 0.45:$$

$1.6 \times 10^{-7} = \dfrac{x^2}{0.45}$, from which $x^2 = 7.2 \times 10^{-8}$ and $x = 2.7 \times 10^{-4}$; assumption valid

$[OH^-] = 2.7 \times 10^{-4}$ M,
$pOH = -\log(2.7 \times 10^{-4}) = 3.57$, and
$pH = 14.00 - 3.57 = 10.43$.

16.33 Follow standard procedures for dealing with equilibrium problems:
(a) The compound is a salt, so major species are NH_4^+, NO_3^-, and H_2O;
(b) The species with acid-base properties are NH_4^+ (a weak acid) and H_2O, so the dominant equilibrium is: $H_2O_{(l)} + NH_4^+{}_{(aq)} \rightleftharpoons NH_3{}_{(aq)} + H_3O^+{}_{(aq)}$
(c) carry out equilibrium calculation to determine $[H_3O^+]$:

Reaction:	H_2O +	$NH_4^+ \rightleftharpoons$	NH_3 +	H_3O^+
Start (M)	---	0.0100	0	0
Change (M)	---	$-x$	$+x$	$+x$
Final (M)	solvent	$0.0100 - x$	x	x

$$pK_a = 9.25, K_a = 5.6 \times 10^{-10} = \frac{[NH_4^+]_{eq}[H_3O^+]_{eq}}{[NH_3]_{eq}} = \frac{x^2}{0.0100 - x}; \text{ Assume } x \ll 0.0100:$$

$5.6 \times 10^{-10} = \dfrac{x^2}{0.0100}$, from which

$x^2 = 5.6 \times 10^{-12}$ and $x = 2.37 \times 10^{-6}$; assumption valid

$[H_3O^+] = 2.37 \times 10^{-6}$ M, and

$pH = -\log(2.37 \times 10^{-6}) = 5.63.$

16.35 $H_2O + SO_2 \rightleftharpoons H_2SO_3$
$H_2SO_3 + H_2O \rightleftharpoons HSO_3^- + H_3O^+$ (determines pH)
$HSO_3^- + H_2O \rightleftharpoons SO_3^{2-} + H_3O^+$

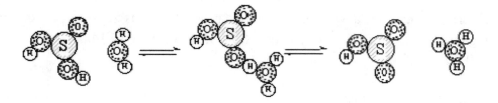

16.37 $FeO(s) + 2 H_3O^+(aq) \rightleftharpoons Fe^{2+}(aq) + 3 H_2O(l)$

16.39 (a) H_2SO_4 is stronger because anions are poorer proton donors than neutral species.

(b)   HClO is stronger because Cl is a more electronegative atom than I. A higher electronegativity means that Cl attracts more of the electron density around it than I, weakening the H-X bond and making it easier to break (hence a better proton donor/acid).

(c) $HClO_2$ is stronger. O atoms are highly electronegative and attract electron density around them. Having two O atoms, $HClO_2$ will have less electron density in the H-X bond than HClO, thus, making the bond easier to break.

16.41   Use arrows of different sizes to show differences in electron density shifts.
(b)

(c)

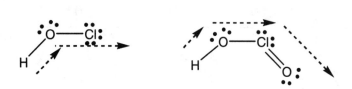

16.43 This problem asks for the concentration of all ionic species present in the solution. Begin by analyzing the chemistry. $NaC_2H_3O_2$ is a salt that dissolves in solution to form Na^+ and $C_2H_3O_2^-$. Acetate is a weak base with equilibrium:

$$H_2O + C_2H_3O_2^- \rightleftharpoons C_2H_3O_2H + OH^- \qquad K_b = 5.6 \times 10^{-10}$$

Every aqueous solution has the water equilibrium:

$$H_2O + H_2O \rightleftharpoons H_3O^+ + OH^- \qquad K_w = 1.0 \times 10^{-14}$$

Thus, the ionic species present in the solution are: Na^+, $C_2H_3O_2^-$, H_3O^+, and OH^-

Na^+ is a spectator ion and will have the same concentration as the initial salt, $[Na^+] = 0.250$ M

For the remaining ions we will need to set up concentration tables, write the equilibrium expressions, and solve for the ionic concentrations.

Reaction:	$H_2O +$	$C_2H_3O_2^- \rightleftharpoons$	$C_2H_3O_2H +$	OH^-
Start (M)	----	0.250	0	0
Change (M)	----	$-x$	$+x$	$+x$
Final (M)	----	$0.250-x$	x	x

The acid-base equilibrium expression is:

$$K_b = 5.6 \times 10^{-10} = \frac{x^2}{0.250 - x} \text{ ; assume that } x \ll 0.250$$

$$5.6 \times 10^{-10} = \frac{x^2}{0.250} \text{ ; for which } x^2 = 1.4 \times 10^{-10} \text{ and } x = 1.2 \times 10^{-5}; \text{ assumption valid}$$

$[C_2H_3O_2H] = [OH^-] = 1.2 \times 10^{-5}$ M

$[C_2H_3O_2^-] = 0.250$ M $- 1.2 \times 10^{-5}$ M $= 0.250$ M

Now use a second concentration table to determine the concentrations of hydronium and hydroxide ions:

Reaction:	$H_2O +$	$H_2O \rightleftharpoons$	$H_3O^+ +$	OH^-
Start (M)	----	----	0	1.2×10^{-5}
Change (M)	----	----	$+x$	$+x$
Final (M)	solvent	solvent	x	$1.2 \times 10^{-5} + x$

$K_w = 1.00 \times 10^{-14} = (x)(1.2 \times 10^{-5} + x)$; assume $x \ll 1.2 \times 10^{-5}$

$1.00 \times 10^{-14} = (x)(1.2 \times 10^{-5})$;

$x = 8.3 \times 10^{-10}$ M $= [H_3O^+]$; assumption valid

$[OH^-] = 1.2 \times 10^{-5}$ M

The ionic concentrations are:
$[Na^+] = [C_2H_3O_2^-] = 0.250$ M
$[OH^-] = 1.2 \times 10^{-5}$ M
$[H_3O^+] = 8.3 \times 10^{-10}$ M

16.45 This problem asks for the concentration of all ionic species present in the solution. Begin by analyzing the chemistry. H_2CO_3 is a diprotic acid with equilibria:

$$H_2O + H_2CO_3 \rightleftharpoons HCO_3^- + H_3O^+ \qquad K_{a1} = 4.5 \times 10^{-7}$$
$$H_2O + HCO_3^- \rightleftharpoons CO_3^{-2} + H_3O^+ \qquad K_{a2} = 4.7 \times 10^{-11}$$

Every aqueous solution has the water equilibrium:

$$H_2O + H_2O \rightleftharpoons H_3O^+ + OH^- \qquad K_w = 1.00 \times 10^{-14}$$

Thus, the ionic species present in the solution are: HCO_3^-, CO_3^{-2}, H_3O^+, and OH^-

Set up concentration tables, write equilibrium expressions, and solve for the ionic concentrations.

Reaction:	$H_2O +$	$H_2CO_3 \rightleftharpoons$	$HCO_3^- +$	H_3O^+
Start(10^{-2} M)	----	1.55	0	0
Change(10^{-2} M)	----	$-x$	$+x$	$+x$
Final(10^{-2} M)	solvent	$1.55-x$	x	x

$$K_{a1} = 4.5 \times 10^{-7} = \frac{x^2}{1.55 \times 10^{-2} - x}; \text{ assume } x \ll 1.55 \times 10^{-2}$$

$$4.5 \times 10^{-7} = \frac{x^2}{1.55 \times 10^{-2}};$$

from which $x^2 = 6.98 \times 10^{-9}$ and $x = 8.4 \times 10^{-5}$; assumption valid

$[H_3O^+] = [HCO_3^-] = 8.4 \times 10^{-5}$ M
$[H_2CO_3] = 1.55 \times 10^{-2}$ M

Now set up a concentration table for the second equilibrium:

Reaction:	$H_2O +$	$HCO_3^- \rightleftharpoons$	$CO_3^{-2} +$	H_3O^+
Start(10^{-5} M)	----	8.4	0	8.4
Change(10^{-5} M)	----	$-x$	$+x$	$+x$
Final(10^{-5} M)	solvent	$8.4-x$	x	$8.4 + x$

$$K_{a2} = 4.7 \times 10^{-11} = \frac{x(8.4 \times 10^{-5} + x)}{8.4 \times 10^{-5} - x} \text{; assume } x << 8.4 \times 10^{-5}$$

$x = 4.7 \times 10^{-11} M = [CO_3^{2-}];$

Now use a third concentration table to determine the ionic concentrations of hydronium and hydroxide ions:

Reaction:	H_2O +	H_2O $\rightleftharpoons$	H_3O^+ +	OH^-
Start(M)	----	----	8.4×10^{-5}	0
Change (M)	----	----	$+ x$	$+ x$
Final (M)	solvent	solvent	$8.4 \times 10^{-5} + x$	x

$K_w = 1.00 \times 10^{-14} = (x)(8.4 \times 10^{-5} + x)$; assume $x << 8.4 \times 10^{-5}$
$1.00 \times 10^{-14} = (x)(8.4 \times 10^{-5})$
$x = 1.2 \times 10^{-10}$ M = $[OH^-]$; assumption valid

ionic concentrations:
$[H_3O^+] = [HCO_3^-] = 8.4 \times 10^{-5} M$
$[CO_3^{2-}] = 4.7 \times 10^{-11} M$
$[OH^-] = 1.2 \times 10^{-10}$ M

16.47 There are two amine groups, one at either end of the molecule, each of which can accept a proton from a water molecule. Convert the line structure to a Lewis structure using the standard procedures, then show the transfer of one proton to each N atom:

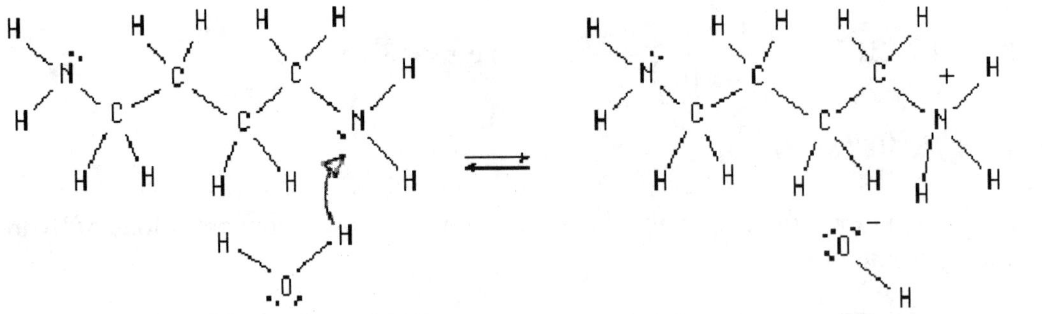

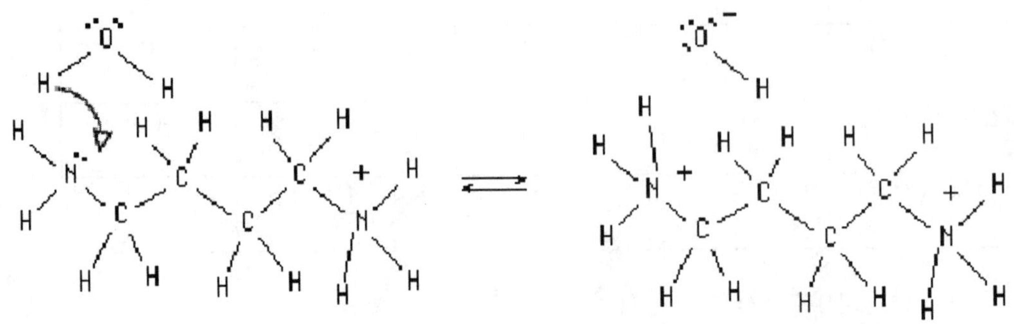

16.49 To determine the pH of a solution, follow the standard procedure for working equilibrium problems:

1.) Major species are H_2O, Na^+, and F^-;

2.) The dominant acid-base equilibrium is $H_2O + F^- \rightleftharpoons HF + OH^-$;

3.) $K_{eq} = K_b = \dfrac{K_w}{K_a}$; from Table 16-3, $K_a = 6.3 \times 10^{-4}$, from which

$$K_b = \frac{1.0 \times 10^{-14}}{6.3 \times 10^{-4}} = 1.6 \times 10^{-11};$$

4.) Concentration table is:

Reaction:	H_2O +	$F^- \rightleftharpoons$	HF +	OH^-
Start (M)	---	0.250	0	0
Change (M)	---	$-x$	$+x$	$+x$
Final (M)	solvent	$0.250 - x$	x	x

$$K_b = 1.6 \times 10^{-11} = \frac{[HF]_{eq}[OH^-]_{eq}}{[F^-]_{eq}} = \frac{x^2}{0.250 - x}; \quad \text{Assume } x \ll 0.250:$$

$1.6 \times 10^{-11} = \dfrac{x^2}{0.250}$, from which $x^2 = 4.0 \times 10^{-12}$ and $x = 2.0 \times 10^{-6}$; assumption valid

$[OH^-] = 2.0 \times 10^{-6}$ M,

pOH $= -\log(2.0 \times 10^{-6}) = 5.70$, and

pH $= 14.00 - 5.70 = 8.30$.

16.51 To determine concentrations of species in a solution, follow the standard procedure:
1.) This is a strong acid. Major species are H_2O, H_3O^+, and HSO_4^-;
2.) The dominant acid-base equilibrium is $H_2O + HSO_4^- \rightleftharpoons SO_4^{2-} + H_3O^+$;
3.) $K_{eq} = K_{a2}$; from Table 16-2, $K_{a2} = 1.0 \times 10^{-2}$;
4.) In this solution, there is a hydronium ion from the strong acid present initially:

Reaction:	H_2O +	$HSO_4^- \rightleftharpoons$	SO_4^{2-} +	H_3O^+
Start (M)	----	2.00	0	2.00
Change (M)	----	$-x$	$+x$	$+x$
Final (M)	solvent	$2.00 - x$	x	$2.00 + x$

$$K_{eq} = 1.0 \times 10^{-2} = \frac{[SO_4^{2-}]_{eq}[H_3O^+]_{eq}}{[HSO_4^-]_{eq}} = \frac{x(2.00 + x)}{2.00 - x}; \quad \text{Assume } x \ll 2.00:$$

$1.0 \times 10^{-2} = \dfrac{(2.00)(x)}{(2.00)}$, from which $x = 1.0 \times 10^{-2}$; assumption valid

$[H_3O^+] = 2.00 + 0.010 = 2.01$ M;

$[SO_4^{2-}] = 1.0 \times 10^{-2}$ M;

$[HSO_4^-] = 2.00 - x = 1.99$ M

16.53 (a) H_2SO_4 is a strong acid, so the major species in solution are H_2O, HSO_4^-, and H_3O^+. The hydrogensulfate ion is a weak acid, so the equilibrium reaction that determines pH is:
$$H_2O + HSO_4^- \rightleftharpoons SO_4^{2-} + H_3O^+;$$
(b) Na_2SO_4 is a salt, so the major species in solution are H_2O, SO_4^{2-}, and Na^+. The sulfate ion is the conjugate base of a weak acid, so the equilibrium reaction that determines pH is:
$$SO_4^{2-} + H_2O \rightleftharpoons HSO_4^- + OH^-;$$
(c) CO_2 associates with H_2O when it dissolves in water, so the major species in solution are CO_2, H_2CO_3, and H_2O. Carbonic acid is a weak acid, so the equilibrium reaction that determines pH is:
$$H_2O + H_2CO_3 \rightleftharpoons HCO_3^- + H_3O^+;$$

(d) NH_4Cl is a salt, so the major species in solution are H_2O, Cl^-, and NH_4^+. The ammonium ion is the conjugate acid of a weak base, so the equilibrium reaction that determines pH is:

$$H_2O + NH_4^+ \rightleftharpoons NH_3 + H_3O^+.$$

16.55 Tabulated equilibrium constants for acid-base reactions always refer to reactions in which H_2O is one of the reactants. The reaction in this problem is the reverse of a base reaction:

$$HPO_4^{2-} \text{ (aq)} + OH^- \text{ (aq)} \rightleftharpoons PO_4^{3-} \text{ (aq)} + H_2O \text{ (l)}$$

Table 16-2 lists K_a values for phosphoric acid:

$$HPO_4^{2-} \text{ (aq)} + H_2O \text{ (l)} \rightleftharpoons PO_4^{3-} \text{ (aq)} + H_3O^+ \text{(aq)} \qquad K_{a3} = 4.8 \times 10^{-13}$$

K_a and K_b for a conjugate acid-base pair are related through $K_a K_b = K_w$:

$$K_b = \frac{1.0 \times 10^{-14}}{4.8 \times 10^{-13}} = 2.1 \times 10^{-2};$$

Thus, $K_{eq} = \dfrac{1}{K_b} = 48$

16.57 (a) An acid-base equilibrium reaction involves proton transfer, in this case from boric acid to water:

(b) To calculate the pH of a solution, follow the standard procedure for equilibrium calculations:

Reaction:	H_2O +	$H_3BO_3 \rightleftharpoons$	$H_2BO_3^-$ +	H_3O^+
Start (M)	----	0.050	0	0
Change (M)	----	$-x$	$+x$	$+x$
Final (M)	solvent	0.050 - x	x	x

$$K_a = 5.4 \times 10^{-10} = \frac{[H_3BO_2^-]_{eq}[H_3O^+]_{eq}}{[H_3BO_3]_{eq}} = \frac{x^2}{0.050 - x}; \text{ Assume } x \ll 0.050:$$

$$5.4 \times 10^{-10} = \frac{x^2}{0.050},$$

from which $x^2 = 2.7 \times 10^{-11}$ and $x = 5.2 \times 10^{-6}$; assumption valid

$[H_3O^+] = 5.2 \times 10^{-6}$ M, and

$pH = -\log(5.2 \times 10^{-6}) = 5.28$

16.59 There is much interesting chemical information provided in the statement of this problem, but the calculation is a straightforward equilibrium determination for a solution of a weak base. Follow the standard procedures to determine the pH. Begin by constructing an equilibrium table, write the equilibrium expression, and solve for the concentration of hydroxide ions.

Reaction:	H_2O +	LSD ⇌	$LSDH^+$ +	OH^-
Start (M)	----	0.55	0	0
Change (M)	----	$-x$	$+x$	$+x$
Final (M)	Solvent	$0.55 - x$	x	x

The equilibrium expression is:

$$K_b = 7.6 \times 10^{-7} = \frac{[LSDH^+]_{eq}[OH^-]_{eq}}{[LSD]_{eq}} = \frac{x^2}{0.55 - x} ; \quad \text{Assume } x \ll 0.55:$$

$7.6 \times 10^{-7} = \dfrac{x^2}{0.55}$, from which $x^2 = 4.2 \times 10^{-7}$ and $x = 6.5 \times 10^{-4}$; assumption valid

$[OH^-] = 6.5 \times 10^{-4}$ M,

$pOH = -\log(6.5 \times 10^{-4}) = 3.19$, and $pH = 14.00 - 3.19 = 10.81$.

16.61 To identify an acid from the pH of its solution, use the pH to calculate the equilibrium constant of the acid. For $pH = 2.71$, $[H_3O^+]_{eq} = 1.95 \times 10^{-3}$ M $= [A^-]$; $[HA]_{eq} = (0.060 \text{ M} - 0.00195 \text{ M}) = 0.058$ M;

$$K_a = \frac{[A^-]_{eq}[H_3O^+]_{eq}}{[HA]_{eq}} = \frac{(1.95 \times 10^{-3})^2}{(0.058)} = 6.6 \times 10^{-5}; \; pK_a = 4.18;$$

The acid is benzoic acid, listed in Appendix E , $pK_a = 4.19$.

16.63 Molecular pictures must show the correct relative numbers of the various species in the solution. From the starting condition (six molecules of oxalic acid), make appropriate changes and then draw new pictures:
(a) Hydroxide ions react with oxalic acid to form water and hydrogen oxalate ions: $H_2C_2O_4 + OH^- \rightleftharpoons H_2O + HC_2O_4^-$ The picture shows 2 molecules of oxalic acid and 4 each of water and hydrogenoxalate:

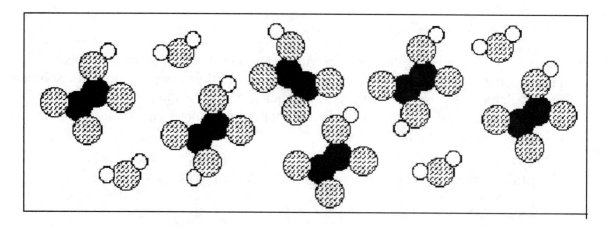

(b) When all oxalic acid has reacted, hydroxide ions react with hydrogen oxalate ions to form water and oxalate ions:

$HC_2O_4^- + OH^- \rightleftharpoons H_2O + C_2O_4^{2-}$. The picture shows 4 hydrogen oxalate ions, 8 water molecules, and 2 oxalate ions:

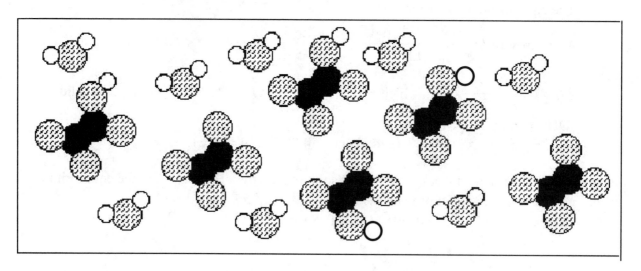

(c) NH₃, a weak base, accepts a proton from oxalic acid, a weak acid:

$H_2C_2O_4 + NH_3 \rightleftharpoons NH_4^+ + HC_2O_4^-$

The picture shows 2 oxalic acid molecules and four each ammonium and hydrogenoxalate ions:

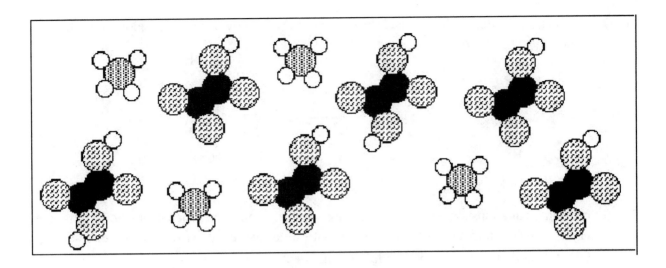

16.65 The chemical reaction that occurs is:

$$P_4O_{10} + 6\,H_2O \rightarrow 4\,H_3PO_4$$

(a) The major species present are H_2O, and H_3PO_4

(b) The minor species are (in order of highest concentration to lowest):
  $H_2PO_4^-$, H_3O^+, HPO_4^{2-}, OH^-, and PO_4^{3-}

(c) The dominant equilibrium that determines the pH is:

$$H_3PO_4 + H_2O \rightleftharpoons H_2PO_4^- + H_3O^+$$

Set up a concentration table, solve for hydronium ion concentration, then calculate the pH. Determine the initial concentration using standard stoichiometric procedures:

$$n(H_3PO_4) = 3.5\ \text{g}\ P_4O_{10}\left(\frac{1\,\text{mol}}{283.88\,\text{g}}\right)\left(\frac{4\,\text{mol}\,H_3PO_4}{1\,\text{mol}\,P_4O_{10}}\right) = 0.0493\ \text{mol};$$

$$[H_3PO_4] = \frac{0.0493\,\text{mol}}{1.50\,\text{L}} = 0.033\ \text{M}$$

Reaction:	$H_2O\ +$	$H_3PO_4 \rightleftharpoons$	$H_2PO_4^-\ +$	H_3O^+
Start (M)	----	0.033	0	0
Change (M)	----	$-x$	$+x$	$+x$
Final (M)	solvent	$0.033 - x$	x	x

Now substitute into the equilibrium constant expression and solve for x:

$$K_{eq} = 0.0069 = \frac{[H_2PO_4^-]_{eq}[H_3O^+]_{eq}}{[H_3PO_4]_{eq}} = \frac{x^2}{0.033\text{-}x}\ ;\ \text{solve by the quadratic equation}$$

$[H_3O^+] = 0.012\ \text{M}$,
$pH = -\log(0.012) = 1.92$

16.67 Proton transfer occurs from the carboxylic acid O–H (shown screened below) and the amino nitrogen atom:

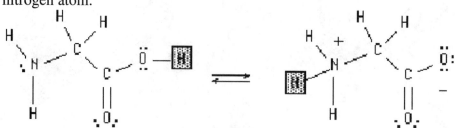

16.69 Net ionic equations show only the reacting species. Remember that strong acids generate H_3O^+ in solution and react to completion with weak bases, and strong bases generate OH^- in solution and react to completion with weak acids:

(a) strong base reacting with weak acid: $OH^- + C_6H_5CO_2H \rightleftharpoons H_2O + C_6H_5CO_2^-$;

(b) strong acid reacting with weak base: $H_3O^+ + (CH_3)_3N \rightleftharpoons H_2O + (CH_3)_3NH^+$;

(c) weak base reacting with weak acid: $SO_4^{2-} + CH_3CO_2H \rightleftharpoons HSO_4^- + CH_3CO_2^-$ HSO_4^-, $pK_a = 1.99$; CH_3CO_2H, $pK_a = 4.75$; HSO_4^- is stronger, so this reaction proceeds to a small extent;

(d) strong base reacting with weak acid: $OH^- + NH_4^+ \rightleftharpoons H_2O + NH_3$;

(e) weak base reacting with weak acid: $HPO_4^{2-} + NH_3 \rightleftharpoons PO_4^{3-} + NH_4^+$; HPO_4^{2-}, $pK_a = 12.32$; NH_4^+, $pK_a = 9.25$; NH_4^+ is stronger, so this reaction proceeds to a small extent.

17.1 A buffer solution must contain both a weak acid and its conjugate weak base. Begin each problem by calculating the initial amounts of acid and base.

(a) The strong base reacts with NH_4^+, a weak acid, to generate the conjugate weak base.

Moles(NH_4^+) = 0.25 M(0.100 L) = 0.025 mol
Moles(OH^-) = 0.25 M(0.150 L) = 0.0375 mol

There is excess strong base (0.0375 mol compared with 0.025 mol), so no weak acid remains after mixing, this solution does not have buffer properties.

(b) Initial amounts:

Moles(NH_4^+) = 0.25 M(0.100 L) = 0.025 mol
Moles(OH^-) = 0.25 M(0.050 L) = 0.0125 mol
The strong base reacts completely with the weak acid to generate the conjugate weak base:

Reaction:	NH_4^+ +	$OH^- \rightleftharpoons$	NH_3 +	H_2O
Start (mol)	0.025	0.0125	0	---
Change (mol)	–0.0125	–0.0125	+0.0125	---
Final (mol)	0.0125	0.0000	0.0125	solvent

Since both acid and conjugate base exist, this solution is an NH_4^+/NH_3 buffer. Use the buffer equation, Equation 17-1, with amounts to determine the pH:

$$pH = pK_a + \log\left\{\frac{(\text{mol } A^-)}{(\text{mol } HA)}\right\}$$

For NH_4^+, $pK_a = 9.25$, so

$$pH = 9.25 + \log\left\{\frac{(0.0125 \text{ mol})}{(0.0125 \text{ mol})}\right\} = 9.25;$$

(c) NH_4^+ is a weak acid and HCl is a strong acid. There is no conjugate base present in high concentration, so this solution does not have buffer properties;

(d) NH_3 is a weak base and HCl is a strong acid. The strong acid reacts completely with the weak base to generate the conjugate weak acid.

Initial amounts:
Moles(NH_3) = 0.25 M(0.100 L) = 0.025 mol
Moles(H_3O^+) = 0.25 M(0.050 L) = 0.0125 mol

Reaction:	NH_3	+	H_3O^+ $\rightleftharpoons$	NH_4^+ +	H_2O
Start (mol)	0.025		0.0125	0	---
Change (mol)	−0.0125		−0.0125	+0.0125	---
Final (mol)	0.0125		0.0000	0.0125	solvent

This is an NH_4^+/NH_3 buffer. Use the buffer equation, Equation 17-1, with amounts:

$$pH = pK_a + \log\left\{\frac{(\text{mol } A^-)}{(\text{mol } HA)}\right\} \quad \text{where } pK_a = 9.25, \text{ so}$$

$$pH = 9.25 + \log\left\{\frac{(0.0125 \text{ mol})}{(0.0125 \text{ mol})}\right\} = 9.25.$$

17.3 The two solutions that are buffered are (b) and (d). The concentration tables for these solutions show that they are identical in acid/base composition, despite being prepared differently. Thus only one calculation is required. The added strong acid reacts completely with the weak base of the buffer to generate the conjugate weak acid.

Reaction:	NH_3	+	H_3O^+ $\rightleftharpoons$	NH_4^+ +	H_2O
Start (mol)	0.0125		0.0050	0.0125	---
Change	−0.0050		−0.0050	+0.0050	---
Final	0.0075		0.0000	0.0175	solvent

Use the buffer equation, Equation 17-1, with the new amounts:

$$pH = pK_a + \log\left\{\frac{(\text{mol } A^-)}{(\text{mol } HA)}\right\} = 9.25 + \log\left\{\frac{(0.0075 \text{ mol})}{(0.0175 \text{ mol})}\right\} = 9.25 - 0.37 = 8.88.$$

17.5 The two solutions that are buffered are (b) and (d). The concentration tables for these solutions show that they are identical in acid/base composition, despite being prepared differently. Thus only one calculation is required. Addition of strong base consumes weak acid, produces weak base, and increases the pH of the buffer. First use the buffer equation to calculate the acid/base ratio in a solution whose pH is greater by 0.10 unit than the initial buffer solution. Then set up a concentration ratio, using x = the amount of added base, and solve for x:
new pH = 9.25 + 0.10 = 9.35;
$[NH_3] = 0.0125 + x$,
$[NH_4^+] = 0.0125 - x$;

$$9.35 = 9.25 + \log\left\{\frac{(0.0125 + x)}{(0.0125 - x)}\right\},$$

$$\log\left\{\frac{(0.0125 + x)}{(0.0125 - x)}\right\} = 0.10,$$

$$\left\{\frac{(0.0125 + x)}{(0.0125 - x)}\right\} = 1.26$$

$(0.0125 + x) = 1.26(0.0125 - x) = 0.0157 - 1.26\,x;$

$0.0032 = 2.26\,x$, and $x = 1.4 \times 10^{-3}$ mol.

17.7 A buffer solution requires a conjugate acid/base pair whose pK_a is within 1 pH unit of the desired pH. For pH = 3.50, the HCO_2H/HCO_2^- system, $pK_a = 3.75$, would be most suitable. Sodium formate, $NaHCO_2$, could be used along with HCl solution which would protonate some formate anions to form the conjugate weak acid. For pH = 12.60, the HPO_4^{2-}/PO_4^{3-} system, $pK_a = 12.32$, would be most suitable. Potassium phosphate, K_3PO_4, could be used along with HCl solution which would protonate some phosphate anions to form the conjugate weak acid.

17.9 This is a NH_3/NH_4^+ buffer. Use the buffer equation, Eqn 17-1, to determine the amount of ammonium chloride needed:

$MM(NH_4Cl) = 14.01$ g/mol $+ 4(1.01$ g/mol$) + 35.45$ g/mol $= 53.50$ g/mol

$$\log\left\{\frac{\text{mol } NH_3}{\text{mol } NH_4^+}\right\} = pH - pK_a = 8.90 - 9.25 = -0.35$$

$$\frac{\text{mol } NH_3}{\text{mol } NH_4^+} = 10^{-0.35} = 0.447$$

$$\text{mol } NH_4^+ = \frac{\text{mol } NH_3}{0.447} = \frac{(1.25 \text{ L})(0.25 \text{ M})}{0.447} = 0.699 \text{ mol} = \text{moles of } NH_4Cl$$

$$m = 0.699 \text{ mol}\left(\frac{53.50 \text{ g}}{1 \text{ mol}}\right) = 37 \text{ g} \quad \text{(note 2 sig figs because there are only two reported for}$$

the molarity of ammonia)

17.11 Addition of a strong acid will lower the pH from 8.90 to 8.65. Use the buffer equation to determine the number of moles of acid required to cause this change:

$$\log\left\{\frac{0.3125 - x}{0.699 + x}\right\} = pH - pK_a = 8.65 - 9.25 = -0.60$$

$$\frac{0.3125 - x}{0.699 + x} = 10^{-0.60} = 0.251; \qquad \text{where } x = \text{moles of acid added}$$

$x = 0.110$ moles HCl

$$V = 0.110 \text{ mol}\left(\frac{1\,L}{2.0\,\text{mol}}\right) = 0.055\,L \text{ or } 55\,mL$$

17.13 To determine the approximate pH at the stoichiometric point of an acid-base titration, examine the equilibrium that determines the pH in the vicinity of the stoichiometric point:
(a) This is a weak base-strong acid titration. At the stoichiometric point, major acid-base species are NH_4^+ and H_2O, and the dominant equilibrium is:
$NH_4^+(aq) + H_2O(l) \rightleftharpoons NH_3(aq) + H_3O^+(aq)$
Hydronium ions are produced in this reaction, so pH < 7 at the stoichiometric point;

(b) This is a strong acid-strong base titration. The only acid-base species present at the stoichiometric point is H_2O, and the dominant equilibrium is:
$H_2O(l) + H_2O(l) \rightleftharpoons OH^-(aq) + H_3O^+(aq)$
Thus pH = 7 at the stoichiometric point.

(c) This is a weak base-strong acid titration. At the stoichiometric point, major acid-base species are CH_3CO_2H and H_2O, and the dominant equilibrium is:
$CH_3CO_2H(aq) + H_2O(l) \rightleftharpoons CH_3CO_2^-(aq) + H_3O^+(aq)$
Hydronium ions are produced in this reaction, so pH < 7 at the stoichiometric point.

17.15 This problem describes an acid-base titration. The major species differ at various points during a titration, so there are different dominant equilibria which must be identified before doing an equilibrium calculation to determine pH.
(a) Before titration begins, the major species are a weak acid, aspirin (HA) and H_2O, and the dominant equilibrium is $HA + H_2O \rightleftharpoons A^- + H_3O^+$, for which $K_a = 3.0 \times 10^{-4}$.

Reaction:	H_2O +	HA $\rightleftharpoons$	A^- +	H_3O^+
Start (M)	---	10^{-2}	0	0
Change (M)	---	$-x$	$+x$	$+x$
Final (M)	solvent	$10^{-2}-x$	x	x

Now substitute into the equilibrium constant expression and solve for x:

$$K_a = 3.0 \times 10^{-4} = \frac{[A^-]_{eq}[H_3O^+]_{eq}}{[HA]_{eq}} = \frac{x^2}{(10^{-2} - x)};$$

$x^2 = (3.0 \times 10^{-4})(10^{-2} - x) = (3.0 \times 10^{-6}) - (3.0 \times 10^{-4})x;$
$x^2 + (3.0 \times 10^{-4})x - (3.0 \times 10^{-6}) = 0;$

$$x = \frac{-b \pm \sqrt{b^2 - 4ac}}{2a} = \frac{-(3.0 \times 10^{-4}) \pm \sqrt{(3.0 \times 10^{-4})^2 - 4(-3.0 \times 10^{-6})}}{2} = 1.6 \times 10^{-3};$$

$[H_3O^+] = 1.6 \times 10^{-3}$ M,
pH = -log(1.6×10^{-3}) = 2.8;

(b) At the stoichiometric point, all the weak acid has been converted into its conjugate base, so the major acid-base species present are A^- and H_2O and the dominant equilibrium is:

$A^- + H_2O \rightleftharpoons HA + OH^-$, $K_{eq} = \dfrac{K_w}{K_a} = \dfrac{1.00 \times 10^{-14}}{3.0 \times 10^{-4}} = 3.3 \times 10^{-11}$:

Reaction:	H_2O +	$A^- \rightleftharpoons$	HA +	OH^-
Start (M)	---	10^{-2}	0	0
Change (M)	---	$-x$	$+x$	$+x$
Final (M)	solvent	$10^{-2} - x$	x	x

Now substitute into the equilibrium constant expression and solve for x:

$K_b = 3.3 \times 10^{-11} = \dfrac{[HA]_{eq}[OH^-]_{eq}}{[A^-]_{eq}} = \dfrac{x^2}{(10^{-2} - x)}$; Assume $x \ll 10^{-2}$;

$x^2 = (10^{-2})(3.3 \times 10^{-11}) = 3.3 \times 10^{-13}$,
from which $x = 5.7 \times 10^{-7}$; the assumption is correct

$[OH^-]_{eq} = 5.7 \times 10^{-7}$ M,
pOH = -log(5.7×10^{-7}) = 6.24, and
pH = 14.00 - pOH = 14.00 - 6.24 = 7.8;

(c) At the midpoint of the titration, half of the weak acid has been converted into its conjugate base, and: pH = pK_a = - log(3.0×10^{-4}) = 3.52.

17.17 The best indicator for this solution is one where pK_{In} = pH$_{stoichiometric point}$; the pH at the stoichiometric point was calculated in 17.15 to be 7.76. The best indicator is phenol red with pK_{In} = 7.9.

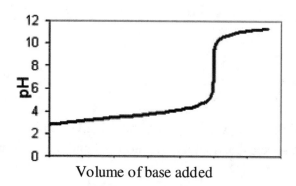

Volume of base added

17.19 This is a titration type problem for a diprotic acid asking for information at the second stoichiometric point. At the second stoichiometric point, all the weak acid (H_2P) has been converted into P^{2-}, so the major acid-base species present are P^{2-} and H_2O and the dominant equilibrium is:

$$P^{2-}(aq) + H_2O(l) \rightleftharpoons HP^-(aq) + OH^-(aq) \; , \; K_{b1}$$

Where,

$$K_{b1} = \frac{K_w}{K_{a2}} = \frac{1.00 \times 10^{-14}}{3.9 \times 10^{-6}} = 2.6 \times 10^{-9}:$$

Begin by determining the initial amount of P^{2-} (we can assume dilution effects to be negligible). Then construct an equilibrium table, write the equilibrium expression, and solve for the hydroxide concentration to determine the pH.
Calculation of initial amount:

$$\text{Moles}(H_2P) = 0.150 \, L \left(\frac{0.015 \, \text{mol}}{1 \, L} \right) = 2.25 \times 10^{-3} \, \text{mol} = \text{moles}(P^{2-}) \; ; \quad \text{(an additional sig.}$$

fig. will be used to avoid round off error, remember to account for this at the end).
We must add twice as many moles of NaOH to neutralize the acid (remember there are two acidic hydrogens):

$$\text{Moles}(OH^-) = 2.25 \times 10^{-3} \, \text{mol} \left(\frac{2 \, \text{mol} \, OH^-}{1 \, \text{mol acid}} \right) = 4.50 \times 10^{-3} \, \text{mol needed}$$

Divide the moles of base by the total volume to obtain the initial concentration of P^{2-}:

$$[P^{2-}] = \left(\frac{2.25 \times 10^{-3} \, \text{mol}}{0.150 \, L} \right) = 1.50 \times 10^{-2} \, M$$

The completed equilibrium table is

Reaction:	H_2O +	P^{2-} $\rightleftharpoons$	HP^- +	OH^-
Start (M)	---	0.0150	0	0
Change (M)	---	$-x$	$+x$	$+x$
Final (M)	Solvent	$0.0150 - x$	x	x

Now substitute into the equilibrium constant expression and solve for x:

$$K_{b1} = 2.6 \times 10^{-9} = \frac{[HP^-]_{eq}[OH^-]_{eq}}{[P^{2-}]_{eq}} = \frac{x^2}{0.0150 - x} =; \text{ assume } x << 0.0150$$

$x^2 = (2.6 \times 10^{-9})(0.0150) = 3.90 \times 10^{-11}$;
$x = 6.24 \times 10^{-6}$;
$[OH^-] = 6.24 \times 10^{-6}$ M,
pOH $= -\log(6.24 \times 10^{-6}) = 5.20$, (there should only be two digits in the mantissa because the initial concentration of H_2P is known to only 2 significant figures)
pH $= 14.00 - 5.20 = 8.80$; a suitable indicator for this titration is thymol blue, $pK_{in} = 8.9$.

Had dilution been taken into account, $[OH^-] = 6.14 \times 10^{-6}$ M, pOH $= 5.21$.

17.21 To write a solubility product equilibrium expression, first determine the chemical formula and stoichiometry of the salt:

(a) $AgCl(s) \rightleftharpoons Ag^+(aq) + Cl^-(aq)$ $K_{sp} = [Ag^+]_{eq}[Cl^-]_{eq}$

(b) $BaSO_4(s) \rightleftharpoons Ba^{2+}(aq) + SO_4^{2-}(aq)$ $K_{sp} = [Ba^{2+}]_{eq}[SO_4^{2-}]_{eq}$

(c) $Fe(OH)_2(s) \rightleftharpoons Fe^{2+}(aq) + 2 OH^-(aq)$ $K_{sp} = [Fe^{2+}]_{eq}[OH^-]_{eq}^2$

(d) $Ca_3(PO_4)_2(s) \rightleftharpoons 3Ca^{2+}(aq) + 2PO_4^{3-}(aq)$ $K_{sp} = [Ca^{2+}]_{eq}^3[PO_4^{3-}]_{eq}^2$.

17.23 "Determine the mass that dissolves" means "calculate the amount present in solution at equilibrium". Initial amounts are zero, so it is easy to complete a concentration table. Calculate molarity from the solubility product expression, then convert to mass using standard stoichiometric methods. Let x be the number of moles of salt dissolving in 1 L of solution.

(a)

Reaction:	$AgCl(s) \rightleftharpoons$	$Ag^+(aq)$ +	$Cl^-(aq)$
Start (M)	---	0	0
Change (M)	---	$+x$	$+x$
Final (M)	solid	x	x

$K_{sp} = 1.8 \times 10^{-10} = [Ag^+]_{eq}[Cl^-]_{eq} = x^2$, from which

$x = 1.34 \times 10^{-5}$ M;

$$m_{AgCl} = 0.475 \text{ L} \left(\frac{1.34 \times 10^{-5} \text{ mol}}{1 \text{ L}} \right) \left(\frac{143.32 \text{ g}}{1 \text{ mol}} \right) = 9.1 \times 10^{-4} \text{ g};$$

(b)

Reaction:	$BaSO_{4(s)}$ ⇌	Ba^{2+}(aq) +	SO_4^{2-}(aq)
Start (M)	---	0	0
Change (M)	---	+x	+x
Final (M)	Solid	x	x

$K_{sp} = 1.1 \times 10^{-10} = [Ba^{2+}]_{eq}[SO_4^{2-}]_{eq} = x^2$, from which

$x = 1.05 \times 10^{-5}$ M;

$$m_{BaSO_4} = 0.475 \text{ L} \left(\frac{1.05 \times 10^{-5} \text{ mol}}{1 \text{ L}} \right) \left(\frac{233.39 \text{ g}}{1 \text{ mol}} \right) = 1.2 \times 10^{-3} \text{ g};$$

(c)

Reaction:	$Fe(OH)_{2(s)}$ ⇌	Fe^{2+}(aq) +	2 OH^-(aq)
Start (M)	---	0	0
Change (M)	---	+x	+2x
Final (M)	Solid	x	2x

$K_{sp} = 4.8 \times 10^{-17} = [Fe^{2+}]_{eq}[OH^-]_{eq}^2 = 4x^3$, from which

$x = 2.3 \times 10^{-6}$ M;

$$m_{Fe(OH)_2} = 0.475 \text{ L} \left(\frac{2.3 \times 10^{-6} \text{ mol}}{1 \text{ L}} \right) \left(\frac{89.87 \text{ g}}{1 \text{ mol}} \right) = 9.8 \times 10^{-5} \text{ g};$$

(d)

Reaction:	$Ca_3(PO_4)_{2(s)}$ ⇌	3 Ca^{2+}(aq)+	2 PO_4^{3-} (aq)
Start (M)	---	0	0
Change (M)	---	+3x	+2x
Final (M)	Solid	3x	2x

$K_{sp} = 2.0 \times 10^{-33} = [Ca^{2+}]_{eq}^3 [PO_4^{3-}]_{eq}^2 = 108x^5$, from which

$x = 1.1 \times 10^{-7}$ M;

$$m_{Ca_3(PO_4)_2} = 0.475\,L \left(\frac{1.1 \times 10^{-7}\,mol}{1\,L} \right) \left(\frac{310.18\,g}{1\,mol} \right) = 1.6 \times 10^{-5}\,g;$$

17.25 To calculate a solubility product from the mass that dissolves in solution, convert to molarity of ions and then evaluate the equilibrium constant expression. Begin by writing the chemical reaction and the equilibrium expression:

$CaC_2O_{4(s)} \rightleftharpoons Ca^{2+}(aq) + C_2O_4^{2-}(aq)$ $\qquad$ $K_{sp} = [Ca^{2+}]_{eq}[C_2O_4^{2-}]_{eq}$

$$n = \frac{m}{MM} = (6.1\,mg) \left(\frac{10^{-3}\,g}{1\,mg} \right) \left(\frac{1\,mol}{128.1\,g} \right) = 4.76 \times 10^{-5}\,mol;$$

Since the stoichiometry is 1:1 the moles of calcium oxalate are the same as the moles of calcium ions and oxalate ions.

$$[Ca^{2+}] = [C_2O_4^{2-}] = \frac{n}{V} = \frac{4.76 \times 10^{-5}\,mol}{1.0\,L} = 4.76 \times 10^{-5}\,mol/L;$$

$K_{sp} = [Ca^{2+}]_{eq}[C_2O_4^{2-}]_{eq} = (4.76 \times 10^{-5})^2 = 2.3 \times 10^{-9}$.

17.27 This problem describes a common ion solution. To determine concentrations of ions at equilibrium, follow the five-step procedure for working equilibrium problems:
1.) Species present initially are Pb^{2+}, NO_3^-, and $PbCl_2$.
2.) Reaction is a solubility process: $PbCl_{2(s)} \rightleftharpoons Pb^{2+}(aq) + 2\,Cl^-(aq)$
3.) $K_{eq} = K_{sp} = [Pb^{2+}]_{eq}[Cl^-]_{eq}^2 = 1.7 \times 10^{-5}$;

4.) Set up and complete a concentration table, letting x be the increase in $[Pb^{2+}]$;

Reaction:	$PbCl_{2\ (s)} \rightleftharpoons$	$Pb^{2+}(aq)\ +$	$2\,Cl^-(aq)$
Start (M)	----	0.650	0
Change (M)	----	$+x$	$+2x$
Final (M)	----	$0.650 + x$	$2x$

5.) Substitute into the equilibrium constant expression and solve for x:
$1.7 \times 10^{-5} = (0.650 + x)(2x)^2$; Assume $x \ll 0.650$; then

$$4x^2 = \frac{1.7 \times 10^{-5}}{0.650}$$

$x^2 = 6.54 \times 10^{-6}$ and
$x = 2.56 \times 10^{-3}$; $\qquad$ $2.56 \times 10^{-3} < 5\%$ of 0.650, so the approximation is valid.

The increase in lead ions is: $\Delta [Pb^{2+}] = 2.56 \times 10^{-3}$ M;

To convert to moles Pb^{2+} that dissolve, multiply by the solution volume:
$n = MV = (2.56 \times 10^{-3}$ M$)(0.750$ L$) = 1.92 \times 10^{-3}$ mol;
The amount of $PbCl_2$ that dissolves equals the amount of increase in Pb^{2+}; convert to mass by multiplying by the molar mass of $PbCl_2$:

$$m = n\,MM = 1.92 \times 10^{-3} \text{ mol} \left(\frac{278.1 \text{ g}}{1 \text{ mol}} \right) = 0.53 \text{ g}; \text{ (note 2 significant figures because } K \text{ is}$$

known to only 2 sig figs).

17.29 The chemical reactions are:

$$\begin{array}{ll} CaSO_3(s) \rightleftharpoons Ca^{2+}(aq) + SO_3{}^{2-}(aq) & K_{sp} \\ \underline{SO_3{}^{2-}(aq) + H_3O^+(l) \rightleftharpoons HSO_3{}^-(aq) + H_2O(l)} & 1/K_{a2} \\ CaSO_3(s) + H_3O^+(aq) \rightleftharpoons Ca^{2+}(aq) + HSO_3{}^-(aq) + H_2O(l) & \end{array}$$

$$K = \frac{[Ca^{2+}][HSO_3^-]}{[H_3O^+]} = \frac{[HSO_3^-]}{[H_3O^+][SO_3^{2-}]} \frac{[Ca^{2+}][SO_3^{2-}]}{1} = \frac{K_{sp}}{K_{a_2}} = \frac{1.0 \times 10^{-4}}{6.3 \times 10^{-8}} = 1.6 \times 10^3$$

17.31 This problem describes a common ion effect solubility reaction. We are asked to determine the final concentration of calcium ions in solution. Begin by analyzing the chemistry. The starting materials are HCl (aq), a strong acid, and $CaSO_3$(s), an insoluble salt. In addition to water, the major species in solution are:
$$H_3O^+, Cl^-, H_2O$$

Hydronium ions will react with calcium sulfite (see problem 17.29):
$$CaSO_3(s) + H_3O^+(aq) \rightleftharpoons Ca^{2+}(aq) + HSO_3{}^-(aq) + H_2O(l)$$

which has the equilibrium expression:
$$K_{eq} = \frac{[Ca^{2+}][HSO_3^-]}{[H_3O^+]} = \frac{K_{sp}}{K_{a2}} = 1.6 \times 10^3$$

Because the K for the reaction is so large, assume the reaction goes to completion and then do the equilibrium calculations:

At completion all of the hydronium reacts to form Ca^{2+} and HSO_3^-:
$[Ca^{2+}] = [HSO_3^-] = 0.125$ M

The complete equilibrium table is:

Reaction:	$CaSO_3$(s)+	H_3O^+(aq) $\rightleftharpoons$	Ca^{2+}(aq) +	HSO_3^-(aq) +	H_2O
Start (M)	----	0	0.125	0.125	----

Change (M)	----	+x	-x	-x	----
Final (M)	----	x	0.125 - x	0.125 - x	----

Finally write the equilibrium expression and solve for the concentration of Ca^{2+} ions:

$$K = 1.6 \times 10^3 = \frac{(0.125 - x)^2}{x}; \text{ assume } x \ll 0.125$$

$$1.6 \times 10^3 = \frac{0.125^2}{x};$$

$x = 9.8 \times 10^{-6}$; assumption is valid
$[Ca^{2+}] = 0.125 - 9.8 \times 10^{-6} = 0.125$ M

17.33 Write the chemical formulas according to the procedures in the text. The coordination number is the number of bonds that the metal atom makes. Both CN^- and NH_3 are monodentate (forms 1 bond with the metal center), ethylenediamine is a bidentate ligand (form two bonds to the metal center).
 (a) $[Fe(CN)_6]^{3-}$; (b) $[Zn(NH_3)_4]^{2+}$; (c) $[V(en)_3]^{3+}$

17.35 This problem describes a common ion effect solubility reaction. We are asked to determine the final ion concentrations in solution. Begin by analyzing the chemistry. The starting materials are NaCl, a salt, and Pb^{2+}. In addition to water, the major species in solution are:

$$Na^+, Cl^-, Pb2+$$

The presence of lead(II) and chloride ions together as major species in solution will result in the reaction:

$$Pb^{2+}(aq) + 4Cl^-(aq) \rightleftharpoons PbCl_4^{2-}(aq)$$

which has the equilibrium expression:

$$K_f = \frac{[PbCl_4^{2-}]}{[Pb^{2+}][Cl^-]^4} = 2.5 \times 10^{15}$$

Because the K_f for the reaction is so large, assume the reaction goes to completion and then do the equilibrium calculations:

The problem gives information about the amounts of both starting materials, so this is a limiting reactant situation. We must calculate the number of moles of each species, construct a table of amounts, and use the results to determine the final solution concentrations.

Calculations of initial amounts:
$n(Na^+) = n(Cl^-) = 0.25$ moles
$n(Pb^{2+}) = 7.5 \times 10^{-3}$ M(1.50 L) = 0.0113 mol

Na$^+$ is a spectator ion and its concentration will be:

$$[Na^+] = \frac{0.25\,mol}{1.50\,L} = 0.167\,M$$

Determine the amounts of each species after assuming the reaction goes to completion:

Reaction:	Pb^{2+} +	4Cl$^-$ ⇌	PbCl$_4^{2+}$
Start (mol)	0.0113	0.25	0
Change (mol)	-0.0113	-4(0.0113)	+ 0.0113
Final (mol)	0	0.205	0.0113

Compute the concentrations by dividing by the volume, 1.50 L:

$$[Cl^-] = \left(\frac{0.205\,mol}{1.50\,L}\right) = 0.137\,M$$

$$[PbCl_4^{2+}] = \left(\frac{0.0113\,mol}{1.50\,L}\right) = 0.0075\,M$$

Use these amounts to complete the concentration table:

Reaction:	Pb^{2+} +	4Cl$^-$ ⇌	PbCl$_4^{2+}$
Start (M)	0	0.137	0.0075
Change (M)	+x	+4x	-x
Equil. (M)	x	0.137 + 4x	0.0075-x

$$K_f = 2.5 \times 10^{15} = \frac{0.0075-x}{x(0.137+4x)^4}\;;\text{ assume } x \ll 0.0075$$

$$x = \frac{0.0075}{(0.137^4)(2.5 \times 10^{15})} = 8.52 \times 10^{-15};\text{ assumption valid}$$

Here are the final concentrations:
[Cl$^-$] = 0.14 M
[Pb^{2+}] = 8.5 × 10^{-15} M
[Na$^+$] = 0.17 M
[PbCl$_4^{2-}$] = 7.5 × 10^{-3} M

(Note all values have two sig. figs. because the initial concentrations are known to two sig. figs.)

17.37 $Mn^{2+} + en \rightleftharpoons [Mn(en)]^{2+}$
$[Mn(en)]^{2+} + en \rightleftharpoons [Mn(en)_2]^{2+}$
$[Mn(en)_2]^{2+} + en \rightleftharpoons [Mn(en)_3]^{2+}$

17.39 This problem describes an equilibrium reaction governed by the common ion effect. We are asked to determine the amount of $CaSO_3$ that will dissolve in the solution. Begin by analyzing the chemistry. The starting materials are $CaSO_3(s)$, an insoluble salt, and nitrilotriacetate. In addition to water, the major species in solution are:
$$H_2O, \; NTA^{3-}$$

The solid salt has a solubility equilibrium:
$$CaSO_3(s) \rightleftharpoons Ca^{2+}(aq) + SO_3^{2-}(aq)$$

The resulting calcium ions can coordinate with the NTA^{3-} to form a complex:
$$Ca^{2+}(aq) + 2NTA^{3-}(aq) \rightleftharpoons [Ca(NTA)_2]^{4-}(aq)$$

Start this problem by determining the overall chemical reaction and the equilibrium expression:

$$
\begin{array}{l}
CaSO_3(s) \rightleftharpoons Ca^{2+}(aq) + SO_3^{2-}(aq) \\
\underline{Ca^{2+}(aq) + 2NTA^{3-}(aq) \rightleftharpoons [Ca(NTA)_2]^{4-}(aq)} \\
CaSO_3(s) + 2NTA^{3-}(aq) \rightleftharpoons [Ca(NTA)_2]^{4-}(aq) + SO_3^{2-}(aq)
\end{array}
$$

$$K = \frac{[[Ca(NTA)_2]^{4-}][SO_3^{2-}]}{[NTA^{3-}]^2} = \frac{[[Ca(NTA)_2]^{4-}]}{[NTA^{3-}]^2[Ca^{2+}]} \frac{[Ca^{2+}][SO_3^{2-}]}{1} = K_{sp}K_f$$
$$K = (1.0 \times 10^{-4})(3.2 \times 10^{11}) = 3.2 \times 10^{7}$$

Because the equilibrium constant is so large we must begin by assuming the reaction has continued to completion and then go to equilibrium. At completion all the ligand will react to form the two products in a 2:1 stoichiometry.

The concentration table is:

Reaction:	$CaSO_3$ (s)+	$2NTA^{3-} \rightleftharpoons$	$[Ca(NTA)_2]^{4-}$ +	SO_3^{2-}(aq)
Start (M)	----	0	0.125/2	0.125/2
Change (M)	----	+2x	-x	-x
Final (M)	----	2x	0.125/2-x	0.125/2-x

$$K = 3.2 \times 10^{7} = \frac{(0.0625 - x)^2}{2x}; \; \text{assume } x \ll 0.0625$$

$3.2 \times 10^7 = \dfrac{(0.0625)^2}{2x}$;

$x = 6.10 \times 10^{-11}$; assumption valid

$[SO_3^{2-}] = 0.0625 - 6.10 \times 10^{-11} = 0.0625$ M

The amount of sulfite ions formed will equal the amount of solid dissolved:

$MM(CaSO_3) = 40.08$ g/mol $+ 32.07$ g/mol $+ 3(16.00$ g/mol$) = 120.1$ g/mol

$n(CaSO_3) = n(SO_3^{2-}) = 0.0625$ M $(0.50$ L$) = 3.13 \times 10^{-2}$ moles

$m(CaSO_3) = 3.13 \times 10^{-2}$ moles $\left(\dfrac{120.1\,\text{g}}{1\,\text{mol}}\right) = 3.8$ g

17.41 A molecular picture of a buffer solution should show molecules/ions of the conjugate acid/base pair in the correct proportions, as determined by the pH of the buffer solution. For formic acid, $pK_a = 3.75$. Use the buffer equation to calculate the base/acid ratio of a formic acid buffer with pH = 4.04:

$$\log\left\{\dfrac{(\text{mol } A^-)}{(\text{mol } HA)}\right\} = pH - pK_a = 4.04 - 3.75 = 0.29;$$

$$\left\{\dfrac{(\text{mol } A^-)}{(\text{mol } HA)}\right\} = 2$$

The picture should show twice as many formate ions as formic acid molecules. The solution is acidic, so there should also be some hydronium ions, but the exact proportion depends on the total concentration of the buffer species. Here is a view showing 6 formate ions, 3 formic acid molecules, and 1 hydronium ion:

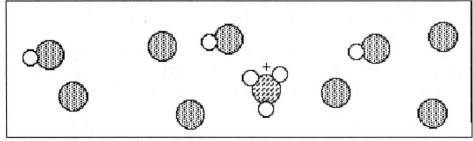

17.43 To determine if all solid dissolves, evaluate Q for all solid dissolving and compare its value with the K_{sp} value. If $Q < K_{sp}$, the spontaneous direction of reaction is to the right even after all solid dissolves, so all solid dissolves, while if $Q > K_{sp}$, not all solid dissolves.

The reaction is $PbCl_2\,(s) \rightleftharpoons Pb^{2+}(aq) + 2\ Cl^-(aq)$,

$K_{eq} = K_{sp} = 2 \times 10^{-5}$ and $Q = [Pb^{2+}][Cl^-]^2$;

Use mole-mass conversions to determine concentrations:

$$n(Pb^{2+}) = \frac{m}{MM} = 0.50 \text{ g} \left(\frac{1 \text{ mol}}{278.1 \text{ g}} \right) = 1.80 \times 10^{-3} \text{ mol};$$

$$[Pb^{2+}] = \frac{1.80 \times 10^{-3} \text{ mol}}{0.300 \text{ L}} = 6.0 \times 10^{-3} \text{ M};$$

$$[Cl^-] = 2[Pb^{2+}] = 1.2 \times 10^{-2} \text{ M};$$

$$Q = (6.0 \times 10^{-3})(1.2 \times 10^{-2})^2 = 8.6 \times 10^{-7}; \quad Q < K_{sp}, \text{ so all the salt dissolves.}$$

17.45 This problem describes the $H_2PO_4^-/HPO_4^{2-}$ buffer solution.

(a) Use the buffer equation to calculate the pH of a buffer solution:

The conjugate acid/base pair is $H_2PO_4^-/HPO_4^{2-}$, $pK_a = pK_{a2} = 7.21$;

$$pH = pK_a + \log\left\{ \frac{[A^-]}{[HA]} \right\} = 7.21 + \log\left\{ \frac{(0.20 \text{ M})}{(0.50 \text{ M})} \right\} = 7.21 - 0.50 = 6.81;$$

(b) Calculate the change in molarity due to the added base, then use the buffer equation to calculate the new pH. The added hydroxide reacts completely with the weak acid and generates the conjugate weak base.

$$n_{NaOH} = \frac{m}{MM} = 0.120 \text{ g} \left(\frac{1 \text{ mol}}{40.0 \text{ g}} \right) = 3.00 \times 10^{-3} \text{ mol};$$

$$\Delta M = \frac{n_{NaOH}}{V} = \frac{3.00 \times 10^{-3} \text{ mol}}{0.15 \text{ L}} = 0.020 \text{ M};$$

$$[HPO_4^{2-}] = 0.20 + 0.020 = 0.22 \text{ M};$$

$$[H_2PO_4^-] = 0.50 - 0.020 = 0.48 \text{ M};$$

$$pH = pK_a + \log\left\{ \frac{[A^-]}{[HA]} \right\} = 7.21 + \log\left\{ \frac{(0.22 \text{ M})}{(0.48 \text{ M})} \right\} = 7.21 - 0.34 = 6.87;$$

$$\Delta pH = 6.87 - 6.81 = 0.06;$$

(c) Adding acid reduces the concentration of weak base, increases the concentration of weak acid, and reduces the pH. Use the buffer equation to calculate the conjugate base/acid ratio at the limit of the pH range. Then calculate the change in molarity that this ratio represents, and convert to moles:

$$pH_{min} = 6.81 - 0.10 = 6.71 = 7.21 + \log\left\{ \frac{[A^-]}{[HA]} \right\};$$

$$\log\left\{ \frac{[A^-]}{[HA]} \right\} = 6.71 - 7.21 = -0.50;$$

$$\left\{ \frac{[A^-]}{[HA]} \right\} = 0.32 = \left\{ \frac{(0.20 - \Delta M)}{(0.50 + \Delta M)} \right\};$$

$$(0.20 - \Delta M) = 0.32(0.50 + \Delta M) = 0.16 + 0.32\Delta M, \text{ so}$$

$0.04 = 1.32\Delta M;$

$\Delta M = 3.0 \times 10^{-2}$ M;

moles of acid neutralized $= \Delta M V = (3.0 \times 10^{-2}$ M$)(0.25$ L$) = 7.5 \times 10^{-3}$ mol.

17.47 To determine if a precipitate forms, evaluate Q for the solution after mixing and compare its value for the value for K_{eq}. If $Q < K_{eq}$, the spontaneous direction of reaction is to the right after the solutions are mixed and a precipitate forms, while if $Q > K_{eq}$ no precipitate forms.

The reaction is: $Ca^{2+}_{(aq)} + SO_4^{2-}$ (aq) $\rightleftarrows$ $CaSO_{4 (s)}$,

$$K_{eq} = \frac{1}{K_{sp}} = \frac{1}{4.9 \times 10^{-5}} = 2.0 \times 10^4 \text{ and } Q = \frac{1}{[Ca^{2+}][SO_4^{2-}]};$$

$$[Ca^{2+}] = 2.00 \times 10^{-2} \text{ M}\left(\frac{350 \text{ mL}}{500 \text{ mL}}\right) = 0.0140 \text{ M};$$

$$[SO_4^{2-}] = 1.50 \times 10^{-2} \text{ M}\left(\frac{150 \text{ mL}}{500 \text{ mL}}\right) = 0.00450 \text{ M};$$

$$Q = \frac{1}{(0.0140)(0.00450)} = 1.59 \times 10^4; \quad Q < K_{eq}, \text{ so the precipitate forms.}$$

17.49 The bromocresol purple indicator with pH $<$ 8.5 has its base color (purple), while with pH $>$ 8.5 has its acid color (yellow):
(a) HCl is a strong acid (pH $\ll$ 5) therefore its color is yellow.
(b) NaOH is a strong base (pH $\gg$ 7) therefore its color is purple.
(c) KCl is a neutral ionic solution (pH $=$7) therefore its color is purple.
(d) NH_3 is a weak base (pH $>$7) therefore its color is purple.

17.51 In a titration of a weak acid by a strong base, pH $= pK_a$ at the midpoint of the titration. Thus, $pK_a = 2.36$ for leucine and $K_a = 4.4 \times 10^{-3}$.

17.53 To calculate an equilibrium constant when experimental data concerning concentrations are available, identify the reaction, then complete an amounts table and substitute into the equilibrium constant expression.

To complete the amounts table, use stoichiometric reasoning and the fact that the change is 2.0×10^{-3} M for Pb^{2+}:

Reaction:	$PbF_{2(s)}$ $\rightleftarrows$	$Pb^{2+}_{(aq)}$ +	2 $F^-_{(aq)}$
Start (M)	---	0	0
Change (M)	---	2.0×10^{-3}	4.0×10^{-3}
Final (M)	solid	2.0×10^{-3}	4.0×10^{-3}

Now substitute into the equilibrium constant expression and evaluate K_{sp}:

$$K_{sp} = [Pb^{2+}]_{eq}[F^-]^2_{eq} = (2.0 \times 10^{-3})(4.0 \times 10^{-3})^2 = 3.2 \times 10^{-8}$$

17.55 The buffer equation is used for calculations involving buffer solutions. Concentrations of acid and conjugate base are needed. The weak acid in the ammonia buffer system is the ammonium ion, $pK_a = 9.25$;

(a) $[NH_3] = 1.00$ M;

$$n(NH_4^+) = 35.0 \text{ g}\left(\frac{1 \text{ mol}}{53.49 \text{ g}}\right) = 0.654 \text{ mol};$$

$$[NH_4^+] = \frac{n}{V} = \frac{0.654 \text{ mol}}{1.00 \text{ L}} = 0.654 \text{ M};$$

$$pH = pK_a + \log\left\{\frac{[A^-]}{[HA]}\right\} = 9.25 + \log\left\{\frac{(1.00 \text{ M})}{(0.654 \text{ M})}\right\} = 9.25 + 0.18 = 9.43;$$

(b) First calculate the ratio of base to acid in the new solution, then use the new ratio to determine the amount of change from the original ratio.
The new pH is $9.43 - 0.05 = 9.38$;

$$\log\left\{\frac{[A^-]}{[HA]}\right\} = pH - pK_a = 9.38 - 9.25 = 0.13;$$

$$\left\{\frac{[A^-]}{[HA]}\right\} = 10^{0.13} = 1.35;$$

Thus $[NH_3] = 1.35[NH_4^+]$; from before, the total concentration is
$[NH_4^+] + [NH_3] = 1.654$ M
$[NH_4^+] + 1.35[NH_4^+] = 1.654$ M, so $2.35[NH_4^+] = 1.654$ M and $[NH_4^+] = 0.704$ M;
The moles of acid required for this change are:
$\Delta n = \Delta M \, V = (0.704 \text{ M} - 0.654 \text{ M})(1.00 \text{ L}) = 0.050 \text{ mol}$;

(c) It is easiest to work with moles in this calculation. In 250 mL of buffer solution, there

are $250 \text{ mL}\left(\dfrac{10^{-3} \text{ L}}{1 \text{ mL}}\right)\left(\dfrac{1.00 \text{ mol}}{1 \text{ L}}\right) = 0.250 \text{ mol NH}_3$ and

$250 \text{ mL}\left(\dfrac{10^{-3} \text{ L}}{1 \text{ mL}}\right)\left(\dfrac{0.654 \text{ mol}}{1 \text{ L}}\right) = 0.163 \text{ mol NH}_4^+$;

The amount of added acid is $5.0 \text{ mL}\left(\dfrac{10^{-3} \text{ L}}{1 \text{ mL}}\right)\left(\dfrac{12.0 \text{ mol}}{1 \text{ L}}\right) = 0.060 \text{ mol}$;

The new amounts are $(0.163 + 0.060) = 0.223 \text{ mol NH}_4^+$ and
$(0.250 - 0.060) = 0.190 \text{ mol NH}_3$;

$$pH = pK_a + \log\left\{\frac{(mol\ NH_3)}{(mol\ NH_4^+)}\right\} = 9.25 + \log\left\{\frac{(0.223\ mol)}{(0.190\ mol)}\right\} = 9.25 + 0.07 = 9.32;$$

17.57 To calculate an equilibrium constant when experimental data concerning concentrations are available, identify the reaction, then complete a concentration table and substitute into the equilibrium constant expression.

The reaction is $Ca_3(PO_4)_{2(s)} \rightleftharpoons 3Ca^{2+}_{(aq)} + 2PO_4^{3-}_{(aq)}$

The solubility in mol/L is $\left(\frac{3.5\ x\ 10^{-5}\ g}{1\ L}\right)\left(\frac{1\ mol}{310.174\ g}\right) = 1.13\ x\ 10^{-7}$ M.

Reaction:	$Ca_3(PO_4)_2 \rightleftharpoons$	$3\ Ca^{2+}\ +$	$2\ PO_4^{3-}$
Start (M)	---	0	0
Change (M)	---	$3(1.13\ x\ 10^{-7})$	$2(1.13\ x\ 10^{-7})$
Final (M)	solid	$3.39\ x\ 10^{-7}$	$2.26\ x\ 10^{-7}$

Now substitute into the equilibrium constant expression and evaluate K_{sp}:

$$K_{sp} = [Ca^{2+}]^3_{eq}[PO_4^{3-}]^2_{eq} = (3.39\ x\ 10^{-7})^3(2.26\ x\ 10^{-7})^2 = 2.0\ x\ 10^{-3}$$

17.59 The major species present are different at various points during a titration, so there are different dominant equilibria which must be identified before doing an equilibrium calculation to determine pH.

(a) Before titration begins, the major species are a weak base, formate (HCO_2^-) and H_2O, and the dominant equilibrium is:

$HCO_2^-_{(aq)} + H_2O_{(l)} \rightleftharpoons HCO_2H_{(aq)} + OH^-_{(aq)}$,

for which $K_b = \dfrac{K_w}{K_a} = \dfrac{1.0\ x\ 10^{-14}}{1.8\ x\ 10^{-4}} = 5.6\ x\ 10^{-11}$.

Reaction:	$H_2O\ +$	$HCO_2^- \rightleftharpoons$	$HCO_2H\ +$	OH^-
Start (M)	---	0.200	0	0
Change (M)	---	$-x$	$+x$	$+x$
Final (M)	solvent	$0.200 - x$	x	x

Now substitute into the equilibrium constant expression and solve for x:

$$K_b = 5.6 \times 10^{-11} = \frac{[HCO_2H]_{eq}[OH^-]_{eq}}{[HCO_2^-]_{eq}} = \frac{x^2}{(0.200\text{-}x)} \text{ ; assume } x << 0.200$$

$x^2 = 1.1 \times 10^{-11}$, from which

$x = 3.3 \times 10^{-6}$;

$[OH^-] = 3.3 \times 10^{-6}$ M,

pOH = -log(3.3 × 10^{-6}) = 5.48 and

pH = 14.00 − 5.48 = 8.52;

(b) At the midpoint of the titration, the buffer equation applies and the concentrations of acid and conjugate base are equal, so pH = pK_a = 3.74;

(c) At the stoichiometric point, all the weak base has been converted into its conjugate acid, so the major acid-base species present are HCO_2H and H_2O and the dominant equilibrium is:

$HCO_2H(aq) + H_2O(l) \rightleftharpoons HCO_2^-(aq) + H_3O^+(aq)$, for which $K_a = 1.8 \times 10^{-4}$.

The initial concentration of HCO_2H is:

$$[HCO_2H] = \frac{0.06 \text{ mol HA}}{0.300 \text{ L} + 0.010 \text{ L}} \text{ ; where 0.010 L is the volume due to the added acid.}$$

The concentration table is:

Reaction:	H$_2$O +	HCO$_2$H $\rightleftharpoons$	HCO$_2^-$ +	OH$^-$
Start (M)	---	0.193	0	0
Change (M)	---	-x	+x	+x
Final (M)	solvent	0.193 - x	x	x

Now substitute into the equilibrium constant expression and solve for x:

$$K_a = 1.8 \times 10^{-4} = \frac{[HCO_2^-]_{eq}[H_3O^+]_{eq}}{[HCO_2H]_{eq}} = \frac{x^2}{(0.193\text{-}x)};$$

$x^2 = (0.193 - x)(1.8 \times 10^{-4}) = (3.5 \times 10^{-5}) - (1.8 \times 10^{-4})x$;

$x^2 + (1.8 \times 10^{-4})x - (3.5 \times 10^{-5}) = 0$;

$$x = \frac{-b \pm \sqrt{b^2 - 4ac}}{2a} = \frac{-(1.8 \times 10^{-4}) \pm \sqrt{(1.8 \times 10^{-4})^2 + 4(3.5 \times 10^{-5})}}{2} = 5.8 \times 10^{-3};$$

$[H_3O^+]_{eq} = 5.8 \times 10^{-3}$ M, and pH = 2.23;

(d) A suitable indicator for this titration must change color around pH = 2. Thymol blue, pK_{In} = 1.75, would be the best choice.

17.61 When a precipitate forms upon mixing aqueous solutions, the equilibrium reaction is the reverse of the solubility reaction. Follow the standard procedure for a reaction with a large equilibrium constant.

Reaction is $Mn^{2+}(aq) + CO_3^{2-}(aq) \rightleftharpoons MnCO_3(s)$, $\qquad K_{eq} = \dfrac{1}{K_{sp}}$

"Initial" concentrations are after mixing but before reaction:

$$[Mn^{2+}] = \left(\frac{0.750\,L}{0.750 + 0.150\,L}\right)(0.0250\,M) = 0.0208\,M$$

$$[CO_3^{2-}] = \left(\frac{0.150\,L}{0.750 + 0.150\,L}\right)(0.500\,M) = 0.0833\,M$$

Begin by assuming all the ions form a solid. The concentrations at completion:
$[Mn^{2+}] = 0.000\,M$
$[CO_3^{2-}] = 0.0625\,M$

Reaction:	$MnCO_3(s) \rightleftharpoons$	$CO_3^{2-}\ +$	Mn^{2+}
Start (M)	----	0.0625	0
Change (M)	----	$+x$	$+x$
Final (M)	Solid	$0.0625 + x$	x

Assume that $x \ll 0.0625$:

$K_{sp} = 2.2 \times 10^{-11} = (x)(0.0625)$, from which $x = \dfrac{2.2 \times 10^{-11}}{0.0625} = 3.5 \times 10^{-10}$

Thus $[Mn^{2+}] = 3.5 \times 10^{-10}\,M$

To calculate the mass of precipitate, use standard stoichiometric methods. The reaction goes virtually to completion, so:

$n_{MnCO_3} = n_{Mn^{2+}} = MV = (0.0250\,mol/L)(0.750\,L) = 0.01875\,mol$;

$m = n\,MM = 0.01875\,mol\left(\dfrac{114.95\,g}{1\,mol}\right) = 2.16\,g.$

17.63 "What mass will dissolve" means "calculate the amount present in solution at equilibrium." Initial amounts are zero, so it is easy to complete a concentration table. Calculate molarity from the solubility product expression, then convert to mass using standard stoichiometric methods. Let x be the number of moles of salt dissolving in 1 L of solution.

Reaction:	$Zn(OH)_{2\,(s)} \rightleftharpoons$	Zn^{2+} +	$2\,OH^-$
Start (M)	----	0	0
Change (M)	----	$+x$	$+2x$
Final (M)	solid	x	$2x$

$pK_{sp} = 16.52$, so $K_{sp} = 3.0 \times 10^{-17} = [Zn^{2+}]_{eq}[OH^-]^2_{eq} = 4\,x^3$ and

$x = 2.0 \times 10^{-6}$;

$m = M\,V\,MM = (2.0 \times 10^{-6}\,M)(1.00\,L)\left(\dfrac{99.40\,g}{1\,mol}\right) = 2.0 \times 10^{-4}\,g$ dissolves.

17.65 The best choice for a buffer solution generally is the weak acid whose pK_a is closest to the desired pH of the buffer solution. For a pH = 4.80 buffer solution, acetic acid/acetate, $pK_a = 4.75$, would be the best choice. To prepare the buffer solution, add the appropriate amounts of sodium acetate (NaAc) and acetic acid (HAc) solution to water and make up to 1.0 L with additional water. Calculate the acetate/acetic acid ratio using the buffer equation:

$\log\left\{\dfrac{[Ac^-]}{[HAc]}\right\} = pH - pK_a = 4.80 - 4.75 = 0.05,\ \left\{\dfrac{[Ac^-]}{[HAc]}\right\} = 10^{0.05} = 1.12$;

Thus, $[Ac^-] = 1.12\,[HAc]$; also, $[Ac^-] + [HAc] = 0.35\,M$,
so $[HAc] = 0.35\,M - [Ac^-]$;
$[Ac^-] = 1.12(0.35\,M - [Ac^-]) = 0.39 - 1.12\,[Ac^-]$, so
$2.12\,[Ac^-] = 0.39$ and $[Ac^-] = 0.18\,M$

$n_{Ac^-} = (0.18\,M)(1.0\,L) = 0.18\,mol$;
$m_{NaAc} = n\,MM = (0.18\,mol)(82.0\,g/mol) = 15\,g$;
$[HAc] = 0.35 - 0.18 = 0.17\,M$;
$n_{HAc} = (0.17\,M)(1.0\,L) = 0.17\,mol$;

$V_{HAc} = \dfrac{n}{M} = 0.17\ mol\left(\dfrac{1\,L}{1.00\,mol}\right) = 0.17\,L$;

Mix together 15 g sodium acetate, 0.17 L 1.0 M acetic acid, and enough water to make 1.0 L.

17.67 "Saturated aqueous solution" identifies this as a solubility equilibrium.

(a) Reaction is $MgF_{2(s)} \rightleftharpoons Mg^{2+}{}_{(aq)} + 2\,F^-{}_{(aq)}$, $K_{sp} = [Mg^{2+}]_{eq}[F^-]^2_{eq}$;

(b) If $[Mg^{2+}]_{eq} = 1.14 \times 10^{-3}\,M$, by stoichiometry $[F^-]_{eq} = 2(1.14 \times 10^{-3}\,M)$;
$K_{sp} = (1.14 \times 10^{-3})[2(1.14 \times 10^{-3})]^2 = 5.93 \times 10^{-9}$;

(c) To estimate the equilibrium constant at a temperature different from 298 K, calculate $\Delta H^o_{reaction}$ and $\Delta S^o_{reaction}$ at 298 K and then use Equations 13-9 and 15-3:

$$\Delta G^{\circ}_{reaction} = \Delta H^{\circ}_{reaction} - T\Delta S^{\circ}_{reaction} \quad \Delta G^{\circ} = -RT \ln K_{eq}$$

$$\Delta H^{\circ}_{reaction} = 1 \text{ mol}(-467.0 \text{ kJ/mol}) + 2 \text{ mol}(-335.4 \text{ kJ/mol})$$
$$- 1 \text{ mol}(-1124.2 \text{ kJ/mol}) = -13.6 \text{ kJ};$$

$$\Delta S^{\circ}_{reaction} = 1 \text{ mol}(-137 \text{ J/molK}) + 2 \text{ mol}(-13.8 \text{ J/molK})$$
$$- 1 \text{ mol}(57.2 \text{ J/molK}) = -222 \text{ J/K};$$

$$\Delta G^{\circ}_{reaction, 373K} = (-13.6 \text{ kJ/mol}) - (373 \text{ K})(-222 \text{ J/mol K})\left(\frac{10^{-3} \text{ kJ}}{1 \text{ J}}\right) = 69.2 \text{ kJ/mol}$$

$$\ln K_{eq} = -\frac{6.92 \times 10^4 \text{ J mol}^{-1}}{(8.314 \text{ J mol}^{-1} \text{ K}^{-1})(373 \text{ K})} = -22.3; \quad K_{eq} = e^{-22.3} = 2.1 \times 10^{-10}$$

16.69 To calculate equilibrium concentrations, set up a concentration table:

Reaction:	HgS(s) ⇌	Hg²⁺(aq) +	S²⁻(aq)
Start (M)	----	0	0
Change (M)	----	+x	+x
Final (M)	solid	x	x

$4.0 \times 10^{-53} = [\text{Hg}^{2+}]_{eq}[\text{S}^{2-}]_{eq} = x^2$, from which $x = 6.3 \times 10^{-27}$

Thus $[\text{Hg}^{2+}]_{eq} = 6.3 \times 10^{-27}$ M (a very small concentration).

To calculate the volume that would be expected to contain a single Hg^{2+} cation,

$V = \dfrac{n}{M}$, with $n = \dfrac{1}{N_A}$:

$$V = \frac{1}{N_A M} = \left(\frac{1 \text{ mol}}{6.022 \times 10^{23} \text{ molecule}}\right)\left(\frac{1 \text{ L}}{6.3 \times 10^{-27} \text{ mol}}\right) = 2.6 \times 10^2 \text{ L}$$

17.71 (a) To determine the ion concentrations in a solution, follow the five step procedure for solving an equilibrium problem:

1.) The species in solution are Ca^{2+}, Cl^-, PO_4^{3-}, and H_2O.

2, 3.) The reaction is formation of $\text{Ca}_3(\text{PO}_4)_2$ precipitate:

$$3 \text{ Ca}^{2+}\text{(aq)} + 2 \text{ PO}_4^{3-}\text{(aq)} \rightleftharpoons \text{Ca}_3(\text{PO}_4)_2\text{(s)} \quad K_{eq} = \frac{1}{K_{sp}} = \frac{1}{[\text{Ca}^{2+}]^3_{eq}[\text{PO}_4^{3-}]^2_{eq}}$$

4.) Initial concentrations are those present after adding solid but before reaction occurs:

$$[\text{Ca}^{2+}] = \frac{(120 \text{ mol})}{(3.00 \times 10^3 \text{ L})} = 4.0 \times 10^{-2} \text{ M}; \quad [\text{PO}_4^{3-}] = 2.2 \times 10^{-3} \text{ M};$$

Take the reaction to completion and then return to equilibrium:

Begin by assuming all the ions form a solid. The concentrations at completion:

$[\text{Ca}^{2+}] = 0.037$ M

$[\text{PO}_4^{3-}] = 0.000$ M

Reaction:	$Ca_3(PO_4)_2$ (s) $\rightleftharpoons$	$3\ Ca^{2+}$ (aq) +	$2\ PO_4^{3-}$ (aq)
Start (M)	----	0.037	0
Change (M)	----	+3x	+2x
Final (M)	solid	0.037 + 3x	2x

5.) Assume that $3x \ll 3.7 \times 10^{-2}$, substitute into the solubility product expression, and solve:

$$K_{sp} = 2.0 \times 10^{-33} = [Ca^{2+}]^3_{eq}[PO_4^{3-}]^2_{eq} = (3.7 \times 10^{-2})^3 (2\,x)^2$$

$$x^2 = \frac{(2.0 \times 10^{-33})}{(3.7 \times 10^{-2})^3 (4)} = 9.9 \times 10^{-30}, \text{ from which } x = 3.1 \times 10^{-15};$$

$$[PO_4^{3-}]_{eq} = 2x = 6.2 \times 10^{-15} \text{ M};$$

(b) Because this reaction goes essentially to completion, the calculation of the mass of precipitate can be done using standard stoichiometric methods:

For phosphate ions, $n = 3.00 \times 10^3 \text{ L} \left(\dfrac{2.2 \times 10^{-3} \text{ mol}}{1 \text{ L}} \right) = 6.6 \text{ mol};$

$$n_{Ca_3(PO_4)_2} = \frac{1}{2} n_{PO_4^{3-}} = 3.3 \text{ mol};$$

$$m = n\ MM = 3.3 \text{ mol} \left(\frac{310.18 \text{ g}}{1 \text{ mol}} \right) = 1.0 \times 10^3 \text{ g}.$$

17.73 To calculate an equilibrium constant when experimental data concerning concentrations are available, identify the reaction, then complete an amounts table and substitute into the equilibrium constant expression.

The reaction is $Sr(IO_3)_2(s) \rightleftharpoons Sr^{2+}$ (aq) + 2 IO_3^- (aq)

At 25 °C, The solubility in mol/L is $\dfrac{0.030 \text{ g}}{(0.100 \text{ L})(437.42 \text{ g/mol})} = 6.86 \times 10^{-4}$ M.

Reaction:	$Sr(IO_3)_2$ (s) $\rightleftharpoons$	Sr^{2+} (aq) +	$2\ IO_3^-$ (aq)
Start (M)	----	0	0
Change (M)	----	$+6.86 \times 10^{-4}$	$+2(6.86 \times 10^{-4})$
Final (M)	solid	6.86×10^{-4}	1.37×10^{-3}

Now substitute into the equilibrium constant expression and evaluate K_{sp}:

$$K_{sp} = [Sr^{2+}]_{eq}[IO_3^-]^2_{eq} = (6.86 \times 10^{-4})(1.37 \times 10^{-3})^2 = 1.29 \times 10^{-9}$$

$\Delta G^o = - RT \ln(K_{eq}) = - (8.314 \times 10^{-3} \text{ kJ/mol K})(298 \text{ K}) \ln(1.29 \times 10^{-9}) = 50.7 \text{ kJ/mol}$

At 100.0 °C, the solubility in mol/L is $\dfrac{0.80 \text{ g}}{(0.100 \text{ L})(437.42 \text{ g/mol})} = 1.83 \times 10^{-2}$ M.

Reaction:	$Sr(IO_3)_2$ (s) $\rightleftharpoons$	Sr^{2+} (aq) +	$2 IO_3^-$ (aq)
Start (M)	----	0	0
Change (M)	----	$+1.83 \times 10^{-2}$	$+2(1.83 \times 10^{-2})$
Final (M)	solid	1.83×10^{-2}	3.66×10^{-3}

Now substitute into the equilibrium constant expression and evaluate K_{sp}:

$K_{sp} = [Sr^{2+}]_{eq}[IO_3^-]^2_{eq} = (1.83 \times 10^{-2})(3.66 \times 10^{-2})^2 = 2.45 \times 10^{-5}$

(Round to two sig. figs: 2.5×10^{-5})

$\Delta G^o = - RT \ln(K_{eq}) = - (8.314 \times 10^{-3} \text{ J/mol K})(373 \text{ K}) \ln (2.45 \times 10^{-5}) = 33 \text{ kJ/mol.}$

17.75 To determine equilibrium concentrations, set up the appropriate concentration table. In this problem, one equilibrium concentration is given and the other two are stoichiometrically related, so the table can be relatively simple (note that the volume of the atmosphere is so large relative to the volume of ground water that the change in CO_2 pressure is negligible):

Reaction:	$CaCO_3(s)$ +	$H_2O(l)$ +	$CO_2(g)$ $\rightleftharpoons$	Ca^{2+} (aq) +	$2 HCO_3^-$ (aq)
Start	----	----	3.2×10^{-4} atm	0 M	0 M
Change	----	----	~ 0	$+x$ M	$+2x$ M
Final	solid	----	3.2×10^{-4} atm	x M	$2x$ M

Substitute equilibrium values into the equilibrium constant expression and solve for x:

$K_{eq} = 1.56 \times 10^{-8} = \dfrac{[Ca^{2+}]_{eq}[HCO_3^-]^2_{eq}}{(p_{CO_2})_{eq}} = \dfrac{(x)(2x)^2}{(3.2 \times 10^{-4})}$

$(1.56 \times 10^{-8})(3.2 \times 10^{-4}) = 4 x^3$, so $x^3 = 1.25 \times 10^{-12}$ and $x = [Ca^{2+}] = 1.1 \times 10^{-4}$ M.

17.77 Salts that are more soluble in acidic solution are those that have anions that have weak conjugate acids. They are: Ag_2CO_3, Ag_2SO_4, and Ag_2S. Those that are independent of pH have anions that are conjugate bases of strong acids: AgBr and AgCl.

17.79 Use the buffer equation to carry out calculations on buffer solutions:

(a) $pH = pK_a + \log\left\{\dfrac{[\text{TRIS}]}{[\text{TRISH}^+]}\right\} = (14.00 - 5.91) + \log\left[\dfrac{(0.30 \text{ M})}{(0.60 \text{ M})}\right] = 7.79;$

(b) Adding HCl provides H_3O^+ ions, which react quantitatively with TRIS to generate TRISH$^+$; the amount of added acid is $n = 5.0 \text{ mL}\left(\dfrac{10^{-3}\text{L}}{1\text{ mL}}\right)\left(\dfrac{12\text{ mol}}{1\text{ L}}\right) = 0.060$ mol;

Using moles in the buffer equation saves us from calculating the dilution effect of adding 5.0 mL of the acid. If the initial solution is 1.0 L of buffer, then the amounts in moles are the same as the molarity of each species;

$$pH = (14.00 - 5.91) + \log\left[\frac{(0.30 - 0.06)\text{ moles}}{(0.60 + 0.06)\text{ moles}}\right] = 8.09 + \log\left[\frac{0.24\text{ mol}}{0.66\text{ mol}}\right] = 7.65.$$

17.81 At different points during a titration, different major species are present, because the titration reaction consumes formic acid and produces formate:

$$HCO_2H(aq) + OH^-(aq) \rightleftharpoons HCO_2^-(aq) + H_2O(l)$$

Point A is before titration begins, when the major species are H_2O and HCO_2H.

Point B is near the midpoint of the titration, when the major species are H_2O, HCO_2H, and HCO_2^-; The dominant equilibrium for both points A and B will be the acid reaction:

$$H_2O(l) + HCO_2H(aq) \rightleftharpoons HCO_2^-(aq) + H_3O^+(aq); \qquad K = \frac{[HCO_2^-][H_3O^+]}{[HCO_2H]}$$

Point C is the stoichiometric point where all formic acid has been consumed, so the major species are H_2O and HCO_2^-. Thus the dominant equilibrium for point C is:

$$H_2O(l) + HCO_2^-(aq) \rightleftharpoons HCO_2H(aq) + OH^-(aq); \qquad K = \frac{[HCO_2H][OH^-]}{[HCO_2^-]}$$

Point D is beyond the stoichiometric point, so excess hydroxide is present and the major species are H_2O, OH^-, and HCO_2^-; The dominant equilibrium for point D is:

$$H_2O(l) + HCO_2^-(aq) \rightleftharpoons HCO_2H(aq) + OH^-(aq); \qquad K = \frac{[HCO_2H][OH^-]}{[HCO_2^-]}$$

The cation of the strong base will also be present as a major species at points B, C, and D.

17.83 Molecular pictures show the correct relative numbers of the various species in the solution. The starting condition shows 9 molecules of acetic acid and 3 of acetate anions. (a) Use the buffer equation to calculate the pH of this solution:

$$pH = pK_a + \log\left\{\frac{[A^-]}{[HA]}\right\} = 4.75 + \log\left[\frac{3}{9}\right] = 4.75 - 0.48 = 4.27.$$

(b) A hydroxide ion reacts with an acetic acid molecule to form a water molecule and an acetate ion, so the new picture shows 8 acetic acid molecules, 4 acetate ions, and 1 H_2O:

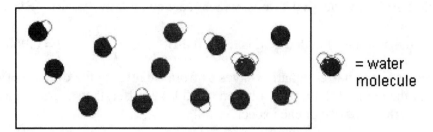

(c) HSO_4^- is a weak acid, so it transfers a proton to an acetate ion:

$$C_2H_3O_2^-{}_{(aq)} + HSO_4^-{}_{(aq)} \rightleftharpoons SO_4^{2-}{}_{(aq)} + HC_2H_3O_2{}_{(aq)}$$

The new picture shows 10 molecules of acetic acid, 2 of acetate anions, and 1 sulfate:

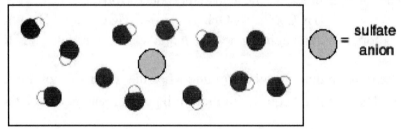

Chapter 18

18.1   Oxidation numbers are determined by applying the rules given in the textbook.
(a) O (most electronegative in this compound) is –2, H is +1, and Fe must be +3 to give overall neutrality;
(b) F (most electronegative) is –1, so N must be +3 to give overall neutrality;
(c) O is –2, H is +1, and C is –2 to give overall neutrality;
(d) ionic compound, with K^+ = +1 and CO_3^{2-}; in the anion, O (more electronegative) is –2 and C is +4 to give the –2 net charge on the ion;
(e) ionic compound; in NH_4^+, H is +1 and N is –3, and in NO_3^-, O is –2 and N is +5;
(f) Cl (more electronegative) is –1, so Ti is +4 to give overall neutrality;
(g) ionic, Pb must be +2 to balance -2 charge on sulfate; O is –2, so S is +6 to give the –2 charge on the anion;
(h) P is 0.

18.3   Identify redox reactions by determining whether or not oxidation numbers change.
(a) no change in oxidation numbers (Br remains –1), so not redox;
(b) redox, because Fe changes from +2 to +3 (O also changes);
(c) no change in oxidation numbers (Fe remains +2), so not redox;
(d) redox, because O in O_2 changes from 0 to –2 (C also changes);
(e) redox, because N changes from 0 to –3 and H changes from 0 to +1.

18.5   Oxidation numbers are determined by applying the rules given in the textbook.
(a) F is –1, so Cl is +5 to give overall neutrality;
(b) Cl is more electronegative, so it is –1;
(c) ionic, ClO_4^- has net charge of -1, and O is –2, so Cl is +7 to give net charge of –1;
(d) Cl is 0 because this is a pure element;
(e) ionic, ClO_2^- has net charge of -1, and O is –2, so Cl is +3 to give net charge of –1;
(f) ionic, ClO^- has net charge of -1, and O is –2, so Cl is +1 to give net charge of –1.

18.7   Determine half-reactions by inspection, making use of oxidation numbers if necessary. Balance each half-reaction following the standard steps in order (balance all but H and O by inspection, balance O by adding H_2O, balance H by adding H_3O^+ cations to the side that is deficient in hydrogen and an equal number of H_2O to the other side, balance charge by adding e^-):
(a) The reactants are Na and H_2O, and one product is H_2; Na^+ must be a second product:
$Na(s) \rightarrow Na^+(aq) + e^-$ and $2 H_3O^+(aq) + 2 e^- \rightarrow H_2(g) + 2 H_2O(l)$;
(b) The reactants are Au and NO_3^-, and the products are $[AuCl_4]^-$ and NO; aqua regia also contains Cl^- anions:
$Au(s) + 4 Cl^-(aq) \rightarrow [AuCl_4]^- (aq) + 3 e^-$ and
$NO_3^-(aq) + 4 H_3O^+(aq) + 3 e^- \rightarrow NO(g) + 6 H_2O(l)$;
(c) The reactants are MnO_4^- and $C_2O_4^{2-}$, and the products are Mn^{2+} and CO_2:
$MnO_4^-(aq) + 8 H_3O^+(aq) + 5 e^- \rightarrow Mn^{2+}(aq) + 12 H_2O(l)$ and
$C_2O_4^{2-}(aq) \rightarrow 2 CO_2(g) + 2 e^-$.
(d) The reactants are C and H_2O, and the products are H_2 and CO:
$C(s) + 3 H_2O(l) \rightarrow CO(g) + 2 H_3O^+(aq) + 2e^-$ and $2 H_3O^+(aq) + 2e^- \rightarrow H_2(g) + 2 H_2O(l)$.

18.9 Balance each half-reaction following the standard steps in order: 1.) balance all but H and O by inspection; 2.) balance O by adding H_2O; 3.) balance H by adding H_3O^+ cations to the side that is deficient in hydrogen and an equal number of H_2O to the other side. If the solution is basic, add a H_2O for each deficient H to the side that is deficient in hydrogen and an equal number of OH^- anions to the other side; 4.) balance charge by adding e^-.

(a) (1) Cu is balanced, (2) add H_2O on left, (3) add 2 H_3O^+ on right and 2 water on the left, (4) add 1 e^- on right:
$Cu^+(aq) + 3\,H_2O(l) \rightarrow CuO(s) + 2\,H_3O^+(aq) + e^-$;

(b) (1) S is balanced, (2) no O present, (3) add 2 H_3O^+ on left and 2 water on the right, (4) add 2 e^- on left:
$S(s) + 2\,H_3O^+(aq) + 2\,e^- \rightarrow H_2S(g) + 2\,H_2O(l)$;

(c) (1) multiply Ag by 2, (2) add H_2O on right, (3) add 2 H_2O on left, and add 2 OH^- to the right side, (4) add 2 e^- on left: $Ag_2O + 2\,H_2O + 2\,e^- \rightarrow 2\,Ag + H_2O + 2\,OH^-$;
Combine and cancel: $Ag_2O(s) + H_2O(l) + 2\,e^- \rightarrow 2\,Ag(s) + 2\,OH^-(aq)$;

(d) (1) I is balanced, (2) add 3 H_2O on left, (3) add 6 H_2O on right, and add 6 OH^- to the left side, (4) add 6 e^- on right: $I^- + 3\,H_2O + 6\,OH^- \rightarrow IO_3^- + 6\,H_2O + 6\,e^-$;
Combine and cancel: $I^-(aq) + 6\,OH^-(aq) \rightarrow IO_3^-(aq) + 3\,H_2O(l) + 6\,e^-$;

(e) (1) I is balanced, (2) add 2 H_2O on right, (3) add 4 H_2O on left, and add 4 OH^- to the right side, (4) add 4 e^- on left: $IO_3^- + 4\,H_2O + 4\,e^- \rightarrow IO^- + 2\,H_2O + 4\,OH^-$;
Combine and cancel: $IO_3^-(aq) + 2\,H_2O(l) + 4\,e^- \rightarrow IO^-(aq) + 4\,OH^-(aq)$;

(f) (1) C is balanced, (2) add H_2O on left, (3) add 4 H_3O^+ on right and 4 H_2O on the left, (4) add 4 e^- on right:
$H_2CO(aq) + 5\,H_2O(l) \rightarrow CO_2(g) + 4\,H_3O^+(aq) + 4\,e^-$.

18.11 To combine half-reactions into a net redox reaction, multiply by appropriate integers so that the electrons will cancel:

(a) $2[Cu^+ + 3\,H_2O \rightarrow CuO + 2\,H_3O^+ + e^-]$
     $+ [S + 2\,H_3O^+ + 2\,e^- \rightarrow H_2S + 2\,H_2O]$
          $2\,Cu^+ + 4\,H_2O + S \rightarrow 2\,CuO + 2\,H_3O^+ + H_2S$;

(b) $3[Ag_2O + H_2O + 2\,e^- \rightarrow 2\,Ag + 2\,OH^-]$
     $+ \quad\quad [I^- + 6\,OH^- \rightarrow IO_3^- + 3\,H_2O + 6\,e^-]$
          $3\,Ag_2O + I^- \rightarrow 6\,Ag + IO_3^-$;

(c) $2[I^- + 6\,OH^- \rightarrow IO_3^- + 3\,H_2O + 6\,e^-]$

$$\underline{+\ 3\ [IO_3^- +\ 2\ H_2O + 4\ e^- \rightarrow IO^- +\ 4\ OH^-\ ]}$$
$$2I^- + IO_3^- \rightarrow 3\ IO^-;$$

(d) $2[S + 2\ H_3O^+ + 2\ e^- \rightarrow\ H_2S + 2\ H_2O]$
$$\underline{+\ \ \ \ \ [H_2CO + 5\ H_2O \rightarrow CO_2 +\ 4\ H_3O^+ + 4\ \ e^-\ ]}$$
$$2S + H_2CO + H_2O \rightarrow 2\ H_2S + CO_2.$$

18.13   The species that loses electrons is oxidized and acts as the reducing agent, while the species that gains electrons is reduced and acts as the oxidizing agent. In this reaction, Cl^- loses electrons (oxidized, reducing agent) to become Cl_2, and Mn in MnO_4^- gains electrons (reduced, oxidizing agent) to become Mn^{2+}.

18.15   Balance redox reactions following the standard procedure. Break into half-reactions, balance each half-reaction using the stepwise technique, then multiply by appropriate integers so the electrons cancel:

(a) $CN^- \rightarrow CNO^-$ (1) C and N already balanced, (2) add 1 H_2O on left, (3) add 2 H_2O on right and add 2 OH^- on the left side, (4) add 2 e^- on right:
$CN^- + H_2O + 2\ OH^- \rightarrow CNO^- + 2\ H_2O + 2\ e^-$;
Combine and cancel: $CN^- + 2\ OH^- \rightarrow CNO^- + H_2O + 2\ e^-$;

$MnO_4^- \rightarrow MnO_2$ (1) Mn already balanced, (2) add 2 H_2O on right, (3) add 4 H_2O on left, and add 4 OH^- on right side, (4) add 3 e^- on left:
$MnO_4^- + 4\ H_2O + 3\ e^- \rightarrow MnO_2 + 2\ H_2O + 4\ OH^-$;
Combine and cancel: $MnO_4^- + 2\ H_2O + 3\ e^- \rightarrow MnO_2 + 4\ OH^-$;

Multiply first reaction by 3, second by 2, and add:
$$3\ [CN^- + 2\ OH^- \rightarrow CNO^- + H_2O + 2\ e^-]$$
$$\underline{+\ 2\ [MnO_4^- + 2\ H_2O + 3\ e^- \rightarrow MnO_2 + 4\ \ OH^-\ ]\ \ \ \ \ \ \ \ \ ;}$$
$$3\ CN^- + 2\ MnO_4^- + H_2O \rightarrow 3\ CNO^- + 2\ MnO_2 + 2\ OH^-;$$

(b) As $\rightarrow HAsO_2$ (1) As already balanced; (2) add 2 H_2O on left, (3) add 3 H_3O^+ on right and 3 H_2O on the left, (4) add 3 e^- on right:
As $+ 5\ H_2O \rightarrow HAsO_2 + 3\ H_3O^+ + 3\ e^-$;

$O_2 \rightarrow H_2O$ (1) no elements other than O and H, (2) add 1 H_2O on right, (3) add 4 H_3O^+ on left, and 4 H_2O on the right (4) add 4 e^- on left:
$O_2 + 4\ H_3O^+ + 4\ e^- \rightarrow 6\ H_2O$;

Multiply first reaction by 4, second by 3, and add:
$$4\ [As + 5\ H_2O \rightarrow HAsO_2 + 3\ H_3O^+ + 3\ e^-]$$
$$\underline{+\ 3\ [O_2 + 4\ H_3O^+ + 4\ e^- \rightarrow 6\ H_2O]\ \ \ \ \ \ \ \ \ \ \ .}$$

$4 \text{ As } + 3 \text{ O}_2 + 2 \text{ H}_2\text{O} \rightarrow 4 \text{ HAsO}_2$;

(c) $\text{Br}^- \rightarrow \text{BrO}_3^-$ (1) Br already balanced, (2) add 3 H_2O on left, (3) add 6 H_2O on right, and add 6 OH^- on the left side, (4) add 6 e^- on right:
$\text{Br}^- + 3 \text{ H}_2\text{O} + 6 \text{ OH}^- \rightarrow \text{BrO}_3^- + 6\text{H}_2\text{O} + 6 \text{ e}^-$;
Combine and cancel: $\text{Br}^- + 6 \text{ OH}^- \rightarrow \text{BrO}_3^- + 3 \text{ H}_2\text{O} + 6 \text{ e}^-$;

$\text{MnO}_4^- \rightarrow \text{MnO}_2$, same as reaction in part (a):
$\text{MnO}_4^- + 2 \text{ H}_2\text{O} + 3 \text{ e}^- \rightarrow \text{MnO}_2 + 4 \text{ OH}^-$;

Multiply second reaction by 2 and add:
$$[\text{Br}^- + 6 \text{ OH}^- \rightarrow \text{BrO}_3^- + 3 \text{ H}_2\text{O} + 6 \text{ e}^-]$$
$$+ 2 [\text{MnO}_4^- + 2 \text{ H}_2\text{O} + 3 \text{ e}^- \rightarrow \text{MnO}_2 + 4 \text{ OH}^-] \qquad .$$
$$\text{Br}^- + 2 \text{ MnO}_4^- + \text{H}_2\text{O} \rightarrow \text{BrO}_3^- + 2 \text{ MnO}_2 + 2 \text{ OH}^- ;$$

(d) $\text{NO}_2 \rightarrow \text{NO}_3^-$ (1) N already balanced; (2) add 1 H_2O on left, (3) add 2 H_3O^+ on right and 2 H_2O on the left, (4) add 1 e^- on right: $\text{NO}_2 + 3\text{H}_2\text{O} \rightarrow \text{NO}_3^- + 2 \text{ H}_3\text{O}^+ + \text{e}^-$;

$\text{NO}_2 \rightarrow \text{NO}$ (1) N already balanced; (2) add 1 H_2O on right, (3) add 2 H_3O^+ on left and 2 H_2O on the right, (4) add 2 e^- on left: $\text{NO}_2 + 2 \text{ H}_3\text{O}^+ + 2 \text{ e}^- \rightarrow \text{NO} + 3 \text{ H}_2\text{O}$;

Multiply first reaction by 2 and add:
$$2 [\text{NO}_2 + 3\text{H}_2\text{O} \rightarrow \text{NO}_3^- + 2 \text{ H}_3\text{O}^+ + \text{e}^-]$$
$$+ [\text{NO}_2 + 2 \text{ H}_3\text{O}^+ + 2 \text{ e}^- \rightarrow \text{NO} + 3 \text{ H}_2\text{O}] \qquad .$$
$$3 \text{ NO}_2 + 3 \text{ H}_2\text{O} \rightarrow 2 \text{ NO}_3^- + \text{NO} + 2 \text{ H}_3\text{O}^+ ;$$

(e) $\text{ClO}_4^- \rightarrow \text{ClO}^-$ (1) Cl already balanced; (2) add 3 H_2O on right, (3) add 6 H_3O^+ on left and 6 H_2O on the right, (4) add 6 e^- on left:
$\text{ClO}_4^- + 6 \text{ H}_3\text{O}^+ + 6 \text{ e}^- \rightarrow \text{ClO}^- + 9 \text{ H}_2\text{O}$;

$\text{Cl}^- \rightarrow \text{Cl}_2$ (1) multiply Cl^- by 2; (2,3) no H or O present, (4) add 2 e^- on right:
$2 \text{ Cl}^- \rightarrow \text{Cl}_2 + 2 \text{ e}^-$;

Multiply second reaction by 3 and add:
$$[\text{ClO}_4^- + 6 \text{ H}_3\text{O}^+ + 6 \text{ e}^- \rightarrow \text{ClO}^- + 9 \text{ H}_2\text{O}]$$
$$+ \qquad 3 [2 \text{ Cl}^- \rightarrow \text{Cl}_2 + 2 \text{ e}^-] \qquad .$$
$$\text{ClO}_4^- + 6 \text{ Cl}^- + 6 \text{ H}_3\text{O}^+ \rightarrow \text{ClO}^- + 9 \text{ H}_2\text{O} + 3 \text{ Cl}_2 ;$$

(f) $\text{AlH}_4^- \rightarrow \text{Al}^{3+}$ (1) Al already balanced, (2) no O present, (3) add 4 H_2O on right, and add 4 OH^- on the left side, (4) add 8 e^- on right:
$\text{AlH}_4^- + 4 \text{ OH}^- \rightarrow \text{Al}^{3+} + 4 \text{ H}_2\text{O} + 8 \text{ e}^-$

$H_2CO \rightarrow CH_3OH$ (1) C already balanced, (2) O already balanced, (3) add 2 H_2O on left, and add 2 OH^- on the right side, (4) add 2 e^- on left:
$H_2CO + 2 H_2O + 2 e^- \rightarrow CH_3OH + 2 OH^-$;

Multiply second reaction by 4 and add:

$$[AlH_4^- + 4 OH^- \rightarrow Al^{3+} + 4 H_2O + 8 e^-]$$
$$+ 4 [H_2CO + 2 H_2O + 2 e^- \rightarrow CH_3OH + 2 OH^-]$$
$$\overline{AlH_4^- + 4 H_2CO + 4 H_2O \rightarrow Al^{3+} + 4 CH_3OH + 4 OH^-}$$

18.17   To determine spontaneity using standard thermodynamic values, calculate $\Delta G^{\circ}_{reaction}$ from tabulated values for ΔG°_f. The reaction is spontaneous if $\Delta G^{\circ}_{reaction}$ is negative:

(a) $\Delta G^{\circ}_{reaction}$ = 2 mol(-129.7 kJ/mol) − 0 kJ/mol = −259.4 kJ; spontaneous;

(b) $\Delta G^{\circ}_{reaction}$ = 2 mol(-58.5 kJ/mol) − 0 kJ/mol = −117.0 kJ; spontaneous;

(c) $\Delta G^{\circ}_{reaction}$ = 1 mol(-300.1 kJ/mol)

                − [1 mol(-53.6 kJ/mol) + 1 mol (0 kJ/mol)] = −246.5 kJ; spontaneous;

(d) $\Delta G^{\circ}_{reaction}$ = [1 mol(-300.1 kJ/mol) + 1 mol(0 kJ/mol)]

                − [1 mol(-100.4 kJ/mol) + 1 mol(0 kJ/mol)] = −199.7 kJ; spontaneous.

18.19   A molecular view of a process occurring at an electrode should show the species involved in charge transfer and indicate the direction of movement of electrons. See Figure 18-5 for an example. At a silver-silver chloride electrode undergoing reduction, solid AgCl gains an electron from the electrode to generate chloride anions in solution and solid Ag:

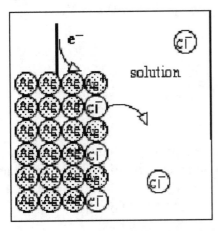 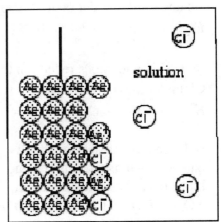

18.21   A passive electrode and a supply of gas are needed to study a redox reaction that includes gases. See Figure 18-9 for a sketch of a hydrogen electrode. A chlorine electrode is exactly analogous:

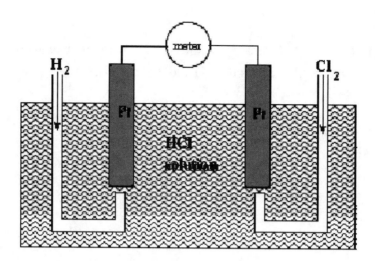

18.23 Active electrodes participate in redox reactions, while passive electrodes only provide or accept electrons. The Ag-AgCl electrode is active, while the Pt electrodes are passive.

18.25 Standard electrode potentials are calculated using Equation 18-1:
$$E^o_{cell} = E^o_{cathode} - E^o_{anode}$$
(a) MnO_4^- undergoes reduction to MnO_2 (cathode), $E^o_{cathode} = 0.595$ V;

CN^- undergoes oxidation to CNO^- (anode), $E^o_{anode} = -0.970$ V;
$E^o_{cell} = 0.595$ V $- (-0.970$ V$) = 1.565$ V.

(b) O_2 undergoes reduction to H_2O (cathode), $E^o_{cathode} = 1.229$ V;
As undergoes oxidation to $HAsO_2$ (anode), $E^o_{anode} = 0.248$ V;
$E^o_{cell} = 1.229 - (0.248) = 0.981$ V

(e) ClO_4^- undergoes reduction to ClO^- (cathode), $E^o_{cathode} = 1.36$ V;
Cl^- undergoes oxidation to Cl_2 (anode), $E^o_{anode} = 1.35827$ V;
$E^o_{cell} = 1.36$ V $- (1.35827$ V$) = 0.002$ V

18.27 Balance redox reactions following the standard procedure. Break into half-reactions, balance each half-reaction using the stepwise technique, then multiply by appropriate integers so the electrons cancel:
$NO \rightarrow N_2O$ (1) multiply NO by 2 to balance N, (2) add 1 H_2O on right, (3) add 2 H_3O^+ on left and 2 H_2O on the right, (4) add 2 e^- on left:
$$2\,NO + 2\,H_3O^+ + 2\,e^- \rightarrow N_2O + 3\,H_2O;$$

$NO \rightarrow NO_3^-$ (1) N already balanced, (2) add 2 H_2O on left, (3) add 4 H_3O^+ on right and 4 H_2O on the left, (4) add 3 e^- on right:
$$NO + 6\,H_2O \rightarrow NO_3^- + 4\,H_3O^+ + 3\,e^-;$$

Multiply first reaction by 3 and second reaction by 2, then add to obtain the overall reaction:

3 [2 NO + 2 H_3O^+ + 2 e^- → N_2O + 3 H_2O]

+ 2 [NO + 6 H_2O → NO_3^- + 4 H_3O^+ + 3 e^-];

8 NO + 3 H_2O → 3 N_2O + 2 NO_3^- + 2 H_3O^+;

Standard electrode potentials are calculated using Equation 18-1:

$$E^o{}_{cell} = E^o{}_{cathode} - E^o{}_{anode}$$

NO undergoes reduction to N_2O (cathode), $E^o{}_{cathode}$ = 1.591 V;

NO undergoes oxidation to NO_3^- (anode), $E^o{}_{anode}$ = 0.957 V;

$E^o{}_{cell}$ = 1.591 – 0.957 = 0.634 V

18.29 The direct way to measure a standard reduction potential is with a cell containing a standard hydrogen electrode as a reference. Because F^- is the conjugate base of a weak acid, the F_2/F^- portion of the cell must be separated from the H_2/H_3O^+ portion, requiring a porous plate. Set up a standard hydrogen electrode on one side and a Pt electrode immersed in a 1 M solution of NaF with F_2 bubbling over the electrode on the other side:

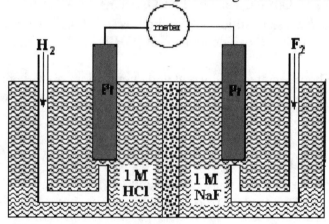

The standard reduction potential of F_2 is + 2.866 volts, so F_2 will be reduced in this cell and the H_2/H^+ electrode will be the anode.

18.31 To act as a reducing agent under standard conditions, a substance must have a more negative standard reduction potential than the substance that is to be reduced. The standard reduction potential of Be is – 1.847 V. Scanning the table, we find the following metals whose standard potentials are more negative than this: Ba(–2.912 V), Ca (–2.868 V), Cs (–3.026 V), Li (–3.0401 V), Mg (–2.37 V), K (–2.931 V), and Na (–2.71 V). All these metals lie in the *s* block of the periodic table and easily lose 1 or 2 electrons.

18.33 The standard free energy change of a reaction is related to the standard reduction potential through Equation 18-3: $\Delta G^o = -nFE^o$

The standard potentials for the reactions were calculated in Problem 18.25,
(Remember that 1 J = 1 V C):

(a)   For this reaction, E°_{cell} = 1.565 V, n = 6:

$$\Delta G^\circ = -(6 \text{ mol})(9.6485 \times 10^4 \text{ C/mol})(1.565 \text{ V})\left(\frac{10^{-3}\text{kJ}}{1\text{J}}\right) = -906.0 \text{ kJ}.$$

(b)   For this reaction, E°_{cell} = 0.981 V, n = 12:

$$\Delta G^\circ = -(12 \text{ mol})(9.6485 \times 10^4 \text{ C/mol})(0.981 \text{ V})\left(\frac{10^{-3}\text{kJ}}{1\text{J}}\right) = -1.14 \times 10^3 \text{ kJ}.$$

(e)   For this reaction, E°_{cell} = 0.002 V, n = 6:

$$\Delta G^\circ = -(6 \text{ mol})(9.6485 \times 10^4 \text{ C/mol})(0.002 \text{ V})\left(\frac{10^{-3}\text{kJ}}{1\text{J}}\right) = -1 \text{ kJ}.$$

18.35   Operating potentials are related to standard cell potentials through Equation 18-6:

$$E = E^\circ - \frac{RT}{nF} \ln Q$$

The reaction for the nickel-cadmium battery is:
$$2 \text{ NiO(OH)}_{(s)} + 2 \text{ H}_2\text{O}_{(l)} + \text{Cd}_{(s)} \rightarrow 2 \text{ Ni(OH)}_{2(s)} + \text{Cd(OH)}_{2(s)}$$
Here, all the substances are solids or pure liquids, so Q = 1, $\ln Q$ = 0, and
$E = E^\circ$ = 1.35 V. Hence, E is independent of [OH⁻].

18.37   The connection between electric current and chemical amounts is provided by Equation
18-7, $n = \dfrac{It}{F}$, where n is moles of electrons flowing.

$$n = 15 \text{ s}\left(\frac{5.9 \text{ C}}{1 \text{ s}}\right)\left(\frac{1 \text{ mol}}{9.6485 \times 10^4 \text{ C}}\right) = 9.2 \times 10^{-4} \text{ mol electrons};$$

The half-reactions in a lead storage battery are:
$$\text{H}_2\text{O} + \text{Pb}_{(s)} + \text{HSO}_4^-{}_{(aq)} \rightarrow \text{PbSO}_4{}_{(s)} + \text{H}_3\text{O}^+{}_{(aq)} + 2 \text{ e}^-$$

$$\text{PbO}_2{}_{(s)} + \text{HSO}_4^-{}_{(aq)} + 3 \text{ H}_3\text{O}^+{}_{(aq)} + 2 \text{ e}^- \rightarrow \text{PbSO}_4{}_{(s)} + 5 \text{ H}_2\text{O}_{(l)}$$

Thus, $n_{Pb} = n_{PbO_2} = 9.2 \times 10^{-4} \text{ mol}\left(\dfrac{1 \text{ mol Pb}}{2 \text{ mol e}^-}\right) = 4.6 \times 10^{-4} \text{ mol};$

Finally convert moles to mass using molar mass:

$$m_{Pb} = 4.6 \times 10^{-4} \text{ mol}\left(\frac{207.2 \text{ g}}{1 \text{ mol}}\right) = 9.5 \times 10^{-2} \text{ g};$$

$$m_{PbO_2} = 4.6 \times 10^{-4} \text{ mol}\left(\frac{239.2 \text{ g}}{1 \text{ mol}}\right) = 0.11 \text{ g}.$$

18.39   The connection between electric current and chemical amounts is provided by Equation
18-7, $n = \dfrac{It}{F}$, where n is moles of electrons flowing. Thus, $t = \dfrac{nF}{I}$;

The reaction being driven by the alternator is :
$$PbSO_4 \text{ (s)} + H_3O^+\text{(aq)} + 2 \text{ e}^- \rightarrow H_2O + Pb \text{ (s)} + HSO_4^- \text{ (aq)}$$

Thus, $n_e = 2\, n_{PbSO_4} = 2(0.850 \text{ g})\left(\dfrac{1 \text{ mol}}{303.3 \text{ g}}\right) = 5.6 \times 10^{-3} \text{ mol};$

$I = (1.750 \text{ A} - 1.350 \text{ A});$

$t = \dfrac{(5.6 \times 10^{-3} \text{mol})(9.6485 \times 10^4 \text{C mol}^{-1})}{0.400 \, \text{C s}^{-1}} = 1.35 \times 10^3 \text{ s.}$

18.41 Balance redox reactions following the standard procedure. Break into half-reactions, balance each half-reaction using the stepwise technique, then multiply by appropriate integers so the electrons cancel:

$Cr_2O_7^{2-} \rightarrow Cr^{3+}$ (1) multiply Cr^{3+} by 2, (2) add 7 H_2O on right, (3) add 14 H_3O^+ on left and add 14 H_2O to the right,   (4) add 6 e^- on the left:

$$Cr_2O_7^{2-} + 14 \, H_3O^+ + 6 \text{ e}^- \rightarrow 2 \, Cr^{3+} + 21 \, H_2O;$$

$CH_3CHO \rightarrow CH_3CO_2H$ (1) C already balanced, (2) add 1 H_2O on left, (3) add 2 H_3O^+ on right and 2 H_2O on the left,   (4) add 2e^- on right:

$$CH_3CHO + 3 \, H_2O \rightarrow CH_3CO_2H + 2 \, H_3O^+\text{(aq)} + 2 \text{ e}^-;$$

Multiply second reaction by 3 and add:

$[Cr_2O_7^{2-} + 14 \, H_3O^+\text{(aq)} + 6 \text{ e}^- \rightarrow 2 \, Cr^{3+} + 21 \, H_2O]$

$+ \quad 3[CH_3CHO + 3 \, H_2O \rightarrow CH_3CO_2H + 2 \, H_3O^+\text{(aq)} + 2 \text{ e}^-];$

$Cr_2O_7^{2-} + 3 \, CH_3CHO + 8 \, H_3O^+\text{(aq)} \rightarrow 2 \, Cr^{3+} + 3 \, CH_3CO_2H + 12 \, H_2O;$

It takes 2 mol of electrons to oxidize 1 mol of CH_3CHO;

$n = 1.00 \text{ g}\left(\dfrac{1 \text{ mol}}{44.1 \text{ g}}\right)\left(\dfrac{2 \text{ mol e}^-}{1 \text{ mol CH}_3\text{CHO}}\right) = 0.0454 \text{ mol of electrons needed;}$

Each mole of $Cr_2O_7^{2-}$ delivers 6 mol of electrons:

$n_{\text{dichromate}} = 0.0454 \text{ mol}\left(\dfrac{1 \text{ mol Cr}_2O_7^{2-}}{6 \text{ mol e}^-}\right) = 7.57 \times 10^{-3} \text{ mol};$

$m = 7.57 \times 10^{-3} \text{ mol}\left(\dfrac{262 \text{ g}}{1 \text{ mol}}\right) = 1.98 \text{ g of sodium dichromate needed.}$

18.43 The Nernst equation is Equation 18-6, $E = E^\circ - \dfrac{RT}{nF} \ln Q$. To determine Q, obtain the balanced redox equation for the standard dry cell:

$2MnO_2 \text{ (s)} + H_2O\text{(l)} + Zn\text{(s)} \rightarrow Mn_2O_3 \text{ (s)} + Zn(OH)_2 \text{ (s)};$

At first glance you may notice that all reactants and products are either solid or liquid and therefore $Q = 1$, $\ln Q = 0$, indicating the potential does not decrease with use. The key to why the voltage must decrease lies in the hydroxide ion concentrations associated with each half reaction:

Cathode: $2 \, MnO_2 \text{ (s)} + H_2O\text{(l)} + 2\text{e}^- \rightarrow Mn_2O_3 \text{ (s)} + 2 \, OH^-\text{(aq)}$

Anode: $Zn(s) + 2 OH^-(aq) \rightarrow Zn(OH)_2 (s) + 2e^-$

From the text we are told that paste is not fluid enough to allow the hydroxide ions generated at the cathode to travel readily to the anode for consumption. Therefore the net reaction can be written as:

$2MnO_2 (s) + H_2O(l) + Zn(s) + 2 OH^-(anode) \rightarrow Mn_2O_3 (s) + Zn(OH)_2(s) + 2 OH^-(cathode)$,

where the hydroxide ions are shown to indicate that the anode and cathode concentrations are different.

Thus, $Q = \dfrac{[OH^-]_{cathode}}{[OH^-]_{anode}}$; as the battery operates, the anode concentration decreases while the cathode concentration increases, $Q > 1$ and $\ln Q > 0$, and the potential of the battery decreases with use.

18.45 When several metals are present, the one that corrodes first is the one with the more negative standard reduction potential. Here are the values for Cr and Fe, from Appendix F:

$Cr^{2+} + 2 e^- \rightarrow Cr$, $E^o = -0.913$ V;
$Fe^{2+} + 2 e^- \rightarrow Fe$, $E^o = -0.447$ V;

Thus, Cr oxidizes more easily and will preferentially corrode.

18.47 Mercury batteries are characterized by a stable potential and compact size but limited current capacity, making them well suited for use where large currents are not needed, such as in pacemakers and cameras. They cannot supply the large current needed to start an automobile engine; moreover, they are irreversible (i.e., not rechargeable) so they would have to be replaced very frequently.

18.49 The connection between electric current and chemical amounts is provided by Equation 18-7, $n = \dfrac{It}{F}$, where n is moles of electrons flowing. Examine the balanced half-reaction to determine the stoichiometric relationship between moles of electrons and moles of chemical species. Here, the reaction is $2 Cl^- \rightarrow Cl_2 + 2 e^-$, so 1 mol of Cl_2 is formed by 2 mol of electrons:

$$t = 200.0 \text{ min}\left(\frac{60 \text{ s}}{1 \text{ min}}\right) = 1.200 \times 10^4 \text{ s};$$

$$n = 1.200 \times 10^4 \text{ s}\left(\frac{4.50 \text{ C}}{1 \text{ s}}\right)\left(\frac{1 \text{ mol e}^-}{9.6485 \times 10^4 \text{ C}}\right)\left(\frac{1 \text{ mol Cl}_2}{2 \text{ mol e}^-}\right) = 0.280 \text{ mol Cl}_2;$$

$$m = n \, MM = 0.280 \text{ mol}\left(\frac{70.906 \text{ g}}{1 \text{ mol}}\right) = 19.9 \text{ g}.$$

18.51 When an external potential is applied to a galvanic cell, the reduction that occurs is the one with the least negative reduction potential, and the oxidation that occurs is the one with the least positive reduction potential. The oxidation reaction is the reverse of the galvanic reduction, $Cu \rightarrow Cu^{2+} + 2 e^-$. However, instead of $Zn^{2+} + 2 e^- \rightarrow Zn$

$(E^\circ = -0.7618$ V), the reduction of H_3O^+ occurs:

$2\ H_3O^+(aq) + 2\ e^- \rightarrow H_2 + 2\ H_2O$ $(E^\circ = 0$ V). In neutral H_2O, the H_3O^+ ion concentration is only 10^{-7} M, but the resulting E is still less negative than E° for Zn^{2+}/Zn.

18.53 (a) To determine where to attach the negative wire from the charger, examine the half-reactions of the battery, which must be run in the reverse direction to recharge:

Cathode: $NiO(OH)\ (s) + H_2O\ (l) + e^- \rightarrow Ni(OH)_2\ (s) + OH^-\ (aq)$

Anode: $Cd\ (s) + 2\ OH^-\ (aq) \rightarrow Cd(OH)_2\ (s) + 2\ e^-$

To recharge, electrons must be supplied to the cadmium electrode, so this is the electrode to which the negative wire (anode) from the charger should be attached. The reaction is: $Cd(OH)_2\ (s) + 2\ e^- \rightarrow Cd\ (s) + 2\ OH^-\ (aq)$.

(b) The connection between electric current and chemical amounts is provided by Equation 18-7, $n = \dfrac{It}{F}$, where n is moles of electrons flowing. Thus $t = \dfrac{nF}{I}$;

$$n_{Cd(OH)_2} = 1.55\ g \left(\frac{1\ mol}{146.4\ g}\right) = 0.0106\ mol;$$

$$n_{e^-} = 2\ n_{Cd(OH)_2} = 0.0212\ mol;$$

$$t = 0.0212\ mol \left(\frac{9.6485 \times 10^4\ C}{1\ mol}\right)\left(\frac{1\ s}{125 \times 10^{-3}\ C}\right) = 1.64 \times 10^4\ s.$$

Convert to hours:

$$1.64 \times 10^4\ s \left(\frac{1\ min}{60\ s}\right)\left(\frac{1\ hr}{60\ min}\right) = 4.56\ hr$$

18.55 Quantity of charge can be calculated once the amount of electron flow has been determined. Use the half-reaction to determine the relationship between amount of chemical change and amount of electron flow:

$$Ag^+ + e^- \rightarrow Ag,\ so\ n_e = n_{Ag} = (12.89\ g - 10.77\ g)\left(\frac{1\ mol}{107.9\ g}\right) = 0.0196\ mol\ e^-;$$

charge $= nF = (0.0196\ mol)(9.6485 \times 10^4\ C/mol) = 1.89 \times 10^3\ C;$

Determine the current by dividing the charge by the time in seconds:

$$t = 15.0\ min \left(\frac{60\ s}{1\ min}\right) = 900\ s;$$

$$i = \frac{charge}{t} = \frac{1.89 \times 10^3\ C}{900\ s} = 2.10\ A$$

18.57 A molecular picture must show the species undergoing redox reactions and the direction of electron flow. In a silver coulometer, both electrodes are silver. Ag is oxidized to Ag⁺ at the anode, and Ag⁺ is reduced to Ag at the cathode.

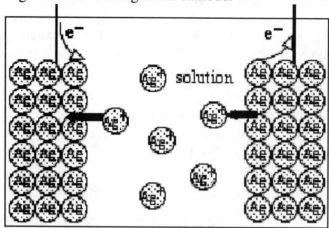

18.59 Standard electrode potentials allow calculation of equilibrium constants if the half-reactions can be combined to give the equilibrium reaction whose constant is desired. In this case, subtract the second reaction from the first and divide by 4:

$$O_2 + 2\,H_2O + 4\,e^- \rightleftharpoons 4\,OH^- \qquad E^\circ = 0.401\ V$$
$$-[O_{2\,(g)} + 4\,H_3O^+_{(aq)} + 4\,e^- \rightleftharpoons 6\,H_2O \qquad E^\circ = 1.229\ V]$$
$$\text{Net: } 8\,H_2O_{(l)} \rightleftharpoons 4\,OH^-_{(aq)} + 4\,H_3O^+_{(aq)}$$

$$E^\circ_{reaction} = 0.401\ V - 1.229\ V = -0.828\ V;$$

Dividing by 4 changes n but not E°:

$$2\,H_2O_{(l)} \rightleftharpoons OH^-_{(aq)} + H_3O^+_{(aq)}\ ,\ n = 1,\ E^\circ = -0.828\ V;$$

Use Equation 18-5, rearranged:

$$\log K_{eq} = \left(\frac{nE^\circ}{5.916 \times 10^{-2}\,V}\right) = \left(\frac{1(-0.828\ V)}{5.916 \times 10^{2}\,V}\right) = -14.00;$$

$$K_{eq} = 1.0 \times 10^{-14} = [OH^-][H^+];$$

18.61 An equilibrium constant can be calculated from standard reduction potentials provided the half-reactions can be combined to give the equilibrium reaction whose constant is desired.

Appendix F provides these standard potentials:

$$Zn^{2+} + 2\,e^- \rightleftharpoons Zn, \qquad\qquad E^\circ = -0.7618\ V;$$
$$Zn[NH]_4^{2+} + 2\,e^- \rightleftharpoons Zn + 4\,NH_3, \qquad E^\circ = -1.04\ V;$$

Subtract the second half-reaction from the first to give the desired equilibrium:
$$E^\circ = (-0.7618\ V) - (-1.04\ V) = +0.28\ V;$$

Now use the relationship between K_{eq} and E° to determine the equilibrium constant:

$$\log K_{eq} = \left(\frac{nE^{o}}{5.916 \times 10^{-2}\,V} \right) = \left(\frac{2(0.28\,V)}{5.916 \times 10^{-2}\,V} \right) = 9.47;$$

$$K_{eq} = 10^{9.47} = 3.0 \times 10^{9}$$

18.63 (a) Balance redox reactions following the standard procedure. Break into half-reactions, balance each half-reaction using the stepwise technique, then multiply by appropriate integers so the electrons cancel:

(a) $H_2 \rightarrow H_2O$ (1) no elements except O and H, (2) add 1 H_2O on left,

(3) add 2 H_2O on right, and 2 OH^- on the left side, (4) add 2 e^- on right:

$$H_2 + 2\,OH^- + H_2O \rightarrow 3\,H_2O + 2\,e^-;$$

Combine and cancel:
$$H_2 + 2\,OH^- \rightarrow 2\,H_2O + 2\,e^-;$$

$Cr(OH)_3 \rightarrow Cr$ (1) Cr already balanced, (2) add 3 H_2O on right, (3) add 3 H_2O on left, and add 3 OH^- on the right sides, (4) add 3 e^- on left:

$$Cr(OH)_3 + 3\,H_2O + 3\,e^- \rightarrow Cr + 3\,H_2O + 3\,OH^-;$$

Combine and cancel:
$$Cr(OH)_3 + 3\,e^- \rightarrow Cr + 3\,OH^-;$$

Multiply first reaction by 3 and second reaction by 2, then add:

$$3\,[H_2 + 2\,OH^- \rightarrow 2\,H_2O + 2\,e^-]$$
$$+ 2\,[Cr(OH)_3 + 3\,e^- \rightarrow Cr + 3\,OH^-]\ ;$$
$$3\,H_2 + 2\,Cr(OH)_3 \rightarrow 6\,H_2O + 2\,Cr;$$

(b) To determine the standard potential, consult values in Appendix F, and subtract E^o for the oxidation from E^o for the reduction:

$2\,H_2O + 2\,e^- \rightarrow H_2 + 2\,OH^-,$ $E^o = -0.828\,V;$

$Cr(OH)_3 + 3\,e^- \rightarrow Cr + 3\,OH^-,$ $E^o = -1.48\,V;$

$$E^o_{cell} = (-1.48\,V) - (-0.828\,V) = -0.65\,V;$$

(c) Calculate ΔG^o from $\Delta G^o = -nF E^o_{cell}$:

$$\Delta G^o = -(6\,mol)(9.6485 \times 10^4\,C/mol)(-0.65\,V)\left(\frac{10^{-3}\,kJ}{1\,J} \right) = 3.8 \times 10^2\,kJ$$

18.65 (a) Balance the half-reaction using the stepwise technique:

$Cr_2O_7^{2-} \rightarrow Cr^{3+}$ (1) multiply Cr^{3+} by 2, (2) add 7 H_2O on right, (3) add 14 H_3O^+ on left and 14 H_2O on the right, (4) add 6 e^- on left:

$$Cr_2O_7^{2-} + 14\,H_3O^+{}_{(aq)} + 6\,e^- \rightarrow 2\,Cr^{3+} + 21\,H_2O;$$

(b) It takes 2 mol K_2CrO_4 to generate 1 mol $Cr_2O_7^{2-}$, which in turn consumes 6 mol of e^-, so 1 mol K_2CrO_4 consumes 3 mol e^- and

$$0.250 \text{ mol}\left(\frac{1 \text{ mol K}_2\text{CrO}_4}{3 \text{ mol e}^-}\right) = 0.0833 \text{ mol K}_2\text{CrO}_4 \text{ is required;}$$

$$m = n \, MM = 0.0833 \text{ mol}\left(\frac{194 \text{ g}}{1 \text{ mol}}\right) = 16.2 \text{ g.}$$

18.67 Identify what is taking place in a galvanic cell by identifying the species present.
(a) On the left, $\quad Ni^{2+} + 2 e^- \rightarrow Ni, \quad E° = -0.257 \text{ V;}$
On the right, $\quad Fe^{2+} + 2 e^- \rightarrow Fe, \quad E° = -0.447 \text{ V;}$

(b) To generate a positive overall potential, the iron half-reaction operates as oxidation:
$Fe + Ni^{2+} \rightarrow Ni + Fe^{2+}, \quad E° = -0.257 \text{ V} - (-0.447 \text{ V}) = 0.190 \text{ V;}$

(c) oxidation occurs at the anode, so the Fe electrode is the anode and Ni is the cathode;
(d) A molecular picture of an electrochemical cell must show the species undergoing redox reactions and the direction of electron flow.

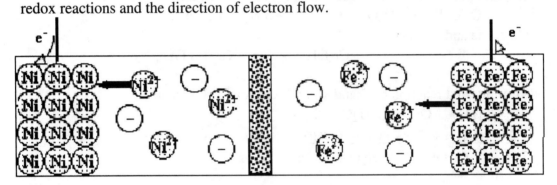

18.69 Use the Nernst equation to determine the concentration at which $E = 0.0$ V:

$$Fe + Ni^{2+} \rightleftharpoons Ni + Fe^{2+}, \quad E° = -0.257 \text{ V} - (-0.447 \text{ V}) = 0.190 \text{ V;}$$

$$E = 0.0 \text{ V} = E° - \frac{(0.0591)}{n} \log\left(\frac{[Fe^{2+}]}{[Ni^{2+}]}\right);$$

$$\log\left(\frac{[Fe^{2+}]}{[Ni^{2+}]}\right) = \left(\frac{nE°}{5.916 \times 10^{-2} \text{ V}}\right);$$

The concentration of reactant must be reduced to reach a potential of 0.0 V:

$$\log\left\{\frac{(1.00)}{[Ni^{2+}]}\right\} = \frac{(2)(0.190 \text{ V})}{(5.916 \times 10^{-2} \text{ V})} = 6.42, \ [Ni^{2+}] = 10^{-6.42} = 3.8 \times 10^{-7} \text{ M.}$$

18.71 Balance redox reactions following the standard procedure. Break into half-reactions, balance each half-reaction using the stepwise technique, then multiply by appropriate integers so the electrons cancel. First balance the half-reaction involving manganese:

$MnO_4^- \rightarrow Mn^{2+}$ (1) Mn is balanced, (2) add 4 H_2O on right, (3) add 8 H_3O^+ on left and 8 H_2O on the right, (4) add 5 e^- on left:

$$MnO_4^- + 8\ H_3O^+(aq) + 5\ e^- \rightarrow Mn^{2+} + 12\ H_2O;$$

(a) $H_2SO_3 \rightarrow HSO_4^-$ (1) S already balanced, (2) add 1 H_2O on left, (3) add 3 H_3O^+ on right and 3 H_2O on the left, (4) add 2 e^- on right:

$$H_2SO_3 + 4\ H_2O \rightarrow HSO_4^- + 3\ H_3O^+(aq) + 2\ e^-;$$

Multiply Mn reaction by 2, S reaction by 5, and add:

$2\ [MnO_4^- + 8\ H_3O^+(aq) + 5\ e^- \rightarrow Mn^{2+} + 12\ H_2O]$
$\underline{\quad\quad + 5\ [H_2SO_3 + 4H_2O \rightarrow HSO_4^- + 3\ H_3O^+(aq) + 2\ e^-]}$
$2\ MnO_4^- + 5\ H_2SO_3 + H_3O^+(aq) \rightarrow 2\ Mn^{2+} + 5\ HSO_4^- + 4\ H_2O;$

(b) $SO_2 \rightarrow HSO_4^-$ (1) S already balanced, (2) add 2 H_2O on left, (3) add 3 H_3O^+ on right and add 3 H_2O on the left, (4) add 2 e^- on right:

$SO_2 + 5\ H_2O \rightarrow HSO_4^- + 3\ H_3O^+(aq) + 2\ e^-;$

Multiply Mn reaction by 2, S reaction by 5, and add:

$2\ [MnO_4^- + 8\ H_3O^+(aq) + 5\ e^- \rightarrow Mn^{2+} + 12\ H_2O]$
$\underline{+ \quad\quad\quad 5\ [SO_2 + 5\ H_2O \rightarrow HSO_4^- + 3\ H_3O^+(aq) + 2\ e^-]}$
$2\ MnO_4^- + 5\ SO_2 + H_2O + H_3O^+(aq) \rightarrow 2\ Mn^{2+} + 5\ HSO_4^-;$

(c) $H_2S \rightarrow HSO_4^-$ (1) S already balanced, (2) add 4 H_2O on left, (3) add 9 H_3O^+ on right and 9 H_2O on the left, (4) add 8 e^- on right:

$H_2S + 13\ H_2O \rightarrow HSO_4^- + 9\ H_3O^+(aq) + 8\ e^-;$

Multiply Mn reaction by 8, S reaction by 5, and add:

$8\ [MnO_4^- + 8\ H_3O^+(aq) + 5\ e^- \rightarrow Mn^{2+} + 12\ H_2O]$
$\underline{+ \quad\quad 5\ [H_2S + 13\ H_2O \rightarrow HSO_4^- + 9\ H_3O^+(aq) + 8\ e^-]}$
$8\ MnO_4^- + 5\ H_2S + 19\ H_3O^+(aq) \rightarrow 8\ Mn^{2+} + 5\ HSO_4^- + 31\ H_2O;$

(d) $H_2S_2O_3 \rightarrow HSO_4^-$ (1) Multiply HSO_4^- by 2, (2) add 5 H_2O on left, (3) add 10 H_3O^+ on right and 10 H_2O on the left, (4) add 8 e^- on right:

$$H_2S_2O_3 + 15\ H_2O \rightarrow 2\ HSO_4^- + 10\ H_3O^+(aq) + 8\ e^-;$$

Multiply Mn reaction by 8, S reaction by 5, and add:

$8\ [MnO_4^- + 8\ H_3O^+(aq) + 5\ e^- \rightarrow Mn^{2+} + 12\ H_2O]$
$\underline{+ \quad\quad 5\ [H_2S_2O_3 + 15\ H_2O \rightarrow 2\ HSO_4^- + 10\ H_3O^+(aq) + 8\ e^-]}$
$8\ MnO_4^- + 5\ H_2S_2O_3 + 14\ H_3O^+(aq) \rightarrow 8\ Mn^{2+} + 10\ HSO_4^- + 21\ H_2O.$

18.73 The connection between electric current and chemical amounts is provided by Equation 18-7, $n = \dfrac{It}{F}$, where n is moles of electrons flowing. Thus $t = \dfrac{nF}{I}$;

The reaction is $Cu^{2+} + 2\,e^- \rightarrow Cu$, so $n = 2\,n_{Cu}$

$$n_{Cu} = 0.250\,L\left(\frac{0.245\,mol}{1\,L}\right) = 0.06125\,mol;\ n_{electron} = 2\,n_{Cu} = 0.1225\,mol;$$

$$t = 0.1225\,mol\left(\frac{9.6485 \times 10^4\,C}{1\,mol}\right)\left(\frac{1\,s}{2.45\,C}\right) = 4.82 \times 10^3\,s.$$

Convert to minutes:

$$t = 4.82 \times 10^3\,s\left(\frac{1\,min}{60\,s}\right) = 80.3\,min$$

18.75 The process in Problem 18.73 is electrodeposition, presumably with Cu serving as both electrodes. Reduction of Cu^{2+} to Cu occurs at the cathode, and oxidation of Cu to Cu^{2+} occurs at the anode:

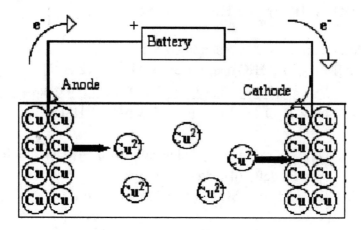

18.77 The better oxidizing agent under standard conditions is the reactant having the more positive reduction potential.
(a) MnO_4^- (1.679 vs. 1.232 V);
(b) O_2 (0.401 vs. 0.109 V);
(c) Sn^{2+} (–0.137 vs. –0.447 V).

18.79 To serve as an oxidizing agent, a substance must have a more positive reduction potential than the substance it is to oxidize: For O_2, $E° = 1.229$ V; all the substances listed in Problem 18.78 (except for Co^{2+}) have less positive reduction potentials, so O_2 can oxidize Cu, Ag, Fe^{2+}, H_2, and I- (not surprisingly, oxygen is a good oxidizing agent!).

18.81 Although the anions are different in the two solutions, the redox process involves only Zn metal and Zn^{2+} cations, so this is a concentration cell.
   (a) $Zn(NO_3)_2$ soln: $Zn^{2+} + 2 e^- \rightarrow Zn$;
   $ZnCl_2$ soln: $Zn \rightarrow Zn^{2+} + 2 e^-$

   (b) The molecular picture shows Zn^{2+} cations depositing on the electrode in the 1.25 M solution and dissolving off the electrode in the 0.250 M solution:

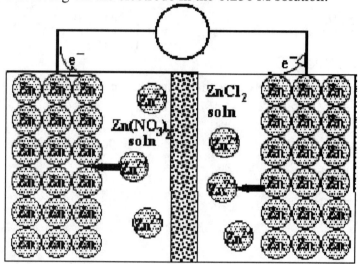

   (c) To calculate a cell potential from concentration measurements, use the Nernst equation. For a concentration cell, $E^{\circ} = 0$ V, so $E = -\dfrac{(0.05916)}{n} \log Q$;
   The reactions are $Zn^{2+} + 2 e^- \rightarrow Zn$, and the concentrations are 0.250 M and 1.25 M:
   Zn^{2+} (1.25 M) $\rightarrow Zn^{2+}$ (0.250 M) , $n = 2$, and $Q = \dfrac{0.250 \text{ M}}{1.25 \text{ M}}$;
   $E = -\dfrac{(0.05916)}{2} \log \dfrac{0.250 \text{ M}}{1.25 \text{ M}} = 0.0207$ V

18.83 The connection between time and amount of electrons is provided by Equation 18-7,
   $n = \dfrac{It}{F}$, from which $t = \dfrac{nF}{I}$. From the half-reaction, $n_{electron} = n_{MnO_2}$. Only 90% of the MnO_2 is available before the battery fails:
   $n_{MnO_2} = 4.0 \text{ g} \left(\dfrac{1 \text{ mol}}{86.94 \text{ g}} \right) \left(\dfrac{90\%}{100\%} \right) = 0.0414$ mol;
   $t = 0.0414 \text{ mol} \left(\dfrac{9.6485 \times 10^4 \text{ C}}{1 \text{ mol}} \right) \left(\dfrac{1 \text{ s}}{0.0048 \text{ C}} \right) = 8.3 \times 10^5$ s;
   Convert to hr: $8.3 \times 10^5 \text{ s} \left(\dfrac{1 \text{ min}}{60 \text{ s}} \right) \left(\dfrac{1 \text{ hr}}{60 \text{ min}} \right) = 2.3 \times 10^2$ hr.

Chapter 18

18.85 Use the Nernst equation to determine a concentration from cell voltages. First determine the net reaction in order to find n, $E°$, and Q.

The reactions are: $Fe^{3+} + e^- \rightleftharpoons Fe^{2+}$, $E° = 0.771$ V;

$Cu^{2+} + 2e^- \rightleftharpoons Cu$, $E° = 0.3419$ V;

Multiply the first reaction by 2 and subtract the second:

$2 Fe^{3+} + Cu \rightleftharpoons 2 Fe^{2+} + Cu^{2+}$, $E° = 0.771$ V $-$ 0.3419 V = 0.429 V;

$$E = E° - \frac{(0.05916)}{n} \log Q = E° - \frac{(0.05916)}{n} \log\left(\frac{[Fe^{2+}]^2[Cu^{2+}]}{[Fe^{3+}]^2}\right);$$

Now substitute values and solve for the unknown concentration:

$$0.00 \text{ V} = 0.429 \text{ V} - \frac{(0.05916)}{2} \log\left\{\frac{(1.00)^2(1.00)}{[Fe^{3+}]^2}\right\};$$

$$-\log\{[Fe^{3+}]^2\} = \frac{2(0.429)}{(0.05916)} = 14.5,$$

$[Fe^{3+}]^2 = 3.2 \times 10^{-15}$;

$[Fe^{3+}] = 5.6 \times 10^{-8}$ M.

18.87 (a) In molten NaCl, the only species present are Na^+ cations and Cl^- anions; the Pt electrodes are passive, so the reactions are: $Na^+ + e^- \rightarrow Na$ and $2 Cl^- \rightarrow Cl_2 + 2 e^-$;
(b) Reduction is driven by electrons supplied at the negative terminal, so the cathode is the Pt electrode connected to the negative pole of the battery, and the anode is the Pt electrode connected to the positive pole of the battery;
(c)

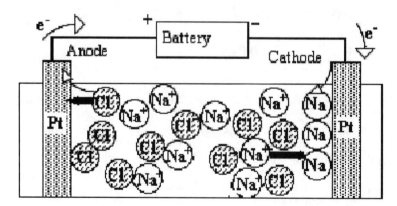

18.89 A sketch of a cell that shows molecular processes should show the molecular species undergoing redox reactions and the direction of electron flow:

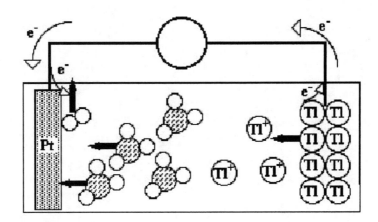

18.91 The standard potential for a redox reaction is easily calculated from the equilibrium constant for the reaction by using Equation 18-5, $E° = \left(\dfrac{5.916 \times 10^{-2} \, V}{n} \right) \log K_{eq}$:

$E° = \frac{1}{2}(5.916 \times 10^{-2} \, V) \log(2.69 \times 10^{12}) = (0.02958)(12.430) = 0.368 \, V.$

19.1 Oxidation states are determined by applying the rules given in Section 18.1. The procedure can be simplified when a polyatomic ion of known charge is present.

(a) CO_3^{2-} has a charge of -2, so Mn must be $+2$ to give overall neutrality;

(b) Cl (more electronegative) is -1, so Mo must be $+5$ to give overall neutrality;

(c) Na is $+1$, so VO_4 has overall charge -3; each O is -2, so V is $+5$ to give net charge of -3;

(d) O is -2, so Au must be $+3$ to give overall neutrality.

(e) H_2O has no charge, SO_4^{2-} has a charge of -2, so Fe must be $+3$ to give overall neutrality;

19.3 Use the periodic table to locate and identify elements from their valence configurations.

(a) There are 6 valence electrons, so the element is in column 6; $3d$ orbital is filling: Cr;

(b) There are full s and d blocks, so the element is in column 12; $4d$ has just filled: Cd;

(c) There are 11 valence electrons, so the element is in column 11; $3d$ orbital is full: Cu;

19.5 The properties of transition metals vary regularly with their valence configurations, so the starting point for predicting relative properties is located in the d block.

(a) Pd is in column 10, Cd is in column 12, both in the $n = 4$ row. Beyond the middle of the d block, melting point decreases with Z because electrons are placed in antibonding orbitals, so Pd melts at higher T;

(b) Cu and Au are both in column 11, but Au has higher Z, so Au has higher density;

(c) Cr is in column 6, Co is in column 9, both in the $n = 3$ row. IE_1 increases with Z across a row because Z_{eff} increases, so Co has the higher IE_1;

19.7 Use the charges of ligands and ions to determine the oxidation states of transition metals in coordination complexes. Because s electrons are always removed first, the count of d electrons is given by the number of valence electrons – the oxidation state.

(a) Each Cl is -1, and NH_3 is neutral, so Ru has oxidation state $+2$. Ru is in column 8 (8 valence electrons), giving d^6;

(b) Each I is -1, and en is neutral, so Cr has oxidation state $+3$. Cr is in column 6 (6 valence electrons), giving d^3;

(c) Each Cl is -1, and trimethylphosphine is neutral, so Pd has oxidation state $+2$. Pd is in column 10 (10 valence electrons), giving d^8;

(d) Each Cl is -1, and NH_3 is neutral, so Ir has oxidation state $+3$. Ir is in column 9 (9 valence electrons), giving d^6;

(e) CO is a neutral ligand, so Ni has oxidation state 0. Ni is in column 10 (10 valence electrons), but here no electrons have been removed, giving s^2d^8. Thus, there are 8 d electrons.

19.9 Compounds that contain coordination complexes are named following the six rules stated in your textbook: name cation first, name ligands in alphabetical order, name metal, add "o" for anions, use Greek prefixes, add "-ate" for anionic complexes, give oxidation number.

(a) hexaammineruthenium(II) chloride;

(b) *trans*-bis(ethylenediamine)diiodochromium(III) iodide;

(c) *cis*-dichlorobis(trimethylphosphine)palladium(II);
(d) *fac*-triamminetrichloroiridium(III);
(e) tetracarbonylnickel(0).

19.11 The structure of a metal complex usually is octahedral (6 ligands), tetrahedral (4 ligands), or square planar (4 ligands).

19.13 The name of a complex contains the information needed to determine its chemical formula. Determine the charge of the complex from charges on the ligands and the oxidation number.
(a) *cis*-[Co(NH$_3$)$_4$ClNO$_2$]$^+$; (b) [PtNH$_3$Cl$_3$]$^-$; (c) *trans*-[Cu(en)$_2$(H$_2$O)$_2$]$^{2+}$; (d)[FeCl$_4$]$^-$.

19.15 (a) *cis*-tetraamminechloronitrocobalt(III) has six ligands and will thus have an octahedral geometry. The *cis* means that the chlorine and nitro ligands will be at 90 degrees with other.

(b) Since platinum has eight *d* electrons in the valence shell for amminetrichloroplatinate(II) it will have a square planar geometry.

(c) In *trans*-diaquabis(ethylenediamine) copper(II) remember that aqua stands for a water ligand and ethylenediamine is a bidentate ligand (it will bond to the copper with both its nitrogens). Because of this it will have an octahedral geometry. The *trans* term means that the water ligands are opposite each other on the complex.

(d) Iron(III) is a 3d metal with 5 valence electrons. Therefore, tetrachloroferrate(III) will have a tetrahedral geometry.

19.17 The crystal field diagram for weak and strong octahedral fields is always the same, with the populations changing depending on how many *d* electrons must be accommodated. The valence configuration provides information about the number of *d* electrons:
(a) Ti^{2+} (column 4 − 2 electrons) is d^2; (b) Cr^{3+} (column 6 − 3 electrons) is d^3.
For these two ions, the low field and high field configurations are the same:

(c) Mn^{2+} (column 7 − 2 electrons) is d^5; (d) Fe^{3+} (column 8 − 3 electrons) is d^5. These two ions have identical diagrams, with low field differing from high field:

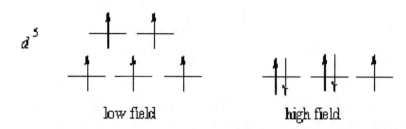

d^5

low field   high field

19.19   The magnetic properties of a complex are determined by its number of d electrons and whether it is high field or low field.
(a) Ir^{3+} (column 9 – 3 electrons) is d^6, and NH_3 is a relatively high-field ligand and all the electrons will be paired; the complex is diamagnetic;
(b) Cr^{2+} (column 6 – 2 electrons) is d^4, and water is a relatively low-field ligand. The complex is paramagnetic with 4 unpaired electrons;
(c) Pt^{2+} (column 10–2 electrons) is d^8, so regardless of field strength this square planar complex is paramagnetic with 2 unpaired electrons;
(d) Pd has d^{10} configuration, so regardless of field strength this complex is diamagnetic.

19.21   The colors of transition metal complexes are generally determined by d-d transitions, but Zr^{4+} (column 4 – 4 electrons) has all its valence electrons removed, so there is no valence electron that can undergo a transition that would absorb visible light.

19.23   Consult your textbook for the chemical reactions of various metallurgical processes.
(a) $CuFeS_2 + 2\ O_2 \rightarrow CuO + FeO + 2\ SO_2$;
(b) $Si + O_2 + CaO \rightarrow CaSiO_3$;
(c) $TiCl_4 + 4\ Na \rightarrow 4\ NaCl + Ti$.

19.25   This is a standard stoichiometry problem. Begin by analyzing the chemistry. The reactants are Cu_2S and air and the given products are Cu metal and SO_2 gas.
The balanced reaction is $Cu_2S + O_2 \rightarrow 2\ Cu + SO_2$.

Start by computing the number of moles of Cu_2S and then use the appropriate mass-mole-mass and P-V-T calculations to determine the amounts of the products:

$$m_{Cu_2S} = 5.60 \times 10^4\ kg\left(\frac{10^3\ g}{1\ kg}\right)\left(\frac{2.37\%}{100\%}\right) = 1.327 \times 10^6\ g;$$

$$n_{Cu_2S} = 1.327 \times 10^6\ g\left(\frac{1\ mol}{159.2\ g}\right) = 8.34 \times 10^3\ mol = n_{SO2};$$

$$m_{Cu} = 8.34 \times 10^3\ mol\left(\frac{2\ mol\ Cu}{1\ mol\ Cu_2S}\right)\left(\frac{63.55g}{1\ mol}\right) = 1.06 \times 10^6\ g;$$

$$P_{total} = 755 \text{ torr}\left(\frac{1 \text{ atm}}{760 \text{ torr}}\right) = 0.993 \text{ atm}$$

$$V_{SO_2} = \frac{nRT}{P} = \frac{(8.34 \times 10^3 \text{ mol})(0.08206 \frac{L \text{ atm}}{\text{mol K}})(273.15 + 23.5 \text{ K})}{0.993 \text{ atm}} = 2.05 \times 10^5 \text{ L}.$$

19.27   Standard free energy changes are calculated using standard free energies of formation found in Appendix D. $\Delta G^o_{reaction} = \Delta G^o_{products} - \Delta G^o_{reactants}$

$ZnO(s) + C(s) \rightarrow Zn(s) + CO(g)$;
$\Delta G^o_{reaction}$ = [1 mol(–137.2 kJ/mol) + 0] – [1mol(–320.5 kJ/mol) + 0] = 183.3 kJ

$ZnO(s) + CO(g) \rightarrow Zn(s) + CO_2(g)$;
$\Delta G^o_{reaction}$ =[1(–394.4 kJ/mol) + 1(0)] – [1(–320.5 kJ/mol) + 1(–137.2 kJ/mol)]
$\Delta G^o_{reaction}$ = 63.3 kJ.

19.29   Consult your textbook, Figure 19-26, for the reactions that convert scheelite into tungsten:
$CaWO_4 + 2 H_3O^+ \rightarrow Ca^{2+} + H_2WO_4 + 2 H_2O$; this reaction dissolves the ore;
$H_2WO_4 + 2 NH_3 \rightarrow (NH_4)_2WO_4 \xrightarrow{\text{heat}} WO_3 + 2 NH_3 + H_2O$; this reaction removes all elements except W and O;
$WO_3 + 3 H_2 \rightarrow W + 3 H_2O$; this reaction reduces the oxide to the pure metal.

19.31   The coinage metals are those that have been used since antiquity for coins: copper, silver, and gold. All are in column 11 of the periodic table. They are characterized by high electrical conductivity, good ductility, and low chemical reactivity, in particular resistance to oxidation. Hence they are used for money (a vanishing use in technologically advanced countries), for electrical wire, for jewelry, and for other decorative objects. See your text for special examples of uses for compounds of these elements.

19.33   Titanium is used as an engineering metal because of its relatively low density, high bond strength, resistance to corrosion, and ability to withstand high temperatures, all of which make it a favored structural material.

19.35   The structural difference between hemoglobin and myoglobin is that the former has four subunits, while the later has just one. As a consequence, hemoglobin has much more complex cooperative chemical behavior than myoglobin does.

19.37   An iron ion bonded to four sulfur atoms from cysteines is in a tetrahedral environment, so the splitting pattern is the 2-3 pattern characteristic of tetrahedral complexes. The iron center loses one electron and is converted from Fe^{2+} to Fe^{3+}:

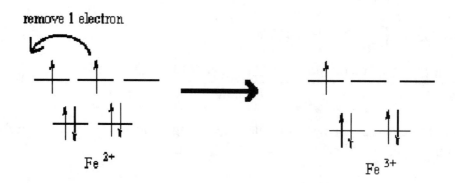

19.39 The number of possible isomers of a complex is determined by its geometry and the number of ligands of each type.

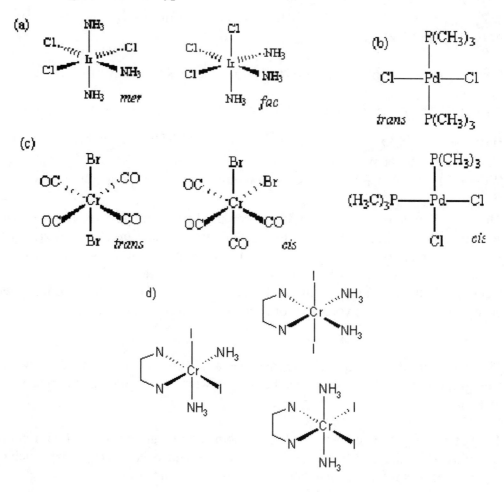

19.41 Bidentate ligands form complexes with two links. Each Fe ion forms an octahedral complex with three oxalate anions. Here is a sketch showing the ligand-metal orientations:

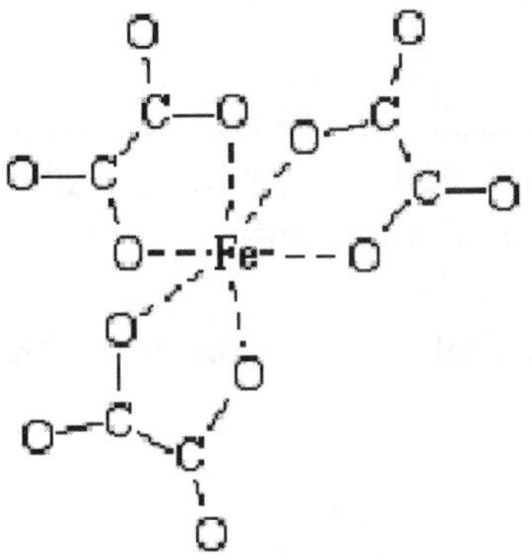

19.43 To determine electron configurations, start from the position of the element in the periodic table and remove s electrons preferentially.
(a) Cr is in column 6, configuration $[Ar]4s^1 3d^5$; Cr^{2+} is $[Ar] 3d^4$; Cr^{3+} is $[Ar] 3d^3$;
(b) V is in column 5, configuration $[Ar]4s^2 3d^3$; V^- is $[Ar]4s^2 3d^4$; V^+ is $[Ar]4s^1 3d^3$; V^{2+} is $[Ar] 3d^3$; V^{3+} is $[Ar] 3d^2$; V^{4+} is $[Ar] 3d^1$; V^{5+} is $[Ar]$;
(c) Ti is in column 4, configuration $[Ar]4s^2 3d^2$; Ti^{2+} is $[Ar] 3d^2$; Ti^{4+} is $[Ar]$.

19.45 Compounds that contain coordination complexes are named following the six rules stated in your textbook: name cation first, name ligands in alphabetical order, name metal, add "o" for anions, use Greek prefixes, add "-ate" for anionic complexes, give oxidation number.
(a) *cis*-tetraaquadichlorochromium(III) chloride; (b) bromopentacarbonylmanganese(I); (c) *cis*-diamminedichloroplatinum(II).

19.47 Superoxide dismutase catalyzes the conversion of superoxide into molecular oxygen and hydrogen peroxide. This reaction occurs at a metal site that contains one Zn^{2+} ion and one Cu^{2+} ion, linked by a histidine ligand that bonds to both metal ions:

$$2\ O_2^- + 2\ H_3O^+ \xrightarrow{\text{SOD}} O_2 + 2H_2O_2 + 2H_2O$$

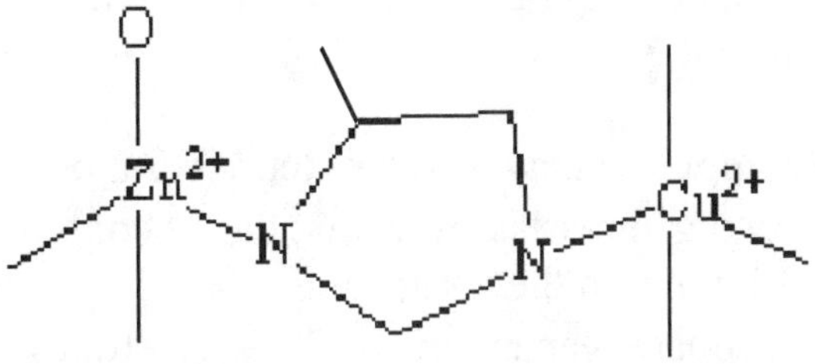

19.49 Cu(II) in water forms an aqua complex. Addition of fluoride produces the insoluble green salt, CuF_2, while addition of chloride produces the bright green tetrachlorocopper(II) complex, $[CuCl_4]^{2-}$.
$$Cu^{2+}(aq) + 2\ F^-(aq) \rightarrow CuF_2\ (s)$$

$$Cu^{2+}(aq) + 4\ Cl^-(aq) \rightarrow [CuCl_4]^{2-}\ (aq)$$

19.51 Silver (column 11 of the periodic table) is difficult to oxidize, so it has good resistance to corrosion and is suitable for jewelry. Vanadium (column 5) is readily oxidized, so it corrodes rapidly and is unsuited to jewelry.

19.53 The *mer* isomer of an octahedral complex has three like ligands arranged in a meridian plane:

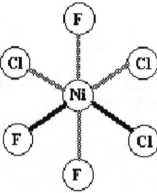

19.55 Silver tarnish is silver sulfide, Ag_2S, formed by reaction with trace amounts of H_2S in the atmosphere:

$$4\ Ag(s)\ + 2\ H_2S(g)\ + O_2(g) \rightarrow\ 2\ Ag_2S(s) + 2\ H_2O(l)$$

19.57 Tetracarbonylnickel(0), $[Ni(CO)_4]$ has a tetrahedral geometry since Ni is in the 1st row transition elements. Because of the strong field ligand, CO, Δ is large, which will result in the compound absorbing UV light and appearing colorless.

Tetracyanozinc(II) $[Zn(CN)_4]^{2-}$ has a tetrahedral geometry (Zn is a first row transition metal) where Zn has a +2 charge (making it a d^{10}). This compound will be colorless because there are no possible *d-d* transitions.

19.59 Brass is an alloy of zinc and copper, and superoxide dismutase contains zinc and copper ions in its reaction center.

19.61 Four-coordinate complexes may be either tetrahedral or square planar. The splitting patterns for these two geometries show that the d^8 configuration can have zero spin in the square planar case but not in the tetrahedral case. Thus the magnetic behavior indicates that $[Ni(CN)_4]^{2-}$ is square planar and $[NiCl_4]^{2-}$ is tetrahedral:

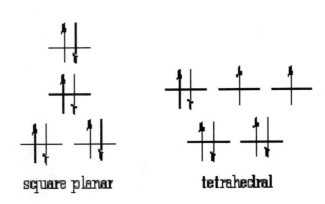

square planar   tetrahedral

19.63 When ferritin is neither empty nor filled to capacity, the protein has the capacity to provide iron as needed for hemoglobin synthesis, or to store iron if an excess is absorbed by the body.

19.65 UV-visible spectroscopy is useful when a compound has an energy gap between the highest occupied and lowest unoccupied orbital that matches the energy of visible light. Because Zn^{2+} has a d^{10} configuration, its d orbitals are completely filled and the lowest unoccupied orbital is quite high in energy. In contrast, Co^{2+} has a d^{7} configuration, giving this cation unfilled d orbitals . Consequently, metalloproteins that contain Co^{2+} absorb visible light, making it possible to study them with uv-visible spectroscopy.

19.67 Use standard reduction potentials from Appendix F to determine which metals can be displaced by Zn. Any metal with a less negative standard potential can be displaced:
$$Zn^{2+} + 2e^- \rightleftharpoons Zn, \quad E° = -0.7618 \text{ V};$$

The following are a few examples of metals which can be displaced by Zn:
$$Co^{2+} + 2e^- \rightleftharpoons Co, \quad E° = -0.28 \text{ V};$$
$$Cu^{2+} + 2e^- \rightleftharpoons Cu, \quad E° = +0.3419 \text{ V};$$
$$Fe^{2+} + 2e^- \rightleftharpoons Fe, \quad E° = -0.447 \text{ V};$$
$$Ni^{2+} + 2e^- \rightleftharpoons Ni, \quad E° = -0.257 \text{ V};$$

When combined with the Zn half-reaction, the overall cell voltage is positive and the process is spontaneous. For example,
$$Zn + Co^{2+} \rightleftharpoons Co + Zn^{2+}, \quad E° = -0.28 \text{ V} - (-0.7618 \text{ V}) = 0.48 \text{ V};$$

20.1 Lewis acids are electron pair acceptors, and Lewis bases are electron pair donors. Thus, Lewis acids are electron-deficient, while Lewis bases have lone pairs to donate.
(a) Lewis acid is Ni, Lewis base is CO;
(b) Lewis acid is $SbCl_3$, Lewis base is Cl^-;
(c) Lewis acid is $AlBr_3$, Lewis base is $P(CH_3)_3$;
(d) Lewis acid is BF_3, Lewis base is ClF_3.

20.3 Construct Lewis structures and determine steric numbers to identify the three-dimensional structure of a molecule.
(a) AlF_3
1. There are 3+3(7) = 24 valence electrons.
2. Three electron pairs are needed for the bonding framework.
3. The remaining electrons are used to place 3 pairs on each outer F atom.

The Lewis structure shows SN = 3 for Al, so the molecule has trigonal planar geometry and there is a vacant $3p$ orbital perpendicular to the molecular plane:

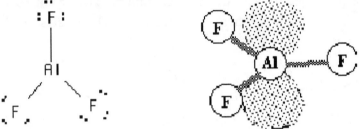

(a) SbF_5
1. There are 5 + 5(7) = 40 valence electrons.
2. Five pairs are used in the bonding framework leaving 40 – 2(5) = 30 electrons.
3. The remaining electrons are used to place 3 electron pairs on each F atom.

The Lewis structure shows SN = 5 for Sb, so the molecule has trigonal bipyramidal geometry. There are vacant $3d$ orbitals that do not participate in dsp^3 hybridization:

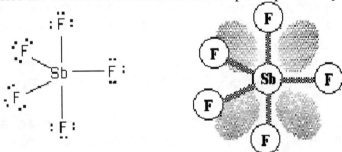

(b) SO_2
1. There are 6 + 2(6) = 18 valence electrons.
2. Two electron pairs are needed for the bonding framework, leaving 18 – 2(2)=14.
3. Place 3 pairs of electrons on each outer O atom.
4. Place the remaining two electrons on the inner S atom.

5. The resulting structure has $FC_S = 6 - 2 - 2 = 2$, move two lone pairs from the outer O atoms to form two double bonds and reduce the formal charge to 0.

The Lewis structure shows $SN = 3$ for S, (so the molecule has a trigonal planar geometry with a lone pair bent) and there is a delocalized π orbital to which an electron pair can be added:

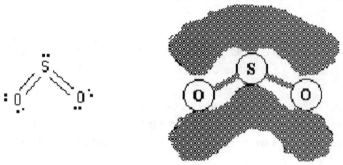

20.5   In the first step, an electron pair from the O atom of H_2O displaces a π bond in Lewis acid-base adduct formation. Then a proton from H_2O migrates to a C–O oxygen atom:

adduct formation   proton transfer

20.7   Polarizability of cations increases with the values of n and Z and decreases with the charge on the ion. V^{3+} (lowest Z, high charge) < Fe^{3+} (high charge) < Fe^{2+} < Pb^{2+} (large n).

20.9   Hard acids have low polarizability, and soft acids have high polarizability.
(a) the hardest acid is BF_3 (both elements from row 2), then BCl_3, and $AlCl_3$ is softest (both elements from row 3);
(b) the hardest acid is Al^{3+} (row 3), then Tl^{3+} (row 6), and Tl^+ (low charge) is softest;
(c) Polarizability increases with n, so hardest is $AlCl_3$, then $AlBr_3$, and AlI_3 is softest.

20.11 When an electronegative O atom bonds to a less electronegative S atom, it withdraws electron density from S, decreasing the polarizability about S and increasing the hardness of the base.

20.13 A metathesis reaction will occur if exchange of partners couples the harder Lewis acid with the harder Lewis base and the softer Lewis acid with the softer Lewis base.
(a) Al^{3+} is harder than Na^+, and Cl^- is harder than I^-, so metathesis occurs, giving $AlCl_3$ (hard-hard) and NaI (soft-soft);

(b) the Lewis acid is Ti^{4+} in each substance, so no reaction occurs;

(c) Ca^{2+} is softer than H^+, and S^{2-} is softer than O^{2-}, so metathesis occurs, giving H_2O (hard-hard) and CaS (soft-soft);

(d) C (in CH_3^-) is a soft base, and Li^+ is a hard acid, so metathesis occurs, giving $(CH_3)_3P$ (soft-soft) and LiCl (hard-hard);

(d) Ag^+ and I^- are both soft and Si^{4+} and Cl^- are both hard, so no reaction occurs.

20.15 Descriptions of bonding always begin with a Lewis structure. As your textbook describes, Al_2Cl_6 contains two "bridging" chlorine atoms. Standard procedures would predict tetrahedral geometry about all inner atoms, but the Al–Cl–Al bond angles of 91° indicate that Cl uses p orbitals. Each Al atom can be described as using sp^3 hybrids to form four σ bonds to four different Cl atoms.

In Lewis acid-base terms, the bridged molecule forms from two $AlCl_3$ units linking together in double adduct formation between the Al Lewis acid atoms and two Cl Lewis base atoms.

20.17 Thallium lies below indium and gallium, so its properties should be similar to those metals: valence of 3, soft Lewis acid. Like its neighbor, Pb, it is toxic.

20.19 Determine the Lewis structure using standard procedures:
$SnCl_4$
 1. There are $4 + 4(7) = 32$ valence electrons.
 2. Four electron pairs are needed for the bonding framework, leaving $32 - 4(2) = 24$.
 3. Place 3 pairs of electrons on each outer Cl atom, leaving $24 - 4(6) = 0$.
 4. No remaining electrons.
 5. The resulting structure has $FC_{Sn} = 4 - 4 = 0$, so the structure is complete.

$SnCl_4$ can function as a Lewis acid because the Sn has empty d orbitals into which it can accept electrons to form more bonds.

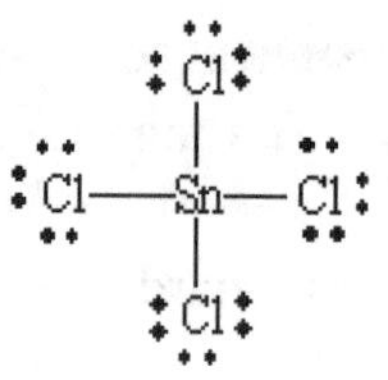

20.21 A BN pair has the same number of valence electrons as a pair of C atoms, so the structure and bonding of $B_3N_3H_6$ is just like that of benzene, a six-membered planar ring with a delocalized set of π bonds. There are $3(3) + 3(5) + 6(1) = 30$ valence electrons, all of which are involved in the bonding network:

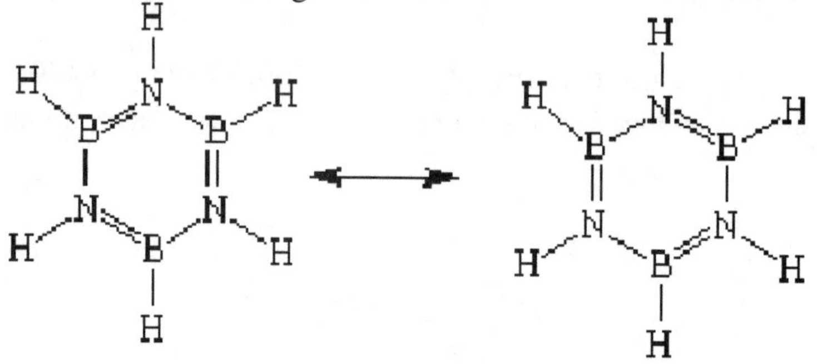

The B and N atoms have bonding and geometry that can be described using sp^2 hybrid orbitals, resulting in 3 σ bonds around each ring atom. In addition, there is a delocalized π bonding network encompassing all six ring atoms and containing 6 electrons.

20.23 The band gap decreases from top to bottom of each column of the periodic table, so Ge has a smaller band gap than Si. In orbital terms, this is because the principal quantum number of the valence orbitals increases, and this makes the valence orbitals larger, leading to less effective overlap and smaller energy difference between bonding and antibonding orbitals.

20.25 Follow the example in section 20.4 of your text on polymer formation.
$$2C_2H_5Cl + Si(Cu) \rightarrow (C_2H_5)_2 SiCl_2 + Cu$$
$$(C_2H_5)_2 SiCl_2 + 2H_2O \rightarrow (C_2H_5)_2 Si(OH)_2 + 2HCl$$
Condensation will eliminate water to give the polymer:
$$2(C_2H_5)_2Si(OH)_2 \rightarrow \text{polymer-linkage} + H_2O$$

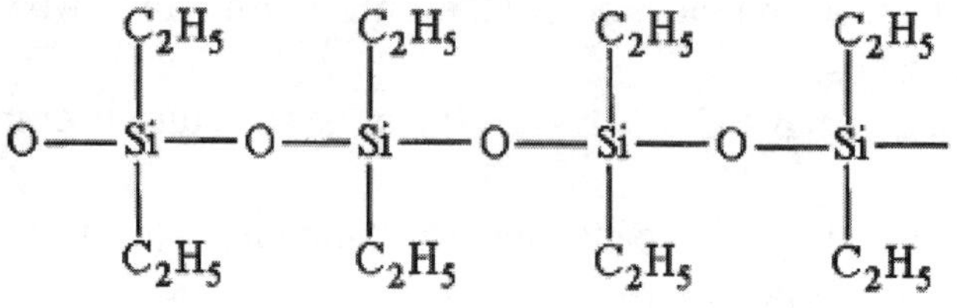

20.27 Polyphosphates form by sequential condensation of PO_3^- units, so the tetraphosphate has chemical formula $P_4O_{13}^{6-}$:

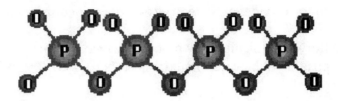

20.29 Phosphoric acid is produced directly from apatite by reaction with sulfuric acid:

$$Ca_5(PO_4)_3X(s) + 5 H_2SO_4 (aq) \rightarrow 3 H_3PO_4 (aq) + 5 CaSO_4 (s) + HX(aq)$$

Where X = (F⁻, OH⁻, or Cl⁻)
This is a Brønsted acid-base reaction in which protons are transferred from sulfuric acid to the phosphate anions. There are no redox reactions in this process.

20.31 Phosphoric acid undergoes condensation reactions with alcohols in which water molecules are eliminated:

20.33 There are three industrial reactions in Section 20.6 in which sulfuric acid acts as a Brønsted acid:

$$CaF_2(s) + H_2SO_4(l) \rightarrow 2 HF(g) + CaSO_4(s)$$
$$Ca_5(PO_4)_3F(s) + 5 H_2SO_4(aq) \rightarrow 3 H_3PO_4(aq) + 5 CaSO_4(s) + HF(aq)$$

20.35 The repeat structure of polyvinylchloride, in common with all polyethylene-type polymers, has an all-carbon backbone. There is a chlorine atom on every other carbon atom:

20.37 The reaction forming $TiCl_4$ from TiO_2 is as follows:

$$TiO_2 \text{ (s)} + C \text{ (s)} + 2 Cl_2 \text{ (g)} \rightarrow TiCl_4 \text{ (l)} + CO_2 \text{ (g)}$$

C goes from zero to +4 oxidation state, so it is oxidized and serves as the reducing agent. Cl goes from zero to -1 oxidation state, so it is reduced and serves as the oxidizing agent.

20.39 This is a straightforward stoichiometry problem, which can be worked by determining the percentage composition of bauxite. Bauxite is AlOOH, MM = 60.00 g/mol;

$$\% \text{ Al} = \frac{26.982 \text{ g/mol}}{60.00 \text{ g/mol}} \times 100\% = 44.97 \%;$$

1 kg of bauxite rock contains $1.00 \text{ kg} \left(\frac{85\%}{100\%} \right) \left(\frac{44.97\%}{100\%} \right) = 0.382 \text{ kg Al};$

If the processing is 75 % efficient, each 1.00 kg of bauxite rock yields (0.382 kg)(0.75) = 0.287 kg Al.

To produce 2500 kg of Al requires:

$$2500 \text{ kg Al} \left(\frac{1 \text{ kg bauxite rock}}{0.287 \text{ kg Al}} \right) = 8.7 \times 10^3 \text{ kg bauxite rock.}$$

20.41 Sulfur has d orbitals available for bond formation, allowing the formation of SF_6, in which the bonding can be described using d^2sp^3 hybrid orbitals on the S atom. Oxygen has no valence d orbitals available. In principle, SBr_6 could also form, but the Br atom is too large for six Br atoms to be accommodated around a central S atom.

20.43 Lewis acids are electron pair acceptors, and Lewis bases are electron pair donors. The As atom in $AsCl_3$ has a lone pair that it donates, giving this compound Lewis base character. The As atoms also can accommodate additional electron pairs by using valence d orbitals, giving this compound Lewis acid character.
$AsCl_3 + BF_3 \rightarrow Cl_3As–BF_3$ ($AsCl_3$ acts as Lewis base).
$AsCl_3 + Cl^- \rightarrow AsCl_4^-$ ($AsCl_3$ acts as Lewis acid).

20.45 Balance a redox half-reaction following the standard steps outlined in Chapter 18 (balance all but H and O by inspection, balance O by adding H_2O, balance H by adding H_3O^+/H_2O, in basic solutions balance H by adding OH^-/H_2O, balance charge by adding electrons):

Begin by balancing the Al reaction:
$Al + 2 H_2O \rightarrow AlO_2^-$; add 4 OH^- on the reactant side and 4 H_2O to the product side to balance H:
$Al + 2 H_2O + 4 OH^- \rightarrow AlO_2^- + 4 H_2O$; cancel 2 H_2O on each side and add 3 electrons to balance charge:
$Al + 4 OH^- \rightarrow AlO_2^- + 2 H_2O + 3 e^-$;

(a) $NO_3^- \rightarrow NH_3 + 3 H_2O$; add 9 OH^- on the product side and 9 H_2O on the reactant to balance H:

$NO_3^- + 9 H_2O \rightarrow NH_3 + 3 H_2O + 9 OH^-$; cancel 3 H_2O on each side and add 8 electrons to balance the charge:

$NO_3^- + 6 H_2O + 8 e^- \rightarrow NH_3 + 9 OH^-$;

Multiply this half-reaction by 3 and the Al half-reaction by 8 to balance the electrons and add:

$8Al + 32 OH^- + 3NO_3^- + 18 H_2O + 24 e^- \rightarrow 3 NH_3 + 27 OH^- + 8 AlO_2^- + 16 H_2O + 24 e^-$;

Canceling duplicated species yields:

$$8 Al + 5 OH^- + 3 NO_3^- + 2 H_2O \rightarrow 3 NH_3 + 8 AlO_2^-;$$

(b) $2 H_2O + 2 e^- \rightarrow H_2 + 2 OH^-$; multiply this half-reaction by 3 and the Al half-reaction by 2, and add:

$$2 Al + 8 OH^- + 6 H_2O + 6 e^- \rightarrow 3 H_2 + 6 OH^- + 2 AlO_2^- + 4 H_2O + 6 e^-;$$

Cancel duplicated species:

$$2 Al + 2 OH^- + 2 H_2O \rightarrow 3 H_2 + 2 AlO_2^-;$$

(c) $SnO_3^{2-} \rightarrow Sn + 3 H_2O$; add 6 OH^- on the product side and 6 H_2O on the reactant side to balance H:

$SnO_3^{2-} + 6 H_2O \rightarrow Sn + 3 H_2O + 6 OH^-$; cancel 3 H_2O on each side and add 4 electrons to balance charge:

$SnO_3^{2-} + 3 H_2O + 4 e^- \rightarrow Sn + 6 OH^-$;

Multiply this half-reaction by 3 and the Al half-reaction by 4, and add:

$4Al + 16OH^- + 3 SnO_3^{2-} + 9 H_2O + 12 e^- \rightarrow 3 Sn + 18 OH^- + 4 AlO_2^- + 8 H_2O + 12 e^-$;

Canceling duplicated species yields:

$$4 Al + 3 SnO_3^{2-} + H_2O \rightarrow 3 Sn + 2 OH^- + 4 AlO_2^-.$$

20.47 Nitrogen, at the top of Group 15, is a non-metal showing a valence of three that forms polar bonds most readily with other non-metals (B, C, O, the halogens). Phosphorus has a similar pattern of reactivity but can also involve *d* orbitals in its bonding. Arsenic and antimony are metalloids with useful semiconductor properties, and Bismuth is metallic..

20.49 The production of aluminum from its ore is described in Section 20.3. First, bauxite ore is treated with strong base to produce soluble $[Al(OH)_4]^-$:

$$Al(O)OH \text{ (s)} + OH^- \text{ (aq)} + H_2O_{(l)} \rightarrow [Al(OH)_4]^- \text{ (aq)}$$

When the solution is diluted with water, aluminum hydroxide precipitates:

$$[Al(OH)_4]^- \text{ (aq)} \rightarrow Al(OH)_3 \text{ (s)} + OH^- \text{ (aq)}$$

Strong heating drives off water, leaving aluminum oxide:

$$2 \, Al(OH)_3 \xrightarrow{1250 \, ^{\circ}C} Al_2O_3 \, (s) + 3 \, H_2O \, (g)$$

Finally, electrolysis of aluminum oxide dissolved in cryolite reduces Al to pure metal:

$$2 \, Al_2O_3(s) \; + \; 3 \, C(s) \; \rightarrow \; 3 \, CO_2 \, (g) \; + \; 4 \, Al \, (l)$$

20.51 Silicon dioxide is first converted into silicon tetrachloride by reaction with molecular chlorine:

$$SiO_2 \, (s) + 2 \, C \, (s) + 2 \, Cl_2 \, (g) \xrightarrow{high \, T} SiCl_4 \, (g) + 2 \, CO \, (g)$$

The $SiCl_4$ is purified by distillation and then reduced by reaction with Mg metal:

$$SiCl_{4(l)} + 2 \, Mg(s) \; \rightarrow \; Si(s) + 2 \, MgCl_{2(s)}$$

20.53 The reaction can be balanced by recognizing the chemical formulas of the substances involved. The reactants are phosphoric acid (H_3PO_4), and fluoroapatite ($Ca_5(PO_4)_3F$); the products are calcium dihydrogen phosphate ($Ca(H_2PO_4)_2$) and HF:
The unbalanced reaction is thus:

$$Ca_5(PO_4)_3F \, (s) \; + H_3PO_4 \, (aq) \rightarrow Ca(H_2PO_4)_2 \, (s) \; + HF \, (aq)$$
$$5 \, Ca + 4 \, P + 16 \, O + F + 3 \, H \rightarrow Ca + 2 \, P + 8 \, O + F + 5 \, H$$

Begin by balancing the Ca by giving $Ca(H_2PO_4)_2$ a coefficient of 5:

$$Ca_5(PO_4)_3F \, (s) \; + H_3PO_4 \, (aq) \rightarrow \textbf{5} \, Ca(H_2PO_4)_2 \, (s) \; + HF \, (aq)$$
$$5 \, Ca + 4 \, P + 16 \, O + F + 3 \, H \rightarrow 5 \, Ca + 10 \, P + 40 \, O + F + 21 \, H$$

Next balance phosphorus by giving H_3PO_4 a coefficient of 7:

$$Ca_5(PO_4)_3F \, (s) \; +\textbf{7} \, H_3PO_4 \, (aq) \rightarrow 5 \, Ca(H_2PO_4)_2 \, (s) \; + HF \, (aq)$$
$$5 \, Ca + 10 \, P + 40 \, O + F + 21 \, H \rightarrow 5 \, Ca + 10 \, P + 40 \, O + F + 21 \, H$$

Note that now all atoms are balanced:

$$Ca_5(PO_4)_3F \, (s) \; + 7 \, H_3PO_4 \, (aq) \rightarrow \; 5 \, Ca(H_2PO_4)_2 \, (s) \; + \; HF \, (aq)$$

Calculate the mass percent phosphorus using standard stoichiometric techniques:

$$MM_{compound} = \{40.078 + 2[(2)(1.008) + 30.974 + (4)(15.999)]\} = 234.05 \text{ g/mol}$$

$$\% \, P = (100\%) \frac{2(MM_P)}{MM_{compound}} = \frac{(2)(30.974 \text{ g/mol})}{234.05 \text{ g/mol}} \times 100\% = 26.47 \, \%.$$

20.55 Gaseous Al_2Cl_6 is in equilibrium with gaseous $AlCl_3$: $Al_2Cl_{6(g)} \rightleftharpoons 2 \, AlCl_{3(g)}$. Le Chatelier's principle predicts that because the forward reaction is endothermic (bonds must be broken), an increase in temperature shifts the position of the equilibrium to the right. Thus as temperature increases, the number of moles of gaseous substance increases, so the pressure increases faster than would be predicted by the ideal gas equation

20.57 Soft metal ions are toxic because they are very soft Lewis acids which react readily with soft Lewis bases such as the sulfur atoms that hold essential enzymes in the conformations that those enzymes require to function properly. When a soft metal ion reacts with an S–S link in an enzyme, the conformation of the enzyme becomes distorted or destroyed and the enzyme can no longer function.

20.59 Mercury is a very soft Lewis acid, while zinc is intermediate in hard/soft character. Thus mercury preferentially associates with sulfur, a soft Lewis base, while zinc forms stable bonds with both soft (sulfur-containing) and hard (oxygen-containing) Lewis bases.

20.61 Polarizability increases with atomic size, and greater polarizability leads to larger dispersion forces. Thus the larger the atoms in a molecule, the larger the intermolecular forces and the easier it is to condense the substance. In the sequence BCl_3, BBr_3, BI_3, the halogens are increasing in size, accounting for the different stable phases at room temperature.

20.63 The structures of red and white phosphorus are shown in Figures 20-11 and 20-12 of your textbook. While both contain tetrahedral P_4 units that are highly strained because of the small bond angles, red phosphorus is a polymeric material made up of long chains of such tetrahedra, while white phosphorus is individual P_4 units. The individual units are much more accessible to other substances, so white phosphorus is the toxic allotrope.

20.65 The reactions can be balanced by inspection:
$2 Al_2O_3 + 6 Cl_2 \rightarrow 4 AlCl_3 + 3 O_2$;
$AlCl_3 + 3 Na \rightarrow Al + 3 NaCl$.

20.67 Compounds with Lewis acid-base properties will undergo metathesis reactions if a transfer of bonding partners associates harder acids with harder bases.
(a) Al^{3+} is quite hard due to its +3 charge, and Cl^- is softer than CH_3:
$AlCl_3 + 3 LiCH_3 \rightarrow Al(CH_3)_3 + 3 LiCl$;
(b) Sulfur is a Lewis acid as well as a Lewis base and forms an adduct with H_2O, a Lewis base: $SO_3 + H_2O \rightarrow H_2SO_4$ (aq)
(c) The antimony atom can bond to one additional F^- anion: $SbF_5 + LiF \rightarrow Li(SbF_6)$;
(d) S is harder than As, and F is harder than Cl, so there is no reaction.

20.69 The reaction consumes 12 H atoms for 4 P atoms, or 3 H atoms for every P atom. Thus the product is H_3PO_4: $P_4O_{10} + 6 H_2O \rightarrow 4 H_3PO_4$.

20.71 Determine the amounts of Al and Na using standard stoichiometric techniques.
$Al^{3+} + 3 e^- \rightleftharpoons Al$
For every 3 moles of charge one mole of aluminum will be refined or:
$$m_{Al} = 1 \text{ mol} \left(\frac{1 \text{ mol Al}}{3 \text{ mol e}^-} \right) \left(\frac{26.98 \text{ g}}{1 \text{ mol Al}} \right) = 8.99 \text{ g}$$

$$Na^+ + e^- \rightleftharpoons Na$$

For every mole of charge one mole of sodium will be refined or 23.0 g.

21.1   The elemental symbol identifies the value of Z, and the left superscript is A. $Z + N = A$:

Part:	(a)	(b)	(c)	(d)	(e)	(f)
Z	3	20	92	52	10	82
A	6	43	238	130	20	205
N	3	23	146	78	10	123.

21.3   Atomic number symbols have the elemental symbol accompanied by a left superscript denoting A and a left subscript denoting Z:

(a) $^4_2$He;   (b) $^{184}_{74}$W;   (c) $^{60}_{28}$Ni ;   (d) $^{26}_{12}$Mg

21.5   Nuclides in the "belt of stability" are stable. Instability occurs if a nuclide has too few neutrons, too many neutrons, or $Z > 83$. In addition, most odd-odd nuclides are unstable: (a) unstable, too few neutrons; (b) unstable, $Z > 83$; (c) stable; (d) unstable, odd-odd.

21.7   Energy releases are calculated using Equation 21-3: $\Delta E = (\Delta m)(8.988 \text{ x } 10^{10} \text{ kJ/g})$:

$$\Delta m = 1.00 \text{ met ton} \left(\frac{10^3 \text{kg}}{1 \text{ met ton}} \right) \left(\frac{10^3 \text{g}}{1 \text{kg}} \right) = 1.00 \text{ x } 10^6 \text{ g};$$

$$\Delta E = 1.00 \text{ x } 10^6 \text{ g} \left(\frac{8.988 \text{ x } 10^{10} \text{kJ}}{1 \text{g}} \right) = 8.99 \text{ x } 10^{16} \text{ kJ}$$

21.9   Binding energy is calculated by determining the mass defect (difference between the mass of the atom and the sum of the masses of its individual components) and converting to energy. When an element has only one stable nuclide, its molar mass is the mass of that nuclide:
$MM_{Cs} = 132.91$ g/mol,
stable nuclide has 55 protons and electrons, and $133 - 55 = 78$ neutrons;
$m_{components} = 55(1.007276 + 0.0005486) + 78(1.008665) = 134.106223$ g/mol;
$\Delta m = 132.91$ g/mol $- 134.106223$ g/mol $= -1.20$ g/mol;

$$\Delta E = \left(\frac{-1.20 \text{ g}}{1 \text{ mol}} \right) \left(\frac{8.988 \text{ x } 10^{10} \text{ kJ}}{1 \text{g}} \right) = -1.079 \text{ x } 10^{11} \text{ kJ/mol};$$

$$\Delta E \text{ (per nucleon)} = \frac{-1.079 \text{ x } 10^{11} \text{kJ/mol}}{133 \text{ nucleons}} = -8.11 \text{ x } 10^8 \text{ kJ mol}^{-1} \text{ nucleon}^{-1}$$

21.11   The coulombic barrier for fusion depends on the product of the nuclear charges and on the radii of the two nuclides. The stability of the product nuclide depends on its proton-neutron ratio: ^{1}H + ^{6}Li $\rightarrow$ ^{7}Be, charge product = (1)(3) = 3 and product has n/p = 0.75; ^{4}He + ^{4}He $\rightarrow$ ^{8}Be, charge product = (2)(2) = 4 and product has n/p = 1.00; Thus the coulombic barrier is greater for the second reaction, but the product nuclide is more stable.

21.13   Symbols and names for nuclear particles appear in Table 21-3 and should be memorized:

	Description	symbol	name
(a)	high energy photon	γ	gamma ray
(b)	mass number 4	α	alpha particle
(c)	positive particle with m_e	$\beta+$	positron.

21.15 To identify the products of nuclear decay processes, make use of the principles of conservation of mass number and charge:

(a) no change in mass or charge, product is $^{125}_{52}\text{Te}$;

(b) Electron capture changes nuclear charge by -1, no change in mass, product is: $^{123}_{51}\text{Sb}$;

(c) β decay changes nuclear charge by $+1$, no change in mass, gamma decay does not change mass or charge, product is $^{127}_{53}\text{I}$.

21.17 Equation 21-4 is used to determine half-lives from radioactive decay data.

$$\text{Rate} = \frac{\Delta N}{\Delta t} = \frac{-N \ln 2}{t_{1/2}}$$

Summarize the known data: rate $= -242$ decays/s; $m = 1.33 \times 10^{-12}$ g;

Convert mass to number of nuclei, using $MM \cong A = 26 + 33 = 59$ g/mol:

$$N = 1.33 \times 10^{-12}\,\text{g}\left(\frac{1\,\text{mol}}{59\,\text{g}}\right)\left(\frac{6.022 \times 10^{23}\,\text{nuclei}}{1\,\text{mol}}\right) = 1.36 \times 10^{10}\,\text{nuclei};$$

$$t_{1/2} = -\frac{(1.36 \times 10^{10}\,\text{nuclei})(\ln 2)}{-242\,\text{decays s}^{-1}} = 3.9 \times 10^7\,\text{s};$$

Convert to a more convenient time unit:

$$3.9 \times 10^7\,\text{s}\left(\frac{1\,\text{min}}{60\,\text{s}}\right)\left(\frac{1\,\text{hr}}{60\,\text{min}}\right)\left(\frac{1\,\text{day}}{24\,\text{hr}}\right)\left(\frac{1\,\text{yr}}{365\,\text{day}}\right) = 1.2\,\text{yr}$$

21.19 Identify products of nuclear decay by noting that α-decay changes A by -4 and Z by -2, while β-decay changes Z by $+1$ while leaving A unchanged:

$^{232}_{90}\text{Th} \rightarrow \alpha + ^{228}_{88}\text{Ra} \rightarrow \beta + ^{228}_{89}\text{Ac} \rightarrow \beta + ^{228}_{90}\text{Th} \rightarrow \alpha + ^{224}_{88}\text{Ra} \rightarrow \alpha + ^{220}_{86}\text{Rn} \rightarrow \alpha + ^{216}_{84}\text{Po}$;

$^{216}_{84}\text{Po} \rightarrow \beta + ^{216}_{85}\text{At} \rightarrow \alpha + ^{212}_{83}\text{Bi} \rightarrow \beta + ^{212}_{84}\text{Po} \rightarrow \alpha + ^{208}_{82}\text{Pb}$

21.21 Identify the compound nucleus and the final products of a nuclear reaction by applying the principles of conservation of mass number and charge:

(a) $^{12}_{6}\text{C} + ^{1}_{0}\text{n} \rightarrow \left[^{13}_{6}\text{C}\right] \rightarrow ^{12}_{5}\text{B} + ^{1}_{1}\text{p}$;

(b) $^{16}_{8}\text{O} + ^{4}_{2}\alpha \rightarrow \left[^{20m}_{10}\text{Ne}\right] \rightarrow ^{20}_{10}\text{Ne} + \gamma$;

(c) $^{247}_{96}\text{Cm} + ^{11}_{5}\text{B} \rightarrow \left[^{258}_{101}\text{Md}\right] \rightarrow ^{255}_{101}\text{Md} + 3\,^{1}_{0}\text{n}$

21.23 Pictures of nuclear reactions should show each nuclide with the appropriate numbers of protons and neutrons. The reaction in this problem is:

$$^{14}_{7}N + ^{4}_{2}\alpha \rightarrow \left[^{18}_{9}F\right] \rightarrow ^{18}_{8}O + ^{0}_{1}\beta^{+}$$

$\bullet$ = proton $\bigcirc$ = neutron

21.25 Stable elements with mass numbers around 135 have N/Z ratios in the range of 1.4. Use $A = N + Z$ to calculate what Z value is likely to be stable:

$$N = A - Z, \quad \frac{135-Z}{Z} = 1.4, \quad 2.4\,Z = 135, \quad Z = 56$$

Thus the most likely high-mass elements resulting from U-235 fission are Cs, Ba, La, Ce.

21.27 Calculate the mass defect from the masses of the individual reactants and products, then convert to energy released:

$$\Delta m = [80.9199 + 151.9233 + 3(1.0087)] - [235.0439 + 1.0087] = -0.1833 \text{ g/mol};$$

$$\Delta E = \left(\frac{-0.1833\text{ g}}{1\text{ mol}}\right)\left(\frac{8.988 \times 10^{10}\text{ kJ}}{1\text{ g}}\right) = -1.648 \times 10^{10}\text{ kJ/mol};$$

This result is somewhat less than the result of the general calculation of Section Exercise 21.4.1 for a net change of 2 neutrons.

21.29 Your description should feature the fact that the core of a nuclear reactor generates radiation that converts any material in its vicinity into radioactive substances. Thus the heat exchanger in immediate contact with the core becomes radioactive and must be separated from the turbine that generates electricity. The primary heat exchanger transfers energy to a secondary heat exchanger, which does not become radioactive and can safely drive the turbine.

21.31 The amount of energy released in a fusion reaction can be calculated from the energy per event and the total amount of matter undergoing fusion. The reactions in this example are:

$$^{2}H + ^{3}H \rightarrow ^{4}He + n, \quad \Delta E = -1.7 \times 10^{9}\text{ kJ/mol}$$
$$^{6}Li + n \rightarrow ^{4}He + ^{3}H, \quad \Delta E = -4.6 \times 10^{8}\text{ kJ/mol}$$

$$n_H = 2.50\text{ g}\left(\frac{1\text{ mol}}{2.014\text{ g}}\right) = 1.241\text{ mol};$$

$$\Delta E = 1.241\text{ mol}\left(\frac{-1.7 \times 10^{9}\text{ kJ}}{1\text{ mol}}\right) = -2.11 \times 10^{9}\text{ kJ};$$

In addition, 0.6207 mol Li react,

$$\Delta E = 0.6207 \text{ mol}\left(\frac{-4.6 \times 10^8 \text{kJ}}{1 \text{ mol}}\right) = -2.86 \times 10^8 \text{ kJ};$$

$\Delta E_{total} = (-2.11 \times 10^9 \text{ kJ}) + (-2.86 \times 10^8 \text{ kJ}) = -2.4 \times 10^9 \text{ kJ}$

(2 sig. figs. because ΔE has 2)

21.33 To determine the speed of a nucleus that can fuse with another, first determine the energy needed to surmount the coulombic barrier, as described by Equation 21-1:

$$E_{coulomb} = \frac{(1.389 \times 10^5 \text{ kJ pm / mol})(Z_1)(Z_2)}{d}$$

Then calculate the speed from the equation for kinetic energy, $E_{kinetic} = \frac{1}{2} mu^2$,

$$u = \sqrt{\frac{2E_{kinetic}}{m}}$$

First determine the nuclide radii using the equation from the problem:

$r_{tritium} = 1.2(3)^{1/3} = 1.7 \times 10^{-3} \text{ pm}$
$r_{deuteron} = 1.2(2)^{1/3} = 1.5 \times 10^{-3} \text{ pm}$

Here are the data:

$Z = 1$ for both nuclides; and $m = 3.01605$ g/mol for tritium:

$$E = \frac{(1.389 \times 10^5 \text{ kJ pm/mol})(1)(1)}{(1.7 \times 10^{-3} \text{pm} + 1.5 \times 10^{-3} \text{pm})} = 4.3 \times 10^7 \text{ kJ/mol};$$

$$u = \left[\frac{(2)(4.3 \times 10^7 \text{ kJ/mol})(10^3 \text{ J/kJ})}{(3.01605 \text{ g/mol})(10^{-3} \text{kg/g})}\right]^{1/2} = 5.3 \times 10^6 \text{ m/s}$$

(Remember that $1 \text{ J} = 1 \text{ kg m}^2/\text{s}^2$, so the units in this calculation cancel to give m/s).

21.35 Your description should include the extremely high energies required to initiate fusion and the difficulties in containing the fusion components at the temperature required for nuclei to have these high energies.

21.37 The characteristics of first-generation stars are described in your text:

Stage	Temperature	Composition
H-burning	4×10^7 K	H, He, e$^-$
He-burning	10^8 K	He, Be, C, O (H depleted)
C-burning	10^9 K	all nuclides from $Z = 6$ (C) up to $Z = 26$ (Fe)

21.39 Your description should include the fact that elements beyond $Z = 26$ are less stable than Fe, so they cannot be generated by fusion of lighter elements.

21.41 The problem asks for the energy released by a radioactive isotope. To determine this, it is necessary to calculate how much material decays during the time period. Determine the amount of the radioactive nuclide that decays in one day using Equation 21-5:

$$\ln\left(\frac{N_o}{N}\right) = \frac{t \ln 2}{t_{1/2}}$$

$$\ln\left(\frac{N_o}{N}\right) = \frac{(1\text{ day})(0.693)}{(8.07\text{ day})} = 0.0859\text{, from which }\left(\frac{N_o}{N}\right) = e^{0.0859} = 1.0897;$$

$$N = \frac{7.45\text{ pg}}{1.0897} = 6.84\text{ pg, and }\Delta N = 7.45\text{ pg} - 6.84\text{ pg} = 0.61\text{ pg;}$$

Now convert to moles and multiply by the energy released per mole to obtain the amount of energy captured by the gland:

$$n = 0.61\text{ pg}\left(\frac{10^{-12}\text{g}}{1\text{ pg}}\right)\left(\frac{1\text{ mol}}{131\text{ g}}\right) = 4.7 \times 10^{-15}\text{ mol;}$$

$$E = 4.7 \times 10^{-15}\text{ mol}\left(\frac{9.36 \times 10^{7}\text{ kJ}}{1\text{ mol}}\right)\left(\frac{10^{3}\text{ J}}{1\text{ kJ}}\right) = 4.4 \times 10^{-4}\text{ J}$$

21.43 Use Equation 21-4 to determine the rate of emission of a radioactive nuclide:

$$N = 7.45\text{ pg}\left(\frac{10^{-12}\text{g}}{1\text{ pg}}\right)\left(\frac{1\text{ mol}}{131\text{ g}}\right)\left(\frac{6.022 \times 10^{23}\text{ nuclei}}{1\text{ mol}}\right) = 3.425 \times 10^{10}\text{ nuclei;}$$

$$t_{1/2} = 8.07\text{ day}\left(\frac{24\text{ hr}}{1\text{ day}}\right)\left(\frac{60\text{ min}}{1\text{ hr}}\right)\left(\frac{60\text{ s}}{1\text{ min}}\right) = 6.97 \times 10^{5}\text{ s}$$

$$\frac{\Delta N}{\Delta t} = \frac{-N\ln 2}{t_{1/2}} = -\frac{(3.425 \times 10^{10}\text{ decays})(0.693)}{6.97 \times 10^{5}\text{ s}} = -3.41 \times 10^{4}\text{ decays/s;}$$

There are 3.41×10^{4} decays/s

21.45 Exposure to radiation results first in damage to those cells that reproduce most quickly, including the white blood cells that are responsible for fighting infection and the mucous membrane lining of the intestinal tract. Thus the early symptoms of radiation exposure include reduced resistance to infection and nausea due to disruption of the digestive tract.

21.47 Dating techniques using radioisotopes are based on Equation 21-5: $\ln\left(\frac{N_o}{N}\right) = \frac{t\ln 2}{t_{1/2}}$.

To obtain an age estimate, the ratio N_o/N must be determined. Convert the mass ratio into the desired numerical ratio. Because both isotopes have the same value for A, their mole ratio and number ratio are the same as their mass ratio:

$$N_o = N_{Sr} + N_{Rb}; \qquad \frac{N_o}{N} = \frac{N_{Sr} + N_{Rb}}{N_{Rb}} = 1 + \frac{N_{Sr}}{N_{Rb}} = 1 + 0.0050 = 1.0050;$$

$$t = \left(\frac{t_{1/2}}{\ln 2}\right)\ln\left(\frac{N_o}{N}\right) = \left(\frac{4.9 \times 10^{11}\text{ yr}}{0.693}\right)\ln(1.0050) = 3.5 \times 10^{9}\text{ yr}$$

21.49 The best radioisotopes for medical imaging have intermediate half-lives, long enough to prepare and administer but short enough to decay in a reasonable length of time, and they decay by processes that do the least damage to body tissue (γ and β^+). Among the isotopes of iodine listed, $A = 123$ would be best because of its 13.3-hr half-life and EC/γ mode of decay.

21.51 To identify the products of nuclear decay processes, make use of the principles of conservation of mass number and charge:

(a) alpha emission changes nuclear charge by -2, mass by -4, product is $^{234}_{90}$Th:

$$^{238}_{92}\text{U} \rightarrow {}^4_2\alpha + {}^{234}_{90}\text{Th}$$

(b) (n,p) changes nuclear charge by -1, no change in mass, product is $^{60}_{27}$Co:

$$^{60}_{28}\text{Ni} + {}^1_0\text{n} \rightarrow {}^1_1\text{p} + {}^{60}_{27}\text{Co}$$

(c) process changes nuclear charge by $+6$, mass by ($+12 - 3$), product is $^{248}_{99}$Es:

$$^{239}_{93}\text{Np} + {}^{12}_6\text{C} \rightarrow 3{}^1_0\text{n} + {}^{248}_{99}\text{Es}$$

(d) (p,α) changes nuclear charge by -1, mass by -3, product is $^{32}_{16}$S:

$$^1_1\text{p} + {}^{35}_{17}\text{Cl} \rightarrow {}^4_2\alpha + {}^{32}_{16}\text{S}$$

(e) β decay changes nuclear charge by $+1$, no change in mass, product is $^{60}_{28}$Ni:

$$^{60}_{27}\text{Co} \rightarrow {}^0_{-1}\beta + {}^{60}_{28}\text{Ni}$$

21.53 Binding energy is calculated by determining the mass defect (difference between the mass of the atom and the sum of the masses of its individual components) and converting to energy. When an element has only one stable nuclide, its molar mass is the mass of that nuclide:

$MM_{\text{Bi}} = 208.980$ g/mol,

The nuclide has 83 protons and electrons, $209 - 83 = 126$ neutrons;

$m_{\text{components}} = 83(1.007276 + 0.0005486) + 126(1.008665) = 210.741232$ g/mol;

$\Delta m = 208.980 - 210.741232 = -1.761$ g/mol;

$$\Delta E = \left(\frac{-1.761\,\text{g}}{1\,\text{mol}}\right)\left(\frac{8.988 \times 10^{10}\,\text{kJ}}{1\,\text{g}}\right) = -1.583 \times 10^{11}\ \text{kJ/mol};$$

$$\Delta E\ (\text{per nucleon}) = \frac{-1.583 \times 10^{11}\,\text{kJ/mol}}{209\ \text{nucleons}} = -7.574 \times 10^8\ \text{kJ mol}^{-1}\ \text{nucleon}^{-1}$$

21.55 Information about nuclear decays can be obtained using Equations 21-4 and 21-5.

$$\ln\left(\frac{N_o}{N}\right) = \frac{t\ln 2}{t_{1/2}}, \text{ from which } \ln N = \ln N_0 - \frac{t\ln 2}{t_{1/2}}$$

$N_o = 5.0$ mg, $t_{1/2} = 138.4$ days, $t = 365$ days;

$$\ln N = \ln(5.0) - \frac{(365 \text{ days})(\ln 2)}{(138.4 \text{ days})} = -0.2182, \quad N = e^{-0.2182} = 0.80 \text{ mg remain};$$

$$\frac{\Delta N}{\Delta t} = \frac{-N \ln 2}{t_{1/2}}; \quad N = 0.80 \text{ mg} \left(\frac{10^{-3} \text{ g}}{1 \text{ mg}} \right) \left(\frac{1 \text{ mol}}{210 \text{ g}} \right) \left(\frac{6.022 \times 10^{23} \text{ nuclei}}{1 \text{ mol}} \right) = 2.3 \times 10^{18} \text{ nuclei}$$

$$t_{1/2} = 138.4 \text{ day} \left(\frac{24 \text{ hr}}{1 \text{ day}} \right) \left(\frac{60 \text{ min}}{1 \text{ hr}} \right) \left(\frac{60 \text{ s}}{1 \text{ min}} \right) = 1.196 \times 10^7 \text{ s}$$

$$\frac{\Delta N}{\Delta t} = \frac{(2.3 \times 10^{18} \text{ decays})(\ln 2)}{1.196 \times 10^7 \text{ s}} = 1.3 \times 10^{11} \text{ decays/s}$$

21.57 Determine the products of nuclear decay by applying conservation of charge and mass number: The other product of neutron decay must be a particle with -1 charge and 0 mass, which is an electron: $n \rightarrow p + e$.

Calculate the decay energy from the mass defect between reactant and products, using masses found in Table 21-1:

$$\Delta m = (1.007276 \text{ g/mol}) + (5.486 \times 10^{-4} \text{ g/mol}) - (1.008665 \text{ g/mol}) = -0.0008404 \text{ g/mol};$$

$$\Delta E = \left(\frac{-0.0008404 \text{ g}}{1 \text{ mol}} \right) \left(\frac{8.988 \times 10^{10} \text{ kJ}}{1 \text{ g}} \right) \left(\frac{10^3 \text{ J}}{1 \text{ kJ}} \right) = -7.554 \times 10^{10} \text{ J/mol};$$

$$E_{\text{kinetic, electron}} = \left(\frac{7.554 \times 10^{10} \text{ J}}{1 \text{ mol}} \right) \left(\frac{1 \text{ mol}}{6.022 \times 10^{23} \text{ electrons}} \right) = 1.25 \times 10^{-13} \text{ J}$$

21.59 Calculations of decay times make use of Equation 21-5:

$$\ln \left(\frac{N_o}{N} \right) = \frac{t \ln 2}{t_{1/2}}, \text{ from which } t = \left(\frac{t_{1/2}}{\ln 2} \right) \ln \left(\frac{N_o}{N} \right)$$

When 1% has decayed, $t = \left(\frac{1622 \text{ yr}}{\ln 2} \right) \ln \left(\frac{N_o}{0.99 N_o} \right) = 24 \text{ yr};$

When 1% remains, $t = \left(\frac{1622 \text{ yr}}{\ln 2} \right) \ln \left(\frac{N_o}{0.01 N_o} \right) = 1.1 \times 10^4 \text{ yr}$

21.61 The n/p ratio of an isotope determines where it lies with respect to the belt of stability. ^{11}C has 6 protons and $(11 - 6) = 5$ neutrons, n/p = 0.833; ^{15}O has 8 protons and $(15 - 8) = 7$ neutrons, n/p = 0.875. Both isotopes are neutron-deficient and lie below (to the right of) the belt of stability. Isotopes that are neutron-deficient decay by positron emission, which increases their n/p ratios: $^{11}_{6}C \rightarrow \, ^{0}_{1}\beta^+ + \, ^{11}_{5}B$ and $^{15}_{8}O \rightarrow \, ^{0}_{1}\beta^+ + \, ^{15}_{7}N$

21.63 Calculations of decay times make use of Equation 21-5:

$$\ln\left(\frac{N_o}{N}\right) = \frac{t\ln 2}{t_{1/2}}, \text{ from which } t = \left(\frac{t_{1/2}}{\ln 2}\right)\ln\left(\frac{N_o}{N}\right)$$

When 1 µg is left, $t = \left(\frac{15\text{ hr}}{\ln 2}\right)\ln\left(\frac{25\text{ µg}}{1\text{ µg}}\right) = 70$ hr.

21.65 Information about nuclear decays can be obtained using Equation 21-5.

$$\ln\left(\frac{N_o}{N}\right) = \frac{t\ln 2}{t_{1/2}}, \text{ from which } \ln N = \ln N_o - \frac{t\ln 2}{t_{1/2}}$$

$N_o = (10^5/\text{s})(30\text{ s}) = 3.0 \times 10^6;$

$$t = 5\text{ hr}\left(\frac{60\text{ min}}{1\text{ hr}}\right)\left(\frac{60\text{ s}}{1\text{ min}}\right) = 1.8 \times 10^4 \text{ s};$$

If $t_{1/2} = 1100$ s, $\ln N = \ln(3.0 \times 10^6) - \dfrac{(1.8 \times 10^4\text{ s})(\ln 2)}{(1100\text{ s})} = 3.57, \quad N = e^{3.57} = 36;$

If $t_{1/2} = 876$ s, $\ln N = \ln(3.0 \times 10^6) - \dfrac{(1.8 \times 10^4\text{ s})(\ln 2)}{(876\text{ s})} = 0.67, \quad N = e^{0.67} = 2.$

21.67 Determine the expected product using conservation of mass number and charge number: $^{208}_{82}\text{Pb} + {}^{48}_{22}\text{Ti} \rightarrow \left({}^{256}_{104}\text{Rf}\right)$. (This compound nucleus probably would decay by emitting neutrons and/or β particles).

21.69 Energy releases are calculated using Equation 21-3: $\Delta E = (\Delta m)(8.988 \times 10^{10} \text{ kJ/g})$:
$\Delta m = (147.9146) + (4.00260) - 151.9205 = -0.0033 \text{ g/mol};$

$$\Delta E = \left(\frac{-0.0033\text{ g}}{1\text{ mol}}\right)\left(\frac{8.988 \times 10^{10}\text{ kJ}}{1\text{ g}}\right) = -2.97 \times 10^8 \text{ kJ/mol};$$

$$\Delta E \text{ (per event)} = \left(\frac{-2.97 \times 10^8\text{ J}}{1\text{ mol}}\right)\left(\frac{1\text{ mol}}{6.022 \times 10^{23}\text{ events}}\right)\left(\frac{10^3\text{ J}}{1\text{ kJ}}\right) = -4.9 \times 10^{-13} \text{ J};$$

Fraction carried off by the α–particle = $\dfrac{3.59 \times 10^{-13}\text{ J}}{4.93 \times 10^{-13}\text{ J}} = 0.73.$

21.71 Nuclides in the "belt of stability" are stable. Instability occurs if a nuclide has too few neutrons, too many neutrons, or $Z > 83$. In addition, most odd-odd nuclides are unstable: (a) too many neutrons; (b) $Z > 83$; (c) odd-odd; (d) too many neutrons.

21.73 In a nuclear reactor, the moderator serves to slow down fast neutrons so they are more efficiently captured by the nuclear fuel. This reduces the amount of fuel required to sustain the reaction. The control rods serve to absorb some of the neutrons, allowing the reactor to operate just below its critical point and generate large quantities of heat without heating up beyond control.

21.75 The precipitate will contain radioactive Na, because once the NaBr is dissolved in solution, its Na^+ cations mix freely with the Na^+ cations of the existing solution. When the solution cools and $NaNO_3$ precipitates, some of the Na^+ cation in the precipitate will be radioactive Na-24.

21.77 Determine the nuclear reaction by applying conservation of mass number and charge number:

$^{10}_{5}B + n \rightarrow \left(^{11}_{5}B\right) \rightarrow ^{7}_{3}Li + \alpha$. This reaction does not pose a significant health hazard, because Li-7 is a stable isotope and the α-particles are easily stopped using an appropriate shield.

21.79 Dating techniques using radioisotopes are based on Equation 21-5: $\ln\left(\dfrac{N_o}{N}\right) = \dfrac{t \ln 2}{t_{1/2}}$.

First, calculate the N_o/N ratio, which is the ratio of counts/g hr for the new and old samples:

Fresh sample: count rate $= \dfrac{18400 \text{ counts}}{(1.00 \text{ g})(20 \text{ hr})} = 920$ counts g^{-1} hr^{-1};

Old sample: count rate $= \dfrac{1020 \text{ counts}}{(0.250 \text{ g})(24 \text{ hr})} = 170$ counts g^{-1} hr^{-1};

$N_o/N = \dfrac{920}{170} = 5.4$

$t = \left(\dfrac{t_{1/2}}{\ln 2}\right)\ln\left(\dfrac{N_o}{N}\right) = \left(\dfrac{5730 \text{ yr}}{0.693}\right) \ln (5.4) = 1.4 \times 10^4$ yr.

21.81 To calculate the useful lifetime of the dating technique use Equation 21-5:

$\ln\left(\dfrac{N_o}{N}\right) = \dfrac{t \ln 2}{t_{1/2}}$, from which $t = \dfrac{t_{1/2}}{\ln 2} \ln\left(\dfrac{N_o}{N}\right)$

$t = \left(\dfrac{5730 \text{ yr}}{0.693}\right) \ln\left(\dfrac{15.3 \text{ counts / g min}}{0.03 \text{ counts / g min}}\right) = 5.2 \times 10^4$ yr

21.83 Determine the products of nuclear decay using conservation of charge number and mass number. Use mass-energy equivalence to determine the mass of a product, given the mass of a reactant and the energy given off in the process:

(a) Sr has $Z = 38$ and Zr has $Z = 40$, so the decay requires 2 β-particles (0 mass number, $-$1 charge number): $^{90}_{38}Sr \rightarrow ^{90}_{39}Y + \beta$; $^{90}_{39}Y \rightarrow ^{90}_{40}Zr + ^{0}_{-1}\beta$;

(b) These nuclear decay processes involve emission of electrons and are exothermic, so there must be a decrease in mass of the isotope in the process. Thus Sr has the larger mass;

(c) $\Delta E = \Delta m(8.988 \times 10^{10}$ kJ/g); net reaction is $^{90}_{38}Sr \rightarrow ^{90}_{40}Zr + 2 ^{0}_{-1}\beta$;

$\Delta m = [89.9043 + 2(0.0005486)] - 89.9073 = -0.0019$ g/mol;

$\Delta E = \left(\dfrac{-0.0019 \text{ g}}{1 \text{ mol}}\right)\left(\dfrac{8.988 \times 10^{10} \text{kJ}}{1 \text{ g}}\right) = -1.7 \times 10^8$ kJ/mol.

21.85 Calculations of decay times make use of Equation 21-5:

$$\ln\left(\frac{N_o}{N}\right) = \frac{t \ln 2}{t_{1/2}}, \text{ from which } t = \left(\frac{t_{1/2}}{\ln 2}\right)\ln\left(\frac{N_o}{N}\right)$$

First calculate the amount initially bound to the thyroid gland:

$$N_o = (0.5 \text{ mg})\left(\frac{10^3 \mu g}{1 \text{ mg}}\right)\left(\frac{45\%}{100\%}\right) = 225 \text{ }\mu g;$$

When 0.1 μg is left, $t = \left(\frac{13.2 \text{ hr}}{\ln 2}\right)\ln\left(\frac{225 \text{ }\mu g}{0.1 \text{ }\mu g}\right) = 1.5 \times 10^2$ hr.

21.87 Around Z = 43, the n/p ratio for stable isotopes is about 1.27, so the isotope that is most likely to be stable has (43)(1.27) = 55 neutrons. This, however, is an odd-odd isotope, so we might expect to find 56 neutrons, $^{99}_{43}$Tc. This, in fact, is the nuclide used widely in nuclear medicine.

21.89 To determine where the oxygen atom in the water molecule comes from, prepare a sample of the alcohol that is enriched in radioactive ^{18}O and run the reaction using this sample. Separate the products and measure the radioactivity of the ester and the water. If the C–OH bond in the alcohol breaks during the condensation, the ^{18}O will appear in the water, while if the C–OH bond in the carboxylic acid breaks during the condensation, the ^{18}O will appear in the ester.